Karl Shuker's ALIEN ZOO
From the pages of *Fortean Times*

Dr Karl Shuker
with forewords by Bob Rickard and David Sutton

Edited by Jonathan Downes
Typeset by Jonathan Downes and an Orange Cat
Cover and Internal Layout by Jon Downes for CFZ Communications
Using Microsoft Word 2000, Microsoft , Publisher 2000, Adobe Photoshop.

First edition published 2010 by CFZ Publications

CFZ PRESS
Myrtle Cottage
Woolfardisworthy
Bideford
North Devon
EX39 5QR

© CFZ MMX

All rights reserved. Without limiting the rights under copyright reserved above, no part of this publication may be reproduced, stored in or introduced into a retrieval system, or transmitted, in any form of by any means (electronic, mechanical, photocopying, recording or otherwise), without the prior written permission of both the copyright owners and the publishers of this book.

ISBN: 978-1-905723-62-1

DEDICATIONS.

To the one and only Charles Hoy Fort, without whom there could never have been a *Fortean Times* or, in turn, an Alien Zoo.

And also to the memory of Jan Williams, who was not only a highly proficient and totally professional cryptozoological researcher but also a good friend and a lovely lady.

Holding a black panther model (Dr Karl Shuker)

LIST OF CONTENTS.

Wanka 2012

7 Foreword by Bob Rickard – The Founder of *Fortean Times*
9 Foreword by David Sutton – The Editor of *Fortean Times*
11 Introduction
15 The Special Years – Carry On Cryptozoology / Absolutely Fabulous Animals
23 Lost Ark: Whatever Happened To Wattie?
           In Search Of Loch Watten's Missing Monster
33 Alien Zoo: 1995-7
51 Lost Ark: The Lovecats - Jungle Cats, Jasper, and a Crossbred Controversy
65 Alien Zoo: 1998
79 Lost Ark: Archangel Feathers – More Than A Divine Flight Of Fancy?
91 Alien Zoo: 1999
105 Lost Ark: Kicking Up A Stink – Barbara Woodhouse's Pouched Skunk
109 Alien Zoo: 2000
125 Lost Ark: Reviewing the Ropen – A New Cryptid From New Guinea
131 Alien Zoo: 2001
145 Lost Ark: Horsing Around In Tibet
149 Alien Zoo: 2002
167 Lost Ark: Out-Foxed By Our High-Flying Cousins?
173 Alien Zoo: 2003
187 Lost Ark: Cz In Oz – An Ark Of New Mystery Beasts Down Under
193 Alien Zoo: 2004
203 Lost Ark: The Art Of Crypto-Twitching
213 Alien Zoo: 2005
225 Lost Ark: Along Came A Spider – Beware The Phantom Flesh-Rotter!
229 Alien Zoo: 2006
241 Lost Ark: Bringing The Winged Cats Down To Earth
257 Alien Zoo: 2007
267 Lost Ark: A Blenny For Your Thoughts – Fishing For A Bite In The Middle East
271 Alien Zoo: 2008
281 Lost Ark: A New Eden In New Guinea
287 Alien Zoo: 2009
303 Lost Ark: A Serpent In Stone – From Adder To Ammonite?
307 Alien Zoo: 2010 - Latest News and Updates
321 Lost Ark: The Monster Of Lake Khaiyr – After Four Decades The Truth At Last...?
325 Alien Zoo/Lost Ark: The Giant Mole of Mesty Croft
329 Lost Ark: Driven Batty By The Orang Bati
337 Reviews: The DVD That Got Away, and a Bevy of Book Reviews
353 UnConvention '94 Lecture: From Panthers to Plesiosaurs
           – Current Affairs In Cryptozoology
365 Index
377 Acknowledgements
378 Dr Shuker's Website and Blogs
379 A Listing of Dr Shuker's Original Articles
381 About The Author

FOREWORD BY BOB RICKARD
THE FOUNDER OF *FORTEAN TIMES*.

Five years after Karl Shuker first started writing for *Fortean Times*, he gave a lecture on cryptozoology to the inaugural session of the journal's 'UnConvention', held in 1994. Instead of presenting a detailed study of a single topic, he chose instead to present an overview of some of the major cases of 'undiscovered animals'. This format has served him well in all his books to date which have become one of the few reliable windows upon the secretive and complex world of cryptozoology. This is made all the more remarkable because Karl is a rare creature himself: an independent professional zoologist.

Karl is also a member of another rare group, the very small but dedicated band who call themselves 'Fortean Zoologists' To some, the term is synonymous with 'cryptozoology', the field of natural history that deals with animals not yet recognised by formal zoology which comprises a spectrum of hypothetical creatures from extinct forms, for which we only have ambiguous fossils, to those for which the evidence is largely narrative, folkloric or legendary. This is sometimes spiced with controversial suggestions that a creature may not actually be extinct or confined to folklore or legend, but living in some remote but shrinking wilderness known only to the accidental explorer or secluded natives.

The term 'Fortean' is derived from the work of Charles Fort (1874-1932), an American writer who devoted his life to scouring the archives of newspapers and scientific journals for reports of anomalies of

all sorts supported by thousands of references laid out in four books, encompassing both the worlds of human experience and of natural occurrences. He made discoveries which have preoccupied his disciples ever since: he coined the word 'teleportation'; he pioneered study of the 'experimenter effect' and of what we now call UFOs; he found associations between disturbed adolescents and poltergeist outbreaks, and between earthquakes and meteorites; and opened up the field of anomalous fossils.

More relevant to this volume is that Fort also foreshadowed the scope of the interests of today's cryptozoologists, both in range (from insects and fishes to wolves and apes); from the unusual involving known creatures (such as falls of frogs and vast

Alien Zoo

swarms of butterflies that takes days to pass) to reports of unusual creatures (sea and lake monsters or 'mystery' beasts); from the mystery of toads entombed in stone to the appearances of animals out-of-place (alligators in England or fish in a newly-dug pond; from ideological debates (acquired versus inherited characteristics, or the fertility of mules) to 'meaningful images' or simulacra in the markings or 'camouflage' of animals; from natural but exceptional teratologies (e.g. two-headed creatures and winged cats) and the absurd (talking dogs) to the incredible navigation feats of long-distance homing and migration, and extraordinary patterns of scarcity and abundance.

Fort also covered contemporary 'flaps' such as the depredations caused by supposed wolves in northern England in and of 'luminous owls' in Norfolk in 1908. He was just as interested in why people imagine things – such as a fluttering black bag caught in a hedge becomes, in the mind's eye, a bear, or a bobbing log on a lake gives rise to a 'monster' – and how these imaginings, in turn, initiate panics, or police hunts for 'mystery big cats'.

Nature defies the attempts of science to categorise and 'normalise' it, and yet it still holds wonders to surprise and delight those who appreciate its mysteries. This selection of Karl's writings from the pages of *Fortean Times* over the last 16 years is an admirable continuation of that tradition.

Bob Rickard

FOREWORD BY DAVID SUTTON
THE EDITOR OF *FORTEAN TIMES*.

Nearly four decades after Bob Rickard founded *Fortean Times*, back in the heady days of 1973, it continues to be the world's foremost journal devoted to the study of strange phenomena and the first port of call for anyone remotely interested in the more mysterious byways of life on our strange planet. It's both gratifying and rather astonishing to see how the magazine has grown from its humble origins to attain a worldwide readership; and equally gratifying and astonishing to realise that Dr Karl Shuker has been our resident cryptozoologist for over half that time, ever since he turned in his first article back in 1989, a round-up of Alien Big Cat sightings (the subject of Karl's first book, *Mystery Cats of the World*, that very same year). By 1997, Karl had a good few more books under his belt, had started writing his regular 'Alien Zoo' and longer 'Lost Ark' columns in *Fortean Times* – the basis for this latest book – and had become widely known and respected not just as *FT's*, but indeed Britain's, leading expert on matters pertaining to animals anomalous, unknown or out of place.

Over his years with us, Karl has detailed many of the amazing new zoological discoveries (and sometimes rediscoveries of species thought to be extinct) made around the world, from the plethora of new species found in Vietnam to the remarkable finds made in the 'New Eden' of a remote region of New Guinea' in 2006. But as well as covering these important discoveries in the field of mainstream zoology, Karl has also been on the trail of rarer and considerably more Fortean beasts. As well as bringing us regular updates about such high-profile cryptids as the orang-pendek or the chupacabras, Karl has also revealed a startling menagerie of quite unfamiliar mystery critters, from the wonderfully named and elusive altamaha-ha, an intriguing water creature from Georgia, USA, to the deadly dodu reported by the Baka and Bantu people of Cameroon; or a whole range of unlikely winged wonders,

from the frightful sounding orang bati and ropen to the rather more homely phenomenon of winged cats.

Another aspect of Karl's work has been his ongoing interest in mapping not just the world of flesh-and-blood crypto-creatures, both real and putative, but in exploring the worlds of history and myth, reminding us that even such legendary creatures as the roc might have zoological inspirations, and investigating such intriguing relics as the fragment of skin said to be that of the serpent that tempted Eve in the Garden of Eden or the elusive 'angel feathers' supposed to reside in the palace of El Escorial, Madrid.

One thing I've come particularly to admire about Karl over the years is his dogged persistence in following up a promising cryptozoological tid-bit or intriguing clue in the hopes that it will yield up something more substantial farther down the line. Even when the trail goes cold, Karl will wait until a new lead emerges – whether from a fresh piece of witness testimony, a letter from one of his many

Alien Zoo

correspondents or a bit of evidence turned up in a forgotten book or archive. As Karl – like a true Fortean – writes: "Some of the most interesting cryptozoological discoveries initially take place not in the field but in the library".

So, even those who have followed Karl's articles over the years will find new gleanings in these pages – updates and additions to the originals and even the odd 'one that got away' in the form of articles and reviews that, for whatever reason, slipped through *FT's* net. And it's particularly good to see in print the lecture Karl gave at Un-Convention back in 1994, a lively snapshot of the subject at the time.

I hope you will enjoy reading these bulletins from the Alien Zoo and the Lost Ark as much as we have enjoyed printing them over the years – and that in a decade's time we'll be looking at a second collection of cryptozoological curios from Dr Shuker's *FT* casebook.

David Sutton

INTRODUCTION.

The level-headed man at home will be tempted to exclaim:
"But this is rubbish, this talk of undiscovered monsters!"
So it may seem - but upon discovery of the Congo's Okapi, scientists gasped:
"This cannot be!" They said the same of the man who first reported the giraffe
and tortured him for being a romancer of the worst description - yet children
feed this "monster" in London Zoo today...

Captain William Hichens – c.1927

Back in the summer of 1989, when my very first *Fortean Times* (*FT*) article was published (in *FT*52) – surveying some recent ABC (alien big cat) events in the UK, including the shooting of an Asian leopard cat *Prionailurus bengalensis* on Dartmoor – I could not have possibly envisaged that more than 20 years later I would still be writing for *FT*, and on a frequent basis. Nevertheless, I am delighted to say that this is indeed how events transpired, yielding not one but two regular series. The first of these, which has been running ever since 1997 and for much of that time as a monthly column (between 2005 and early 2010 it switched temporarily to bimonthly status along with most of *FT*'s other regular columns before reverting to its former monthly status in May 2010), is of course my Alien Zoo, which presents a round-up of cryptozoological news stories not covered in larger, headline reports elsewhere in *FT*.

I have also been penning for much the same span of time a roughly quarterly series of longer, Forum-style cryptozoology articles usually carrying my very own 'Lost Ark' banner, and which were even accompanied for a time by a wonderful, specially-commissioned Hunt Emerson 'Lost Ark' cartoon image of me. This is reproduced in the present book by kind permission of Hunt himself.

Alien Zoo

As you can imagine, in over 20 years, and especially with regard to the Alien Zoo column, I've covered a very appreciable range and number of topics, with many of them never before (and some never again) documented in the mainstream fortean literature, if indeed in print anywhere at all. Browsing through them some time ago, it came to mind that a compilation volume, concentrating upon the Alien Zoo reports but interspersed with longer Lost Ark articles, would be a very effective way of gathering together this weighty corpus of work and making it readily accessible to cryptozoological enthusiasts and other long-term followers of my *FT* writings. Consequently, after Dennis Publishing, the publishers of *Fortean Times*, and its current editor, David Sutton, very kindly gave me permission to do so, I decided to prepare just such a volume at the earliest opportunity.

Once I began doing so, however, it soon became abundantly clear that there was far too much material to fit into a single volume. Yet as I also felt that it would not be a satisfactory plan to attempt to eke it out into several volumes, some drastic yet at the same time selective editing would need to be employed here. But what to keep in and what to leave out? That, as Shakespeare never said, was the question. Obviously the bulk of the material needed to be drawn from the Alien Zoo columns, because that had always been my primary focus and drive when contemplating the production of a book based upon my *FT* writings - but with such a huge quantity to hand, what criteria should I impose upon my selection for inclusion in this book?

After due consideration, I decided to concentrate upon items relating specifically to cryptozoology and out-of-place (oop) animals, thereby excluding for the most part those items dealing with new and redis-covered animals and those dealing with freak animals. The reasons for their omission were that the for-mer category of creatures was already documented (and in much greater detail) within my books *The Lost Ark* (1993), *The New Zoo* (2002), and the forthcoming third version in this series (currently enti-tled *New and Rediscovered Animals of the World*, and scheduled for publication in late 2010); and the latter category of creatures constitutes the subject of another forthcoming book of mine, *The Anoma-larium of Doctor Shuker* (scheduled for publication in 2011).

(The only exceptions to this ruling that I have allowed, and which do therefore appear in this book, are items referring to new and rediscovered animals of local, rather than global, interest, which therefore do not appear in my above-mentioned books but whose stories are nonetheless interesting; and some re-ports of freak creatures that have cryptozoological as well as teratological significance.)

With those categories omitted, all of the cryptozoological and oop Alien Zoo items could be included within a single volume, together with a representative selection of Lost Ark articles, plus a number of book reviews penned by me that have been published in *FT* over the years, as well as the frequently-requested but hitherto long-misplaced transcript to my cryptozoology lecture that I gave at the very first *FT* UnConvention (held in June 1994), and the present book is the result. In terms of the material's text, however, I decided use my original submitted text rather than the version actually published in *FT*. This is because space requirements in certain issues of the magazine had meant that my items' text had sometimes needed to be edited down in order to fit the space allocated to it in those particular issues.

Consequently, what you will be reading in this volume are the unedited versions - which in some cases include Alien Zoo items that were edited down or were omitted completely (or which, again for space considerations, I wrote but did not submit), never appearing at all in the published Alien Zoo columns, or which occasionally were amalgamated into longer news reports appearing elsewhere in *FT*. More-over, in order to impose a well-defined chronological continuity upon the Alien Zoo columns, not only have I arranged them in separate yearly chapters, but within each chapter I have also arranged the items chronologically, by way of the dates of their cited sources.

Alien Zoo

In earlier books of mine that have compiled selections of my previously-published articles from various magazines, I have always sought to update the material if significant new findings or developments had occurred in relation to their subjects since their original publication. Conversely, Alien Zoo has always been a current cryptozoology news column, i.e. documenting and summarising the news stories using only the information available at the time of each column's preparation. Consequently, in the vast majority of cases it would not be appropriate to update the items here in this volume, because to do so would destroy their 'captured in time' format and appeal (in those few exceptions where it did seem necessary to offer an update, I have done so by inserting some extra information in square brackets within or at the end of the reports in question). Having said that, and as will be seen here, if some news stories reappeared in later media reports with new follow-up details, they were often duly returned to and updated accordingly in later Alien Zoo columns anyway. In contrast, as the Lost Ark articles were less tied-in to contemporary news reports, those included here have been updated as need be and as normal for articles of mine in compilation volumes, unless specifically stated otherwise.

Finally: a couple of caveats. Firstly: Although the internet has become an unparalleled source of information in modern times, it does have one major drawback. Whereas information printed in a book, periodical, magazine, or newspaper comfortingly remains there for all eternity, information in online websites has a disquieting tendency to disappear altogether without warning or change from one address (url) to another. Consequently, please note that some urls used and cited as sources for Alien Zoo news items (especially many of the earlier ones) at the time of my writing them no longer exist and hence are cited here only for historical interest. In cases where the url has merely changed, I have sought wherever possible here to provide not only the one that I originally used but also the new, current version (inserted in square brackets after the original one).

Secondly: Readers will notice that the system of measurements used in the Alien Zoo columns fluctuates back and forth between imperial and metric, rather than remaining consistent. This is because, as Alien Zoo is a news column, I always cite those measurements that were given in the original news reports (thus avoiding the need for conversions – which sometimes lead to approximations having to be used in place of the original, precise measurements).

I have thoroughly enjoyed writing for *FT*, and I very much hope to continue in this happy vein indefinitely. Hence it will come as no surprise to learn that I have derived equal joy from preparing this volume, which has rekindled so many memories for me while revisiting my earlier years of writing and research on such a vast range of cryptozoological and associated subjects. Consequently, I hope that it will be a source of interest and pleasant recollections for you too - and if, like so many others, you have shared the years with me as readers of my Alien Zoo column, Lost Ark articles, and book reviews, I thank you most sincerely for your very welcome company then, now, and in the future.

Meanwhile, if the weather outside is wet and windy, or even if it's fine but you've nowhere exciting to go, how would you like to visit a mesmerising realm of monsters and mystery beasts from the comfort and safety of your very own armchair?

Where can you meet alien big cats and snake-headed dogs, resurrected thylacines and death worms from the Gobi, nightgrowlers and goatsuckers, lake monsters a-plenty and sabre-tooths alive-o, paradise parrots and rainbow serpents, glowing lizards and donkey-eared deer, sachamamas and curupiras, whale-chomping sea monsters and murderous jellyfish, giant rats and New York sqrats, rock-painted mermaids and unicorn bones, river dragons and elephant birds, blood-sweating horses and squids from the swamps, man-beasts, müshmurghs, and mapinguaries, didis, dodos, dodus, and dobhar-chús, orang pendeks and albatwitchers, Nessie teeth, yeti hairs, and archangel plumes, pouched skunks and pig po-

Alien Zoo

Oriental river dragon lampstand
(Dr Karl Shuker)

nies, duendes and cureloms, horse-eels and globsters, flying snakes and cats with wings, scratch monsters and shell monsters, mystery quails from New Zealand and mini-men from Maine, tigers in Tanzania and lions in Chile, sex-mad super-otters and cow-snatching tiger trees, seal mothers, ghoul cats, mouse-whales, and other Icelandic exotica, the lost songbirds of Audubon and new species from New Guinea, mokele-mbembes, kuil kaaxs, nittaewos, banakons, Saharan crocodiles, Birdzilla...and who knows what else too?

In Karl Shuker's Alien Zoo, that's where! So what are you waiting for? Its gates are still open, so let's go inside - right now!

Ex umbris et imaginibus in veritatem - 'Out of shadows and phantasms into the truth'.

On headstone of the
Venerable John Henry
Cardinal Newman

THE SPECIAL YEARS.

He who travels far afield beholds things which lie beyond the bounds of belief; and when he returns to tell of them, he is not believed, but is dismissed as a liar, for the ignorant throng will refuse to accept his word, but must needs see with their own eyes, touch with their own hands.

Ludovico Ariosto – *Orlando Furioso*

The following two articles were specifically commissioned from me by *FT* for two *FT* special issues. The first of these issues, published in 2003, celebrated *FT*'s first 30 years via a series of thematic review articles; mine focussed upon the past three decades of cryptozoology, and in its published version it was combined by *FT* with an article by fellow cryptozoologist Loren Coleman in which he provided his own survey of this subject. The second issue, published in 2006, concentrated heavily upon cryptozoological subjects, with my article surveying the most significant crypto-events of 2005. Consequently, on account of their time-specific nature, I am reproducing each of my two articles here in its original submitted, unedited form, without any updating.

CARRY ON CRYPTOZOOLOGY
- 30 YEARS OF MYSTERY BEASTS WORLDWIDE.

Thirty years is a long time in any field of investigation and discovery - and cryptozoology is certainly no exception. Ably demonstrating this is that when the first issue of *Fortean Times* appeared in 1973, such remarkable beasts as the megamouth shark, Vu Quang ox, dingiso tree kangaroo, mimic octopus, Chao Phraya giant stingray, Indonesian coelacanth, two entirely new phyla of invertebrates (loriciferans and cycliophorans), giant muntjac, and at least two new beaked whales, not to mention the gigantic tube-dwelling vestimentiferan worms and the other bizarre members of the ocean-floor vent world fauna, were all still unknown to science. And as for the cryptozoological litera-

Alien Zoo

ture - many of today's standard books on the subject, by the likes of the Bords, Coleman, Costello, Eberhart, Krantz, Mackal, and yours truly, had yet to be conceived, let alone published. The greatest exception was of course Dr Bernard Heuvelmans's *On the Track of Unknown Animals* (referred to hereafter as *OTTOUA*), the most extensive single source of data on mystery beasts available at that time. Indeed, flicking through its pages and noting the current status of its contents' elusive beasts is a good way of assessing the progress of cryptozoological endeavour during the past three decades.

Sadly, none of *OTTOUA*'s major stars has been discovered. The yeti, tatzelworm, Madagascan giant lemurs, New Zealand moas, Queensland tiger, Nandi bear, bunyip, kongamato, and giant anaconda, for instance, all still linger in zoological obscurity. Even the African pygmy elephant, known from living specimens and superb video footage, is still denied official taxonomic recognition by some authorities. The best hope is for Sumatra's orang pendek or 'short man', whose scientific recognition is quite literally a hair's breadth away, thanks to hair samples whose DNA does not correspond with that of any tested species. In marked contrast, de Loys's South American 'ape', in spite (or, more probably, because) of the existence of a controversial close-up photo of a dead, propped-up specimen, is nowadays discounted even by many cryptozoologists as a sham.

Conversely, some of *OTTOUA*'s lesser stars have risen rapidly to 'A'-list status in the cryptozoological world since the early 1970s, most notably the mokele-mbembe. And the sasquatch or bigfoot, which did not even rate a mention in *OTTOUA* and was still little-known outside North America when *Fortean*

A selection of water monsters (Richard Svensson)

Times made its debut, is unquestionably one of the world's most famous mystery beasts today, As for sea serpents and lake monsters: the former nowadays have a much-reduced profile in comparison with media enthusiasm for them 30 years ago, while interest (and certainly belief) in Nessie probably peaked during the mid-1970s when it was formally dubbed *Nessiteras rhombopteryx* by Peter Scott and Robert Rines, and today has been somewhat eclipsed by Canada's Ogopogo and the bioacoustically-adept Champ in the USA as perhaps more credible cryptids.

Back on dry land, meanwhile, the vast media coverage and ostensible ubiquity of ABCs (alien big cats) in Britain initiated by the Exmoor Beast during the early 1980s could scarcely have been predicted a mere decade earlier (Surrey puma flaps of the 1960s notwithstanding), and has undoubtedly absorbed at least some of the public interest hitherto focused upon Nessie.

Most significant of all, however, is that during the past 30 years there has been an ever-increasing public, and even scientific, recognition that extraordinary new animals still can be (and are being) discovered. With the zoological revelations from Indo-China in the 1990s, and the regular unfurling of new South American monkeys over the past decade, plus an array of remarkable maritime finds, and a steady procession of serious expeditions into the world's remotest localities in search of such diverse and dramatic crypto-fauna as the orang pendek, Mongolian death worm, ivory-billed woodpecker, giant squid, Chinese wild-man, mokele-mbembe, Lake Seljord monster, and living dodos, it is evident that there will be plenty of cryptozoological events for *Fortean Times* to document during the next 30 years, and well beyond.

Certainly, as expressed by Heuvelmans himself, the great days of zoology are not done. Or, to put it another way, carry on cryptozoology!

ABSOLUTELY FABULOUS ANIMALS.

Doesn't time fly when you're having fun? Reading through Chris Moiser's ninki nanka article [in a selection of crypto features that followed this present one of mine in this *FT* issue], in which he refers to the carcase of the mysterious Gambian sea serpent washed up on Bungalow Beach back in the 1980s, I was shocked to realise that it is 20 years ago since my first full-length cryptozoological article was published - a two-part account of that self-same beached cryptid in a now long-defunct British magazine called *The Unknown* (my article was later republished in updated, expanded form by *FT*). It hardly needs stating that in the two decades that have passed since then, an awful lot has happened in the challenging world of cryptozoology, which has seen countless expeditions, discoveries, rediscoveries, new theories proposed, and old theories rejected.

One thing about this subject is that it is never dull or stagnant. Even a single year, as exemplified by 2005, can yield a diverse array of headlines - which also ably demonstrate the many fronts on which cryptozoology continues to advance. Take new discoveries and rediscoveries, for instance. True, the real stars of the subject, such as Nessie, bigfoot, and mokele-mbembe, continue to elude us, but the past 12 months or so have not been without sufficient surprises to encourage hope that even greater revelations may yet be made.

New discoveries since 2005 began have included a new snubfin dolphin from Australia, a new right whale, a major new mangabey monkey from Tanzania, an entirely new taxonomic family of rodents [later shown not to be new but resurrected] courtesy of the bizarre Laotian rock rat or kha-nyou

Alien Zoo

Laonastes aenigmamus, a novel species of woolly lemur named after Monty Python star John Cleese, a deadly new box jelly, and (yawn) even more new species of South American monkey. Rediscoveries were headed by the spectacular resurrection of North America's ivory-billed woodpecker *Campephilus principalis* (notwithstanding a few dissenting voices claiming that the observed bird in question was nothing more than a pileated woodpecker *Dryocopus pileatus* - personally, the evidence obtained for the existence of at least one living ivory-bill is too convincing for that to be likely), plus, as exclusively revealed in Loren Coleman's excellent cryptomundo website, a plausible sighting of the equally elusive, and even bigger, imperial woodpecker *C. imperialis*. And even as I write this piece (February 2006), the newspapers are full of reports concerning a dramatic tally of new and rediscovered species (including birds of paradise and bowerbirds) uncovered within a mist-shrouded 'lost world' high up in the remote Foja Mountains of Irian Jaya in western New Guinea [see my Lost Ark chapter 'A New Eden?' in this present book for my coverage of these finds].

Artist reconstruction of the orang pendek, based upon eyewitness descriptions (Tim Morris)

Of course, excluding those rare but magical instances when a notable new discovery is made entirely by chance, with no prior planning or search (e.g. the khanyou), cryptids will only be unveiled if a conscious effort is made to track them down. Hence it is good to know that the interest and willingness to seek such creatures, which in recent years has increased very considerably (and has been reflected accordingly in enhanced, serious media coverage), continues apace. And so in 2005, we saw expeditions to Mongolia in search of its formidable death worm, to Malaysia in search of its very own version of bigfoot, to the Himalayas for yeti data by a fact-finding team from Disney, and plans drawn up for quests in 2006 to Central Africa in pursuit of the mokele-mbembe and other Congolese cryptids, Chris Moiser's ninki-nanka investigations in West Africa, and forays after man-beasts in South America.

Reaffirming an opinion aired half a century ago by Dr Bernard Heuvelmans, I have little doubt that if any major cryptozoological discoveries remain to be made on land, they will occur in the remote, mountainous rainforests and swamplands of regions like tropical Africa, Madagascar, Asia, South America, New Guinea, and northwestern Australia. So it is very encouraging to know that there are still researchers and investigators out there willing to pit themselves against the inhospitable conditions of such localities in order to

Alien Zoo

expose their secrets. Moreover, it is no longer being left entirely to enthusiastic, self-funded laymen to do the hard work. Fully-equipped, sizeable scientific explorations in search of new species – at one time every bit as scarce as the creatures themselves! - are nowadays becoming ever more frequent, and successful.

Of the wide assortment of cryptids currently being sought, I still feel that the likeliest to be revealed and formally confirmed by science as a major new species is the Sumatran orang pendek - the subject of two expeditions by the CFZ as discussed here in Richard Freeman's article, as well as an ongoing WWF-associated search featuring veteran orang pendek seeker Debbie Martyr. Needless to say, the revelation in 2004 that the nearby island of Flores was home until at least as recently as 12,000 years ago to a species of dwarf human, Flores man *Homo floresiensis* (popularly dubbed the hobbit), and that there are even modern reports of such beings (ebu gogo) still existing here, have done the orang pendek's credibility no harm at all - creating a significant precedent for a 'short man' (the translation of 'orang pendek') elsewhere in the Sundas. Furthermore, even before the bones of *H. floresiensis* had been discovered, unique, unclassifiable hair samples and footprints purported to be from the orang pendek had been obtained, giving its status as a valid undiscovered species considerable weight.

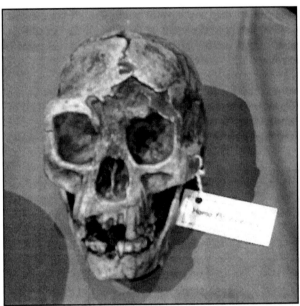

Life-size skull replica of *Homo floresiensis* (Dr Karl Shuker)

Of course, for every convincing piece of cryptozoological evidence procured, many much more enigmatic, ambiguous examples are also proffered, and 2005 once again did not disappoint. So take your pick from a motley collection of crypto-curiosities that included some controversial unpublished photos purporting to show the back of a living thylacine or Tasmanian wolf *Thylacinus cynocephalus*; a highly controversial 'tooth' said to have been found embedded in a half-eaten deer carcase at Loch Ness; an intriguing video of an alleged bigfoot filmed at Nelson River in Manitoba; a carving from Cameroon ostensibly depicting the horned swamp-dwelling emela-ntouka or killer of elephants; and photos in India's Kerala State of pygmy elephants known locally as kallaana.

More satisfactory was the procurement of a DNA sample confirming the long-claimed but hitherto-unproved existence of at least one moose *Alces alces* existing in the forests of New Zealand; the vindication of a long-held belief of mine that the European eagle owl *Bubo bubo* is now breeding in the UK; the first filming of a living giant squid (by Japanese scientists off the Ogasawara Islands); and the unexpected photographing of a mysterious civet-like beast in Borneo that may prove to be a virtually unknown species called Hose's palm civet *Diplogale hosei*.

There is little doubt that DNA techniques offer the way forward in determining the zoological identity of biological material that previously would – and did – remain unclassifiable and contentious. Globsters, for example, which were once the bane of marine biologists and the delight of monster devotees,

Alien Zoo

can now be readily identified via their DNA in spite of their bizarre morphological appearance. In addition, motion-sensor cameras, which the Bornean mystery beast triggered, may well snap some further surprises in the remote jungles in which they are being set up by field researchers investigating these regions' biodiversity. Indeed, this, I feel, offers the best hope for obtaining conclusive evidence for the survival of the thylacine and certain other highly reclusive, medium-sized quadrupedal beasts.

Having said that, searches for cryptids, analysing physical remains claimed to be from cryptids, and photographing cryptids all rely on the – very big – assumption that these creatures are real. But one of the major problems facing any cryptozoological investigation is determining whether the mystery beast in question is indeed real, as opposed to wholly mythical, or even paranormal.

The articles presented here span this entire spectrum – from the indisputably zoological orang pendek and the semi(?)-mythical ninki nanka to what appears to be a wholly folkloristic water monster of Italy's Lake Maggiore (echoing in many ways the long-running saga of China's Lake Tianchi monster, represented by a further flurry of sightings in 2005), and reports of Australian yowies that feature certain decidedly preternatural aspects.

Some cryptozoologists frown upon the merest suggestion that any cryptid could exhibit paranormal behaviour, but this is a facet of cryptozoology that has surfaced time and again through the decades in relation to a range of different creatures, and is a subject as worthy of serious attention and recognition as the creatures – or creature-like entities? – themselves.

Lastly, it is a pleasure to see so many worthy new additions to the cryptozoological literature in the form of books, articles, websites, and other sources of information, for 2005 was certainly rich in all of these. As I stated in a recent *FT* book review, cryptozoology seems to have entered a new golden age of published literature, with more information on hand than ever before for the enthusiastic newcomer and the knowledgeable veteran alike.

With my copy of *Fabulous Animals* (Dr Karl Shuker)

Back when my interest in cryptozoology began, over 30 years ago, there were far fewer readily accessible sources available, but there was one slim book (which I still have) and its accompanying television series that whetted my appetite just as much in those days as any of the major crypto tomes around today would do now – which is why I am so pleased to see Martin Gately's article, recalling to a new generation, and bringing back fond memories to readers from mine, the sheer delight and fascination that was David Attenborough's *Fabulous Animals*.

Why this enchanting series was never repeated or released on video (or, as yet, DVD) never ceases to amaze me, because it was a true

Alien Zoo

landmark in cryptozoological television. BBC Archives, take note – and re-issue!

Little did I know when watching that pioneering crypto-show 30 years ago that a decade later I would be making my own first contribution to its subject, and that two decades after that I'd be writing this introduction to the following selection of articles heralding another promising crypto year. But that is the joy of cryptozoology – and long may it continue!

[NB – since the original article constituting this chapter was written, the common consensus regarding the Bornean mystery mammal is that it is not a civet of any kind but is much more likely to be a species of large flying squirrel, *Aeromys thomasi*, already documented by science, although not a well-known species.]

Alongside a life-size model of megalodon shark jaws at Lyme Regis, Dorset (Dr Karl Shuker)

Plesiosaur carving in Nothe Gardens, Weymouth, Dorset (Mark North)

Holding my copy of *The Monster Trap and Other True Mysteries* alongside an inquisitive, extra-large *Velociraptor* (Dr Karl Shuker)

The Lost Ark

WHATEVER HAPPENED TO WATTIE? IN SEARCH OF LOCH WATTEN'S MISSING MONSTER.

It was a bright spring day, and steely sunshine glinted over the mountains of Caithness when Colonel Arthur Trimble first saw the monster of Loch Watten. The monster's eyes were slits in a huge squat head, and its body, which loomed under the rippling water, appeared at least 20 feet wide. It observed him for several seconds. He even had time to take a photograph of it.

John Macklin – 'The Trap He Set Was For A Monster...',
Leicester Mercury, 28 March 1966.

I first learnt about the existence of winged cats (see pp. 241-254) – which subsequently became an investigative passion of mine - when, as a teenager, I read a fascinating little book by prolific author Peter Haining entitled *The Monster Trap and Other True Mysteries* (1976). That same book introduced me to a couple of other subjects that I have since pursued in depth too – the Green Children, and the mysterious mini-mummy of Wyoming.

Ironically, however, the chapter that interested me most of all (and which gave its title to the entire book) was also the one that has mystified me most of all – because, over 30 years later, and in spite of the fact that it is potentially of immense cryptozoological significance, its subject has resisted every attempt made by me to uncover any additional details regarding it. Consequently, I feel that it is now time to give this whole perplexing matter a long-overdue public airing.

THE (VERY) MYSTERIOUS CASE OF THE MONSTER TRAP

The setting for the truly extraordinary episode documented in this chapter is Loch Watten – a Scottish freshwater lake in Caithness's River Wick drainage system. Its grim tale as given in Haining's book (in which the chapters' stories, although all allegedly true, are written up in a dramatised, novel-like style) can be summarised as follows.

Loch Watten (Flaxton/Wikipedia)

According to Haining, the incident in question took place some 10 years before the flap of Nessie sightings in 1933, and featured local estate owner Colonel Arthur Trimble (who had retired in 1922 from the British army). It all began on the morning of 21 April 1923, when Trimble was walking his spaniel, Bruce, by the lochside, not far from his estate. He had a camera with him, as it was a pleasant morning and he hoped to take some photographs. After reaching his usual point for turning back, Trimble called to Bruce, who had run some distance further ahead, and after waiting for him to come back, Trimble looked out across the loch – where, in Haining's words:

> Something dark and looming had suddenly appeared on the surface of the water.

Alien Zoo

> The Colonel squinted his eyes and raised his hand to half-shade his face. The form was clearer now. It looked like a kind of neck with a huge flat head.
>
> Keeping quite still, he looked harder and could see that it was indeed a head and neck, and that there were slit eyes staring directly at him. Below the surface of the water he could make out the shape of an immense body, at least twenty feet wide.
>
> Colonel Trimble could hardly believe the evidence of his senses. It seemed like some huge water monster.

Haining stated that although the monster was less than a hundred yards away, thanks to his years of army discipline Trimble did not panic, and lifted his camera. Just as he was about to take a photograph, however, his dog Bruce spied the monster and immediately ran towards it, barking loudly. Startled, the monster disappeared beneath the water almost at once, but at that same instant Trimble succeeded in snapping a single photo, although he had no idea whether he had actually captured the beast's image. When he and Bruce arrived back home and he told his housekeeper, a local woman called Mrs Doris Dougal, what had happened, she confirmed that he had seen the loch's legendary 'serpent', and suggested that he report his sighting.

That same day, Trimble took his camera's film to the local chemist shop for developing, and when he collected his photos two days later he was delighted to discover that although the picture snapped by him at the loch was slightly blurred, it did indeed depict the monster's head and neck above the water surface. Consequently, that afternoon he penned an account of his sighting for London's *Times* newspaper, enclosed with it a copy of his photograph, and posted it a few hours later. From then on, Trimble visited the loch daily, in the hope of seeing and photographing the monster again, but leaving Bruce at home to ensure that he didn't cause any disturbance this time if the monster should reappear.

Unbeknownst to Trimble, however, on 1 May, while he was once again at the loch, Bruce managed to sneak out, and when Trimble returned home later that day he was met by Mrs Dougal with the disturbing news that Bruce was missing. The two of them spent some time searching for the dog locally, but to no avail – until Trimble saw a man approaching from the direction of the loch. The man was Trimble's nearest neighbour, the local doctor Robert McArdish, who told Trimble that he had spied Bruce swimming in the loch – but just as the doctor had been about to call out to him, he had seen a flurry in the water, as if something else was also there, and then the dog disappeared, after which the waves settled again, but with no sign of Bruce.

Enraged by the apparent killing of his dog by the monster, the following day Trimble set about building an extraordinary 'monster trap', consisting of 50 fathoms of rope attached to an enormous sharpened spike of steel that had been shaped into a massive hook. Trimble baited this hook with a large piece of freshly-purchased horsemeat, and after rowing into the middle of the loch in his dinghy he lowered the hook into the water, attached a marker buoy to the end of the rope, and dropped it overboard. Then he rowed back to shore, and returned home.

The next morning Trimble went out to inspect the trap, but it had not been touched, so he repositioned it elsewhere in the loch, and came back home. This procedure was repeated up until the evening of 4 May, when he informed Mrs Dougal that he was going out to the loch again, even though it was almost dark. Just on 9.30 pm, after looking outside to see whether he was returning as he was late, Mrs Dougal suddenly heard a single loud, terrified scream, from the direction of the loch. Racing outside to the gardener's cottage close by, she hammered on his door, explained what had happened, and the two of them ran fearfully to the loch. There, in some reeds at the lochside, was the half-submerged body of Trimble, and as they looked down at it, they saw to their horror that his chest had been pierced by the giant hook, which was still attached to the rope. And as they stood there, they heard something:

> ...something that turned their blood to ice – and haunted them for the rest of their days.

> It was a sound which came from the loch. The sound of something large that splashed as it swam away from the shore...

And with that dramatic little flourish, there endeth Haining's tale of the Loch Watten monster (let's call it Wattie, for short).

WHITHER WATTIE?

Needless to say, one would imagine that such an episode, far more sensational than anything that even Nessie can lay claim to, would have subsequently featured in every major (and minor!) cryptozoology publication as a matter of course, as famous – or infamous – as the story of the Surgeon's Photograph and other endlessly rehashed and recycled cryptozoological histories. Yet nothing could be further from the truth.

Indeed, I have yet to discover a single mention of the Wattie history anywhere – I know of no book, periodical paper, magazine article, newspaper report, or website that contains even the briefest reference to it. Moreover, the only acknowledged claim to fame of Loch Watten, other than having been formally designated as an SAC (Special Area of Conservation), is that it is a good body of water for fly-fishing for brown trout. In stark contrast, any celebrity status as a monster-haunted lake is conspicuous only by its absence. So how can such anomalies be explained?

Let's look at some background information, beginning with a few additional details supplied by Haining himself. In his book's introduction, he stated that when selecting stories to be covered by him in it, he didn't want to repeat ones that lots of other writers had already utilised. Instead, he decided:

> I would use stories that had particularly fascinated me in which I had done considerable research, if not actually visited the places in question.

Furthermore, in the opening to the 'Monster Trap' chapter itself, he stated that although Nessie was certainly the most famous Scottish monster, she was not the only one, noting that there were stories of water horses and serpents from many other Highland lochs, and then commenting:

> One particular monster story has always fascinated me, but amid all the fuss about 'Nessie' it rarely gets mentioned.

For 'rarely', substitute 'never'!

Alien Zoo

Yet according to Haining's book, local people claim that there have been stories of a monster, which they term 'the serpent', in Loch Watten for many years, but no documentary records of actual sightings prior to Trimble's ultimately fatal incident. Is this true? Never having visited the loch myself, which is only 14 miles from John O'Groats in the far northeast of Scotland, I have no idea whether there is any verbal tradition of a monster here (though I have yet to communicate with anyone versed in Scottish mythology or cryptozoology who has ever heard of such tales). However, I would have expected at least some documentation of it, were such a tradition to exist. After all, as Haining correctly pointed out, there are accounts of monsters for a number of other lochs – including Ness, Morar, Oich, Lochy, Shiel, Arkaig, Lomond, Beiste, Quoich, and Trieg.

Another anomaly concerns Loch Watten itself. Despite being the second largest of Caithness's lochs, it is under three miles (4.65 km) long, less than a mile (1.6 km) across at its widest point, and boasts an average depth of only 10-12 ft (2.5-3.0 m) – a very far cry from the immeasurably greater size of Loch Ness, Loch Morar, and other notable bodies of Scottish freshwater associated with monster traditions. If the kind of huge reptilian monster (at least 20 ft *wide* – so how *long* was it?!) allegedly encountered by Trimble were truly real, it would surely require a much more substantial aquatic domain than Watten.

Nor do these inconsistencies constitute the full extent of my concern for the validity of Wattie as a bona fide cryptid. When I first attempted to research this subject, back in the early 1990s, I wrote on two separate occasions to Haining, having obtained his correct address, but I never received a reply to my requests for information, and as he died in 2007 this most direct line of investigation is no longer an option. In addition, I met with a succession of dead-ends when attempting to uncover any Trimble-related leads (not even trawling through death registers and army records online elicited any evidence for his supposed former existence). I also searched meticulously through the relevant period of back issues for *The Times*, but did not find any published letter or photo by Trimble.

In short, the only known source of information (to me, at least) concerning Wattie is Haining's book, and, therefore, Haining himself – which to my mind is the most disturbing aspect of all concerning this mystifying tale. The reason why I say this is that some of Haining's other publications have already attracted considerable controversy in relation to the validity – or otherwise – of their claims.

SPRING-HEELED JACK, SWEENEY TODD, AND WOOLPIT'S GREEN CHILDREN – THREE REASONS FOR WATTIE WORRIES

For example: in a detailed paper on Spring Heeled Jack (*Fortean Studies*, vol. 3, 1996), Mike Dash revealed that he was unable to obtain independent corroboration of various accounts and details that had been published by Haining in his book on this subject (*The Legend and Bizarre Crimes of Spring Heeled Jack*, 1977). And even an engraving claimed by Haining to show the recovery from a marsh of one of Jack's victims – a victim, incidentally, undocumented by anyone else – in reality showed no such thing.

Moreover, when Mike Dash wrote to him asking for sources, Haining replied that he was unable to supply any because all of his research material had been loaned to a film scriptwriter who had subsequently vanished. Not surprisingly, perhaps, in his paper's annotated bibliography, Mike made the following comments regarding Haining's book:

> The only full-length work on the subject is a curious hodge-podge of the

accurate, the overtly-dramatised and the invented...it repeats many existing errors, creates new ones, and is so single-mindedly determined to fit evidence to the theory that Jack was the Marquis of Waterford that it does not flinch from introducing made-up evidence to support this case.

Spring Heeled Jack (Richard Svensson)

Equally controversial are Haining's books on Sweeney Todd, Fleet Street's homicidal hair-snipper. Although Todd is widely assumed to be an entirely fictitious character spawned by the Penny Dreadfuls of Victorian times, Haining published two book-length treatments, respectively entitled *The Mystery and Horrible Murders of Sweeney Todd, the Demon Barber of Fleet Street* (1979) and *Sweeney Todd: The Real Story of the Demon Barber of Fleet Street* (1993), in which he alleged that such a person had actually existed. However, this claim has attracted much criticism, for a variety of reasons, including those summarised succinctly in Wikipedia's entry for Haining (as accessed by me on 2 July 2009):

> In two controversial books, Haining argued that Sweeney Todd was a real historical figure who committed his crimes around 1800, was tried in December 1801, and was hanged in January 1802. However, other researchers who have tried to verify his citations find nothing in these sources to back Haining's claims. A check of the website 'Old Bailey' for "Associated Re-

cords 1674-1834" for an alleged trial in December 1801 and hanging of Sweeney Todd for January 1802 show no reference; in fact the only murder trial for this period is that of a Governor/Lt Col. Joseph Wall who was hanged 28 January 1802 for killing a Benjamin Armstrong 10 July 1782 in "Goree" Africa and the discharge of a Humphrey White in January 1802.

In short, there are notable precedents when faced with questioning the reliability of claims made by Haining in the absence of any independent sources of evidence to examine. Even in another chapter of *The Monster Trap,* documenting the Green Children, it is curious to note that the famous, historically-recorded incident of the Woolpit Green Children receives no mention whatsoever. Instead, Haining devotes the entire chapter to an exceptionally similar version allegedly occurring several centuries later in Spain – a version subsequently revealed by other researchers to be a complete fabrication, by person (s) unknown, directly inspired by the Woolpit episode.

And I hardly need point out that Haining's description of Trimble's supposed photo – slightly blurred but showing a head and neck – is more than a little reminiscent of the Surgeon's Photograph of Nessie. Also worth remembering is that aside from his non-fiction books, Haining was a well-respected, extremely knowledgeable anthologist of horror and mystery short stories of fiction.

WAS THERE EVER A WATTIE? OVER TO YOU!

It gives me no pleasure whatsoever in questioning the legitimacy of the Wattie affair as documented by Haining, especially as the book in which it appears is one that has been instrumental in introducing to me various other subjects that have since become significant in my own researches – and I would therefore be delighted if my concerns regarding this case could be convincingly dismissed. Yet it is clear that the omens for Wattie's validity are not good.

Nevertheless, it would be rash to deny this tantalising tale out of hand without having first given an opportunity for it to be investigated publicly. So here, gentle readers, is where you come in. If there is indeed anyone out there with direct or indirect, integral or background information relating in any way to monsters reported from Loch Watten, and to the Trimble incident in particular, I'd love to hear from you.

Similarly, if Haining's research files have been preserved, any details of where and whether they can be accessed would be very welcome. After all, if we are to believe his claim that all of the subjects in his book were ones in relation to which he had conducted considerable research, these archives undoubtedly offer the most likely source of primary and additional data concerning this most monstrous of Scottish crypto-mysteries.

FOUR WATTIE UPDATES

The publication by *FT* in its September 2009-dated issue (but on sale in August) of my Wattie article forming the basis of this chapter has triggered four notable responses to date.
The first of these was an email of 28 August 2009 that I received from Rod Williams of Talgarth, Wales:

Alien Zoo

I am a regular reader of *Fortean Times* and your item on Wattie was interesting but feel that it was a concocted tale by Peter Haining.

I have read Hugh Miller's *Scenes and Legends of the North of Scotland* and also Samuel Smiles's biography of Robert Dick [*Robert Dick, Baker of Thurso, Geologist and Botanist*, 1878], baker, biologist (botanist mostly) and geologist. A man who walked many miles at night over large parts of Caithness.

I cannot recollect in either book mention of Loch Watten or a/its monster. Both men were not above mentioning curious tales, particularly Miller who was well into hauntings and weird happenings; apart from being a quarryman turned geologist he seemed to thrive on such tales.

I may have missed any reference of course but the book of Miller's can be read on line for free should you wish to check it out.

Not sure of Dick's biography being on line but probably is.

George Borrow's *Wild Wales* (circa 1854) mentions 'crocodiles' in Welsh lakes or rather stories of these mythical beasts and enquires of people on his journey whether they knew of any local legends relating to these little lakes and crocs.

Again I don't remember any specific stories as it has been many years since reading the book. I think I need to re-read it sometime.

Our local lake Llangorse Lake has large pike in it and one chap told me that when he was wind surfing and was stood in the lake (shallow in many places) something large brushed his leg.

Quite apart from confirming the absence of Wattie information from some literary sources new to me, Rod's email is also of value for the interesting snippets of information concerning Wales's mystifying water 'crocodiles', which I've read about in a number of publications and which deserve a detailed examination in their own right.

The second response was a letter penned by German cryptozoologist Ulrich Magin, which was published by *FT* in November 2009. In his letter, Ulrich revealed that Haining's account of Wattie was almost identical to a tale included by French fiction writer George Langelaan in *Les Faits Maudits* (not *Maufits* – as erroneously titled in Ulrich's letter) or 'Cursed Facts' - a book of forteana published in 1967, containing an eclectic mixture of retold press clippings and fictional stories. Langelaan claimed that his source for that particular tale was a *Times* news report from May 1932, but a search for it undertaken by Ulrich failed to unearth any such report.

The third, and most significant, response was a letter that I received from *FT* on 31 March 2010, which had been written to me on 23 March by Lance Shirley of Cornwall and was accompanied by a remark-

Alien Zoo

able enclosure – a photocopy of an article that had been published in the *Leicester Mercury* newspaper on 28 March 1966 in what appears to have been a regular, long-running series of articles published under a 'Stranger Than Fiction' banner. Written by a John Macklin, the article was entitled 'The trap he set was for a monster...but it was the colonel who died'. Reading it through, I discovered that its content and wording were so similar to Peter Haining's chapter 'The Monster Trap' that it seemed highly likely either that Haining had directly copied Macklin's account or that he and Macklin were one and the same person.

As I learnt from Mike Dash, Haining is known to have written under various pen-names as well as his own, so could John Macklin be yet another one? After receiving Lance's letter and enclosure, I googled John Macklin on the internet, and discovered that just like Haining, he is/was a prolific author, and, again just like Haining, has authored many popular-format compilation books of supposedly true mysteries. Just another coincidence?

In his letter to me, Lance mentioned that he and his family had lived in Caithness, near to Loch Watten, from 1966 to 1976, during his childhood. While still living there in the early 1970s, he had read the *Leicester Mercury* article, which had belonged to his mother (it had been forwarded to her for its interest value from her father, who lived in Loughborough and always bought this newspaper), and was excited to think that such a creature may live so close to them. Whenever they passed the loch in the car, they always scanned the surface, just in case they could catch sight of the monster. Upon reading my *FT* Wattie article in September 2009, Lance realised that Haining's account matched what he could still recall from that newspaper cutting from long ago. Moreover, while subsequently clearing out the loft in the family home, he was delighted to rediscover it - yellowed with age but still intact - stored inside a biscuit tin crammed with other cuttings (including another John Macklin 'Stranger Than Fiction' article from the *Leicester Mercury*, this time dating from 1969 and documenting a ghostly occurrence in Hoy Harbour).

As Lance points out, what is so interesting is that the *Leicester Mercury* article predates not only Haining's book (by 10 years) but also that of George Langelaan (by a year). Consequently, it now seems that Langelaan did not originate this tale after all. Regardless of who did do so, however, no independent, substantiating evidence for its veracity or the existence in Loch Watten of a mysterious creature has ever come forward. Consequently, in my opinion the most reasonable conclusion remains that Wattie is a complete invention.

Irrespective of this, after receiving Lance's letter I lost no time in pursuing the Macklin line of investigation further. My ultimate goal was the procurement of some current contact details if he is still alive (in which case, of course, he and Haining could not be the same person!); or, if he is dead, uncovering as much biographical information concerning him as possible, in the hope of determining conclusively whether or not John Macklin was indeed merely another pen-name of Peter Haining.

In April 2010, I emailed an enquiry to Macklin regarding Wattie via Sterling, the American publisher of the most recent Macklin book that I have yet been able to trace (a children's book of true ghost stories, published by Sterling in 2006). So far, however, I have yet to receive any response from him.

Moreover, to me it seemed undeniably thought-provoking that whereas Macklin and Haining are/were both extremely prolific authors who wrote on extremely similar subjects, none of Macklin's works are cited in the bibliographies of any of Haining's books accessed by me (or vice-versa). Equally, I have been unable to trace any indication that Macklin has published any books or articles in the years following Haining's death. And whereas photos of Haining are readily obtainable by googling his name,

Google is currently (as of May 2010) unable to locate a single photograph of Macklin. Also, whereas Haining has a detailed entry in Wikipedia, Macklin (despite being a comparably prolific – and hence successful - author) has no entry whatsoever.

So were Haining and Macklin the same person, with Wattie merely the figment of an inordinately prolific writer's fertile imagination? I was soon to discover the answer, which provided yet another unexpected surprise.

On 8 May, I received the fourth response to my enquiry for Wattie information. This time it was a highly illuminating email from none other than *FT*'s own Paul Sieveking, who informed me that John Macklin was indeed a pseudonym – but not of Peter Haining! Instead, it was one of many pen-names used by another author of popular-format writings on mysteries – Tony James. The plot thickens! So did Tony James originate the storyline for the Wattie tale, or is there an even earlier version out there somewhere that he had read? If anyone has current contact information for James, I'd like to hear from you!

Meanwhile, my sincere thanks go to Rod Williams, Ulrich Magin, and especially Lance Shirley and Paul Sieveking for shining some important light upon this increasingly complicated mystery, and I am intrigued to see if any new developments will occur in the future. After all, as a certain cult television series used to proclaim, the truth is out there – it's finding it that's the problem!

REFERENCES

DASH, Mike (2010). Pers. comms, 1 & 2 April.
HAINING, Peter (1976). *The Monster Trap and Other True Mysteries*. Armada (London).
LANGELAAN, George (1967). *Les Faits Maudits*. Encyclopédie Planète (Paris).
MACKLIN, John (1966). The trap he set was for a monster...but it was the colonel who died. *Leicester Mercury*, 28 March, p. 7.
MAGIN, Ulrich (2009). Wattie. *Fortean Times*, no. 255 (November): 69.
SHIRLEY, Lance (2010). Pers. comms, 23 March & 2 April.
SIEVEKING, Paul (2010). Pers. comm., 8 May.
WILLIAMS, Rod (2009). Pers. comm., 28 August.

ALIEN ZOO

1995-7

Please note: My very first Alien Zoo column was faxed to *Fortean Times* on 3 March 1997; however, in the first couple of columns I purposefully included a few items from a year or so earlier, because of their particularly interesting nature yet previous lack of widespread publicity, thus explaining why the first six items here date from 1995 and 1996.

GIANT RAT WITH PAWS OF CAT? Belated news has reached me of a peculiar-sounding beast captured some time in early 1995 on the outskirts of Beijing, China, by two soldiers belonging to the Beijing paramilitary armed police while on patrol. According to the Xinhua News Agency, this still-unidentified animal was a huge, toothy, rat-like creature measuring 2 ft long and weighing 8 lb, with "paws like a cat". It also sported a tail "as thick as a man's thumb". ***Whiteboard News*, 23 May 1995**

IN THE SHADOW OF THE NIGHTGROWLER. Another fascinating snippet of belated news: According to inhabitants of Goldsborough Valley, south of Cairns in Queensland, Australia, this region is currently being prowled by a mysterious, exclusively nocturnal creature whose mere presence in the vicinity is enough to terrify the local dogs, and which earns its name from its deep, ferocious growls.

People who have heard these are convinced that they are not of canine origin, thereby ruling out dogs and dingoes, but are unquestionably feline, reminding them of a tiger's sounds. Speculation is rife that the elusive nightgrowler may be one and the same as the mystifying yarri or Queensland tiger, which is banded with black and white stripes, has a feline head and tusk-like teeth, and is believed by some cryptozoologists to be a surviving member of the officially extinct lineage of thylacoleonids or marsupial lions. A team of investigators, including local naturalists and a ranger, and led by Pat Shepherd, is keen to capture a nightgrowler, in the hope of finally unmasking its cryptic identity. ***Queensland Sunday Mail*, 12 November 1995**

Alien Zoo

Representation of Queensland tiger (William Rebsamen)

A SURFEIT OF CIVETS? What may well be the latest significant cryptozoological revelations from Vietnam have been reported by Dr Pham Nhat of the Forestry University, Xuan Mai, in the journal *Lam Nghiep*. According to Pham Nhat, two highly unusual civet specimens have been documented from Lao Cai Province in the far northwest of Vietnam. Could they comprise a species or subspecies new to science? While on the subject of strange civets: in June 1986, tropical agriculturalist Tyson Hughes visited the little-explored Indonesian island of Seram (Ceram) in the Moluccas group. During his work there for the VSO, he received reports from the local inhabitants concerning several mystifying, unidentifiable beasts.

The most tangible evidence obtained by him for any of these, however, was the furry tail of a mysterious mammal, about 18.5 in long and encircled by a series of dark rings. According to the natives, it was from a beast that was half-dog and half-cat, and when he presented it to the WWF's Indonesian branch for examination, officials suggested that it may be from an unknown cat. Judging from the natives' description of its former owner, however, I consider a viverrid, specifically a civet, identity rather more likely, because the description is very reminiscent of the giant palm civet *Macrogalidia musschenbroekii*, indigenous to the nearby island of Sulawesi. ***Lam Nghiep*, 2/1995, p. 22 (1995); Bill Gibbons, pers. comm., 29 July 1993.**

Alien Zoo

PARLEZ-VOUS PUMA? A large puma-like ABC (alien big cat) was reported so frequently during spring 1996 in the Forêt de Chizé that this highly popular French tourist area south of Niort had to be closed to all visitors on safety grounds. Beaters, live and dead bait, and even a female puma all failed to entice it from its 13,000-acre forest hideaway. ***The Times*, 18 May 1996.**

WHIPPING UP A JUMPER? For decades, a mysterious belligerent snake said to be able to jump into the air has been reported from the environs of Sarajevo and elsewhere in the former Yugoslavia. Its identity has never been ascertained, but eminent British zoologist Professor John L. Cloudsley-Thompson has now suggested that it may be *Hierophis viridiflavus*, the western whip snake, which is a very fast-moving and aggressive species. ***British Herpetological Society Bulletin*, no. 55, 1996.**

SNAKE-HEADED DOG? On 24 May 1996, a bizarre dog-like beast 4-5 ft in total length, but with a sleek serpentine head, red reptilian eyes, a slender 24-30-in neck, shaggy black fur, long hind limbs, shorter forelimbs, and no tail at all, ran out in front of the car driven by Sheila Charles as she was taking her son Shane to school in Magalia, California. She swerved to avoid hitting the creature, and veered out of control into a canyon. Fortunately, no-one was badly injured, and the mysterious beast's reality was later confirmed by the driver of the car following her, who had also seen it. ***UFO News World Report*, 1996.**

Snake-headed dog, based upon eyewitness description

ONE EEL OF A DISCOVERY? In June 1996, a TV news bulletin broadcast in Oregon allegedly carried a report describing the recovery from waters somewhere off the American Pacific Northwest of a dead giant eel-like animal. It measured 25 ft long, and possessed genuine bones rather than mere cartilage. Could this have any relevance to sightings of the elongate sea monster often reported from this region and known as Caddy or Cadborosaurus?

IN QUEST OF THE GIANT SQUID. Despite being among the world's lengthiest invertebrates, measuring over 60 ft long, the southern giant squid *Architeuthis longimanus* is known entirely from dead specimens: it has never been seen in the living state. In February 1997, however, an international

Alien Zoo

scientific team, led by Dr Clyde Roper from the U.S. National Museum in Washington, plans to launch a six-week quest for sightings of this elusive marine monster. The team will be exploring the deep trenches off South Island, New Zealand, armed with underwater video cameras, and plenty of fish bait. ***Daily Telegraph*, 30 January 1997.**

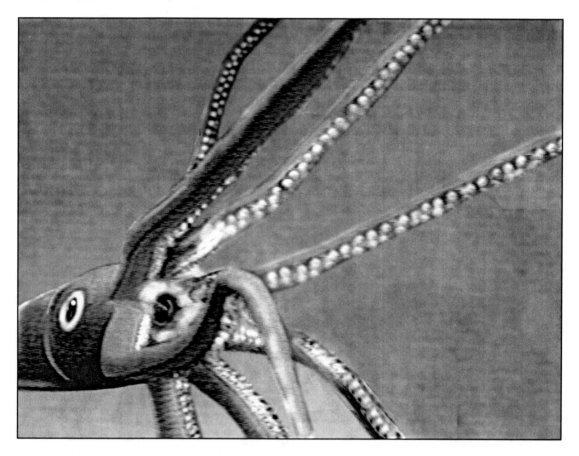

Giant squid painting (William Rebsamen)

A MYSTERY FROM MASBATE. In February 1997, the central Philippine province of Masbate was the focus of intense cryptozoological interest, following the news that the carcase of a mysterious sea monster had been washed ashore here on Christmas Day 1996 in the coastal town of Claveria. According to press reports, the carcase was of a 26-ft-long eel-like creature with a turtle-like head. A photograph of its preserved skull, vertebrae, and limbs was published in the *Philippine Star*. This photo (which I have not seen so far), together with portions of its dried flesh, was presented to various unnamed experts for examination, but no-one could identify it. According to Philippines University zoologist Dr Perry Ong: "Judging from what I see now, it's an eel-like fish. It must be an ancestral or primitive fish. It had fins. But if it's a fish, where are the ribs? It is not a mammal". Curiously, however, its head apparently possessed a blowhole-like orifice, resembling that of a dolphin, though it lacked the latter's characteristic elongate beak (rostrum). Perhaps the most mystifying aspect of all concerned the alleged suggestion by some 'scientists' that its bones should be carbon-dated "to determine if the animal is prehistoric". Yet as the creature evidently died only very recently, as evinced, for instance, by the presence of dried flesh still attached to its carcase, it is clearly not prehistoric. So why on earth would any scientist propose carbon-dating it?! *South China Morning Post*, **25 February 1997.**

HOW DEAD IS THE DODO? The Plain Champagne is a rarely-visited area of rainforest stretching down to the coast on the island of Mauritius. And it is here, at dawn and dusk, where some of its occasional modern-day human visitors have claimed to have spied strange birds bearing more than a passing resemblance to this island's most famous deceased species - the dodo *Raphus cucullatus*, which officially became extinct in 1681. Needless to say, the possibility that dodos have survived undetected by science into the present day, even in such a secluded locality as this one, is exceedingly slim, to say the least. Nevertheless, it needs to be checked out, just in case - a worthy task soon to be undertaken by cryptozoological explorer Bill Gibbons, leader of the 'Operation Congo' expeditions seeking the dinosaurian mokele-mbembe, who now plans to visit this mysterious plain in July 1997. **Bill Gibbons, pers. comm., March 1997.**

YOU CAN'T KEEP A GOOD THYLACINE DOWN. The thylacine or Tasmanian wolf *Thylacinus cynocephalus* officially died out in New Guinea around 2000 years ago...but no-one seems to have mentioned this to the thylacine. In late March 1997, reports emerged from Irian Jaya, western New Guinea, claiming that a creature closely resembling a thylacine had been attacking the livestock of villagers from Oksibil and Okbibab in the Jayawijaya regency. These claims were examined, and accepted, by officials sent to this area by its regent, J.B. Wenas; the villagers were too scared of the animal to attempt catching it. A striped, dog-headed beast with a long stiff tail and a mouth with a huge gape (all thylacine characteristics) has been reported before from Irian Jaya, where it is termed the dobsegna. In view of the recent discovery here of a striking new species of black-and-white tree kangaroo called the dingiso *Dendrolagus mbaiso*, it would be unwise to rule out entirely the possible persistence of the thylacine (supposedly extinct everywhere since 1936). *Melbourne Herald-Sun*, **25 March 1997.**

A captive thylacine (photo origin unknown)

MEGA-BUSHBABY, OR MAINLAND LEMUR? During June 1985, in the centre of Senegal's Casamance Forest, wildlife enthusiast Owen Burnham spied a mysterious creature resembling a giant

Alien Zoo

Are these mysterious mammals an extra-large form of bushbaby, or an undiscovered species of mainland lemur?

bushbaby. It was the size of a half-grown cat, with pale grey fur, and was accompanied by two or three babies. Several years later, a similar creature was reported from the Ivory Coast. Moreover, in 1994, an assistant of bushbaby taxonomist Dr Simon K. Bearder, from Oxford Brookes University, encountered a strange creature in Cameroon superficially reminiscent of a giant bushbaby. Although the creature was captured, measured, weighed, and photographed, it was not retained, and hair samples obtained from it for DNA analysis proved insufficient in quantity. In 1996, Dr Bearder visited Cameroon and saw one of these creatures himself, but although he was able to photograph it he was not able to capture it. Cameroon's mysterious 'giant bushbaby' is the size of a cat, with grey-brown fur, small pointed ears, and a potto-like body. However, its face is lemur-like in profile, and its tail is long and thick, longer even than that of the newly-named false potto *Pseudopotto martini*. It doesn't jump like a potto, but walks instead. I recently spoke on the phone to Dr Bearder, and he opined that it may conceivably be a mainland African lemur. Cameroon is well-known as a region containing many endemic animal and plant species of a relict, primitive nature. Consequently, the survival here of a mainland representative of the lemurs (hitherto believed to have persisted into modern times exclusively on Madagascar, having become extinct elsewhere) is by no means impossible, and is a very exciting prospect. **Dr Simon K. Bearder, pers. comm., 26 March 1997.**

HAVE THE PEAK DISTRICT WALLABIES HOPPED OFF? Over the years, several naturalised populations of wallabies, descended from zoo escapees/releases, have established themselves in parts of Britain (the most recent is a colony on an island in Loch Lomond), and according to police wildlife liaison officers around the country, there has been a rise in wallaby sightings during recent times. This is believed to be due to the warm climate enjoyed by the UK lately, but it may have come too late for the most famous British colony. The Peak District's population of red-necked wallabies *Macropus rufogriseus* has experienced great fluctuations in fortunes and numbers since its establishment in the 1930s. At one stage, it may have boasted as many as 60 individuals, but as there does not appear to have been any conclusive sighting in the past two years, some zoologists and local farmers fear that the wallabies might have died out. *Wolverhampton Express and Star*, **5 May 1997;** *Daily Telegraph*, **12 May 1997.**

EAR, WHAT'S THIS DEER? Undeniably intriguing is a story that possibly appeared on a CBS news programme in or around June 1997 of a deer observed in the Florida suburbs. What made it so distinc-

Alien Zoo

tive, however, were its peculiarly large ears, which apparently resembled those of a donkey! Despite attempts by baffled animal control officers, however, this anomaly eluded identification and capture. **Brad LaGrange, cz@onelist.com 24 June 1997.**

HAS MANIPOGO BEEN MURDERED? Equally tantalising are rumours that sometime during early summer 1997, a 45-ft-long serpentine monster with a horse-like head was shot in Canada's Lake Manitoba, secretly smuggled away, and either sold to a reclusive purchaser for $200,000, or placed under guard with the Royal Canadian Mounted Police.

This lake has long been claimed to harbour such a creature, nicknamed Manipogo, which some cryptozoologists consider may be a surviving species of zeuglodont whale, traditionally reconstructed as being exceedingly elongate in shape. The Mounties themselves, inundated with enquiries, have been investigating the stories, but no evidence in support of such claims has been revealed.

Was a creature like this shot in Lake Manitoba? (Richard Svensson)

ON THE TRAIL OF THE CURUPIRA. June 1997 is the planned month of departure for a cryptozoological expedition to the Brazilian Amazon rainforests, led by French ethnologist Dr François-Xavier

Pelletier and continuing a search initiated by him in 1996. His quarry is a mysterious form of giant monkey, referred to locally as the curupira. According to native descriptions, the curupira is just over 4 ft high, very hairy, with a notable fringe or mane of hair around its neck, a flattened nose, very large mouth and feet, crooked toes, and a head like that of a chimpanzee. It is predominantly arboreal, jumping from branch to branch, thus leaving few traces on the ground below, and it has been seen feeding upon bananas. **Institut Virtuel de Cryptozoologie, http://perso.wanadoo.fr/cryptozool, June 1997 [now http://pagesperso-orange.fr/cryptozoo/].**

'O Curupira' – painting by Manoel Santiago

A VERTICAL CHALLENGE FOR SEA SERPENTS? One of the most popular 'orthodox' identities for sea serpents, especially some of the elongate, serpentiform versions, is the oarfish *Regalecus glesne*. The world's longest bony fish, it has traditionally been thought to swim horizontally, propelled via sinuous lateral undulations.

During a recent dive off Nassau in the Bahamas, however, Brian Skerry was fortunate enough not only to encounter a living oarfish at close range but also to photograph it, and he was amazed to discover that it did not swim horizontally at all. Instead, it held its long thin body totally upright and perfectly rigid, with its pelvic rays splayed out to its sides to yield a cruciform outline, and it seemed to propel itself entirely via movements of its dorsal fin.

Oarfishes may swim horizontally too, but until now no-one had ever suspected that this serpentine species could orient itself and move through the water in this strange, perpendicular fashion. ***BBC Wildlife*, June 1997.**

Alien Zoo

Oarfish depicted in more traditional horizontal mode

ARE MAINLAND THYLACINES NATIVE OR NATURALISED? Although the Tasmanian wolf or thylacine *Thylacinus cynocephalus* supposedly died out on mainland Australia about 2300 years ago, in modern times there have been many sightings here of thylacine-like beasts, particularly in the Gippsland region, including Wilsons Promontory. Does this mean that the thylacine has persisted here, undiscovered by science, for over two millennia?

Perhaps there is a simpler, alternative thylacine-featuring explanation. Wilsons Promontory was reserved as a national park in 1898, but during the next 40 years at least 23 different species of native animal from elsewhere in Australia were released here by misguided naturalists. Hence it is possible that thylacines too, which survived on Tasmania until at least 1936, were imported and released in Gippsland. Needless to say, if that is indeed true, and if they duly survived, bred, and thrived in this area, eventually establishing naturalised populations, that would explain the present-day plethora of unconfirmed thylacine reports here. ***Brisbane Sun-Herald*, 1 June 1997.**

EYE-TO-EYE(?) WITH THE LAKE VAN MONSTER. In June 1997, television news programmes around the world and also the internet screened a brief video purporting to show the monster of Lake Van in eastern Turkey. The video depicts an ill-defined hump moving across the water before disap-

Alien Zoo

pearing beneath the surface, followed by a closer but unclear shot of presumably the same hump, revealing what some observers have dubbed "an eye". However, it has not been possible to attempt any degree of positive identification of the filmed object. The video was shot by 26-year-old Unal Kozak, a teaching assistant at Van University. Kozak claims to have filmed the beast on three separate occasions, he estimates its total length at just under 50 ft, and has even written a book on this freshwater cryptid, which contains drawings reconstructing its likely appearance as based upon the testimony of some 1000 eyewitnesses. **CNN Interactive, World News, 12 June 1997.**

(INTER)NETTING THE WATER MONSTERS. One of the joys of cryptozoology on the internet is discovering a surprising number of cryptids not previously documented by more traditional means. Gryttie is a good example, with its own homepage, no less. Depicted as a serpentine Ogopogo-lookalike, Gryttie is the monster of Lake Gryttjen, located between Hudiksvall and Ljusdal in central Sweden. Among the identities on offer for it is a type of relict sea-cow. **http://www.algonet.se/~sbr/ gryttie/index.html accessed July 1997 [no longer online].**

NOT TO BE LAUGHED AT? THE ALTAMAHA-HA. Equally intriguing is the so-called altamaha-ha. This elusive water beast reputedly frequents the Altamaha river at Darien, a small fishing town in Georgia, USA, and is represented online by Ann R. Davis's website, containing several pages of eye-witness reports. The most recent, dating from 6 July 1997, featured a sighting by Jim and Mary Marshall of "a smallish head carried low with three definite humps of body undulating in an effortless rhythm...grey-brown and smooth, 10-12 feet in length, 10-12 inches in diameter". They were certain that it was not an alligator, manatee, or otter. **http://www.gate.net/~anndavis/altahaha.htm accessed July 1997 [no longer online].**

IN A MESS OVER MESSIE. Nor should we overlook the unforgettably-named Messie, a reputedly aggressive monster inhabiting Lake Murray at Irmo, South Carolina. In October 1996, an unidentified caller to the radio station WNOK FM 104.7 claimed that while fishing on Shell Island, he saw a 2-ft-long fin surface in the lake. He threw his line out in front of it - and to his great surprise, the monster broke his rod in half and swam away with it, rolling in the water before disappearing. Previous eyewitnesses have speculated that Messie may be a huge sturgeon, already confirmed as the identity of several sizeable North American lake monsters. **http://www.geocities.com/CapitolHill/1171/irmo019.html accessed July 1997 [no longer online].**

MULTIPLYING THE MIGO. Known locally as the migo, the monster of Lake Dakataua in New Britain, east of New Guinea, was initially hailed as a major addition to cryptozoology after being filmed by a Japanese TV team in 1994. At that time, Prof. Roy P. Mackal, who accompanied the team as their scientific advisor, speculated that it may be a surviving species of serpentine whale called a zeuglodont. A few months after their first visit to the lake, however, the team returned, and this time obtained a second video, shot at much closer range.

Having scrutinised this, Mackal now considers that the migo is more likely to be of crocodilian origin - three times over. He believes the 'monster' in the video to be not one but three separate specimens of the saltwater crocodile *Crocodylus porosus*, filmed in the act of mating. The monster's 'head' is the female crocodile. The hump just behind it is apparently the back of the tracking male crocodile. And the second, rear hump is probably the back of a second male. If Mackal's deductions are correct, as they seem to be, the new video is highly significant, because this species has never previously been observed in the mating phase, let alone filmed in it. **http://www.omnimag.com/antimatter/more_antimatter/ hump.html accessed July 1997 [no longer online].**

Alien Zoo

Reconstruction of zeuglodont (Tim Morris)

JOVIAL NEWS FROM JAVA? Zoologists are celebrating some optimistic news lately received from Java. Sightings of two officially extinct endemics, the Javan tiger *Panthera tigris sondaica* and the Javan hawk-eagle *Nisaetus bartelsi*, have been reported by a team of mountaineers in East Java's Bromo Tengger Semeru National Park. Encouraged by their welcome news, this Indonesian island's Ministry of Forestry has pledged a cash reward of 3 million rupiahs to anyone who can substantiate these sightings with recent photos of the two potential resurrectees. *Antara News Agency*, **1 July 1997.**

FROM BIGFOOT TO BIG BIRDS? A curious snippet that apparently featured a while ago on the internet (possibly in the ongoing Virtual Bigfoot Conference website) has been brought to my attention by palaeontologist Darren Naish. As far as he can recall, the snippet claimed that several sightings had been made, the most recent during 1975, of a 7-ft-tall bird in the Mount Adams area of Washington State, USA, and which has been likened to a giant brown bird, called the pach-an-a-ho' (variously translated as 'crooked-beak bird' or 'rough-looking bird'), from traditional Yakima legends.

In addition, a party of American Indians apparently visited a certain American museum not long ago, and became very excited when they saw a life-sized reconstruction of a giant flightless prehistoric predatory bird from North America called *Diatryma*, because they claimed that this was the pach-an-a-ho'. Sources have informed me that issue #20 (August 1992) of the Western Bigfoot Society's newsletter, *The Track Record*, may include details concerning all of this, but I have not seen a copy of that issue so far.

[In April 2010, I learnt from Chad Arment that in fact this issue does not contain any mention of such a

Alien Zoo

bird.]

As for *Diatryma* reconstructions, the only one that I am aware of in the USA features two adults and a chick, and is at the Californian Academy of Sciences. If any readers can shed further light on this mystifying case, I'd greatly welcome any details. **Request made on 30 July 1997.**

A philatelic *Diatryma*

BEAST OF GÉVAUDAN UNMASKED AT LAST? During the early 1760s, a rapacious carnivorous beast of undiscovered identity was blamed for the serial killing of a number of people, particularly children, in a village-speckled district of Lozère, southeastern France, called Gévaudan. Finally, a creature claimed to be this elusive Beast of Gévaudan was shot dead by local hunter Jean Chastel at Mount Chauvet, after which the killings ceased. Tragically, however, its corpse was supposedly 'lost', and its identity has thus remained a mystery...until now. In summer 1997, taxidermist Franz Julien from France's National Museum of Natural History, in Paris, sensationally revealed that the Beast had actually been preserved as a stuffed specimen in the museum until 1819, and during that time it was examined and conclusively identified - as a striped hyaena! As this species, *Hyaena hyaena*, is not native to Europe, it was evidently an escapee or deliberate release from captivity - and it is a nothing if not interesting coincidence that hunter Chastel's own son, Antoine, actually possessed a striped hyaena in his personal menagerie. **Michel Raynal/Institut Virtuel de Cryptozoologie http://perso.wanadoo.fr/cryptozool, August 1997 [now http://pagesperso-orange.fr/cryptozoo/].**

Engraving from 1765 depicting the Beast of Gévaudan

THYLACINES IN NEW GUINEA? Since I last mentioned the possible survival in New Guinea of the Tasmanian wolf or thylacine, reports of striped dog-like beasts bearing a notable similarity to this

Alien Zoo

supposedly extinct marsupial continue to emerge - and not just reports, it would seem. In August 1997, media accounts carried a claim attributed to WWF spokesman Ron Lilley that thylacines had been reported roaming the Pass Valley in the Abenaho district of Irian Jaya (western New Guinea), and that villagers had reputedly killed three of them. News of these cryptozoologically-priceless carcases' fate, however, remains unknown - a tragedy not only for science but also for the villagers, bearing in mind that a local Irian Jaya government head official, U.B. Wenas, has reportedly offered a R2 million bonus (US$ 1000) to anyone procuring a bona fide thylacine specimen here. ***Daily Post*, 20 August 1997; *The Australian*, 20 August 1997.**

OUT OF AFRICA. It is every entomologist's dream to discover an unlikely species of insect in his own back garden - and this is precisely what happened last year to John Holloway, a member of Europe's largest insect charity, Butterfly Conservation. There on his geraniums in his Lewes, East Sussex, garden were over half a dozen specimens of a tiny butterfly known as the geranium bronze *Cacyreus marshalli* - a species native to South Africa! What made Holloway's find even more significant was that when he looked closer, he found that these specimens had laid a large quantity of eggs on his plants - the first record ever of this species breeding in Britain. It seems likely that they reached Britain upon geraniums (the natural food-plant of this species' caterpillars) imported here from overseas, as they are far too small and slow-flying to have flown here themselves. ***Daily Mail*, 2 October 1997, *Independent*, 6 October 1997.**

CALLING OUT FOR KOKAKOS. The wattlebirds comprise a small taxonomic family (Callaeidae) indigenous to New Zealand. Of the three known species, one, the huia *Heteralocha acutirostris*, has been deemed extinct since 1906; the other two, the kokako (wattled crow) *Callaeas cinerea* and the tieke (saddleback) *Philesturnus carunculatus*, are both very rare. Indeed, the last confirmed sighting of the South Island subspecies of kokako *C. c. cinerea* (resembling a plump steely-grey crow but differentiated from its North Island counterpart *C. c. wilsoni* by possessing blue-and-orange rather than wholly blue facial wattles) occurred 30 years ago, since when it too has been feared extinct. However, in early 1996 a single kokako-like feather was found in an expanse of cut-over forest in South Island's West Coast area, inducing a Timberlands West Coast team to initiate an ongoing search led by Nelson ornithologist Rhys Buckingham for conclusive evidence that the bird still exists.

An unconfirmed West Coast sighting took place in November 1996 above a Glenroy River terrace, and searches uncovered unusual moss grubbing characteristic of kokakos, suggesting that there could be specimens alive and well in the Grey and Maruia valley forests. Most recently, kokako-like calls have been heard and recorded during October 1997 in a patch of forest near Murchison by Lloyd Robbins, whose considerable knowledge and experience with the North Island kokako makes him an excellent 'earwitness'. Sponsored by the WWF and other wildlife organisa-

Both subspecies of the kokako (South Island ssp is behind North Island ssp), painted by John Gerrard Keulemans

tions, Buckingham's team is now using automatic camera surveillance and digital audio tapes to collect what they hope will be conclusive proof to verify the South Island kokako's rediscovery. *Nelson Mail*, 7 October 1997; *Christchurch Press*, 7, 8 October 1997; *Dominion* (NZ), 11 November 1997.

PACKING A TRUNK TO SEEK THE ELEPHANT BIRD. Barry Ingram, a conservation officer with Leicester City Council, is planning to search for one of the largest birds ever known - the aptly-named great elephant bird *Aepyornis maximus* of Madagascar. Standing 9-10 ft tall, weighing a colossal 1000 lb, and famed for its gargantuan eggs boasting a 2-gallon fluid capacity, it was flightless and probably resembled a sturdy ostrich, but on account of its huge size it is believed to have partly inspired the Arabian Nights legends of the roc – a gargantuan mythical bird of prey capable of carrying off elephants in its great claws. Formerly inhabiting the great swampy forests in this island's interior, *A. maximus* was known to the natives as the vorompatra, but is generally believed to have died out by the 18th Century when the swamps dried out and the forests were felled. Nevertheless, Madagascar is one of the world's largest islands, and new species of large lemur have been discovered here in recent years, so Ingram is keen to look into reported modern-day sightings of the great elephant bird...just in case. *Bradgate Mail*, 6 November 1997.

Alongside egg of great elephant bird (Dr Karl Shuker)

ANOTHER ROC OF AGES (PAST). While on the subject of the roc, in recent years a second major candidate for its inspiration has come forward. In 1994, based upon subfossil finds on Madagascar, an enormous species of now-extinct bird of prey was formally described – *Stephanoaetus mahery*, the Malagasy crowned eagle. Believed to have died out by the early 16th Century, this fearsome predator is thought to have fed upon giant lemurs (including specimens weighing as much as 26.5 lb) and possibly even upon the great elephant bird itself. If such a spectacular eagle had been observed by early European sailors visiting the island, it is easy to understand how exaggerated retellings of these sightings could have ultimately given rise to the roc legend. **Alan Feduccia (1996).** ***The Origin and Evolution of Birds.*** **Yale University Press (New Haven) – consulted 6 November 1997.**

A skeleton of the great elephant bird in Paris's National Museum of Natural History (Wikipedia)

A BOG-BAGGED BUG. One inch long and resembling a giant golden-haired bee, the Maid of Kent beetle *Emus hirtus* has been believed extinct in Britain since 1966, and was last reported in Kent back in 1951. At the beginning of November 1997, however, the entomological history books were dramatically rewritten - albeit in a public lavatory. For this is where a Maid of Kent beetle was spotted by Adam Roland, an assistant warden at a bird sanctuary in Elmley on Kent's Isle of Sheppey. Roland spied the beetle crawling through the entrance to the sanctuary's public toilets, and swiftly rescued the unexpected visitor, maintaining it in a warm tank to shelter it from the prevailing frost before releasing it back into the wild when the weather warmed up. *Sunday Telegraph*, **9 November 1997.**

SACHAMAMA - A SNAKE IN A SHELL? In 1997, Czech explorer Arnost Vasícek returned home from Peru, after searching for one of cryptozoology's most obscure mystery beasts - the sachamama, or

Alien Zoo

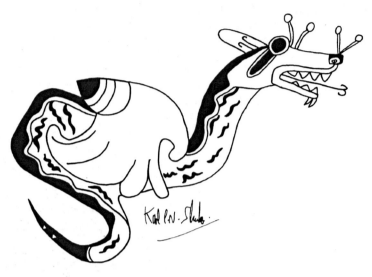

Sachamama-like beast on 16th-Century Peruvian pottery
(Dr Karl Shuker)

shelled snake. According to local eyewitnesses, this bizarre creature resembles a giant black anaconda-like snake, but possesses a large conch-like shell. Intriguingly, although he did not spy the beast itself, Vasícek did document some 16th-Century examples of Peruvian pottery depicting a very similar animal, which also sported a forked tongue and two pairs of snail-like, bulbous-tipped feelers - one pair at the base of its snout, and the other pair at its snout's tip. **Blesk Magazin, 14 November 1997.**

SKULLS, STRIPES, AND SPECKLES. For several years, zoologist Dr Peter Hocking from Lima Natural History Museum has been investigating native reports of four different types of Peruvian mystery cat. One of these is the so-called speckled tiger, said to be as big as the jaguar (known colloquially as the 'tiger' in much of South America), but patterned with a unique speckling of tiny dots, and equipped with very large, almost tusk-like canine teeth. Another is the 'striped tiger', also as big as the jaguar, but striped like a genuine Asian tiger. In 1994, Hocking succeeded in obtaining a female skull for each of these two mystery cats, and sent photos of them to a number of American zoologists. Comments revealed so far have been mixed - whereas one (un-named) expert claimed that the speckled tiger skull appeared to represent a completely new species, others have declined to pass any unequivocal opinion as to their possible taxonomic identity. **Angel Morant Forés, pers. comm., 18 December 1997; Matthew Bille, pers. comm., 18 December 1997.**

MERELY A MARA? A mystifying rodent was brought to my attention recently. According to Toni Williams, a strange beast has been spotted lately in rural Virginia that resembles a huge rodent, estimated at 20-30 lb in weight, and which, although quadrupedal, uses its hind feet to move in a hopping fashion.

Although it has been likened to an American beaver *Castor canadensis*, it apparently lacks the beaver's characteristic flat tail and slick coat. Talk of huge rodents readily brings to mind the South American capybara *Hydrochoerus hydrochaeris*, but this is much larger, weighing up to 100 lb. Somewhat smaller, up to 35 lb or so, is the coypu *Myocastor coypus*, again native to South America, but which can also be found in a naturalised state in parts of the USA, including Louisiana, due to fur-farm escapees establishing a thriving population here. However, the coypu has a very long, conspicuous tail. Much more likely is that this mystery beast is simply based upon an exaggerated account of a mara (Patagonian cavy) *Dolichotis patagonum*, which is indeed quadrupedal but given to hopping movements with its hind legs, has only a tiny tail, and is often kept in wildlife parks. **Matthew Bille, pers. comm., 18 December 1997.**

Alien Zoo

Maras or Patagonian cavies

MYSTERY TRACKS IN MONGOLIA. Chad Arment's cryptozoology website contains some interesting recent news from Mongolia. In December 1997, Chad was contacted by a Russian zoologist seeking sponsors for a planned expedition to northern Mongolia to investigate what might be responsible for a series of mysterious "turtle-like" tracks that appeared annually between 1985 and 1991 on the sandy bank of one of this region's lakes, and which were seen on one occasion by a Russian geologist. As noted by Chad, seals have been reported from some of Mongolia's inland lakes. Could these be responsible? **http://www.herper.com/Mongturtle.html December 1997 [no longer online; see now http://www.strangeark.com/].**

The Ludlow jungle cat (Dr Karl Shuker)

The Lost Ark

THE LOVECATS –
JUNGLE CATS, JASPER,
AND A CROSSBRED CONTROVERSY.

Older natural history guides sometimes mention 'giant feral cats'. These were often said to be the second-generation offspring of a domestic cat gone wild; huge and long-legged in comparison to the norm. Dr Maurice Burton refers to them in Wild Animals of the British Isles, *stating that gamekeepers tend to keep quiet about such animals, and bury them as soon as possible. They don't take measurements, but describe them as 'enormous', 'big as a dog' and 'twice as big as an ordinary cat'.*

Recent studies of feral cats show that such animals tend to be smaller than normal domestics, rather than larger; and specimens of such outsized cats seem not to exist in collections. It is often suggested that people mistake large feral cats for 'pumas' and 'panthers'. Could mistaken identity have worked the other way?

In 1962, a fourteen year old boy shot an outsized cat on his father's smallholding in Shropshire. There were many feral cats in the area, but all of normal size, and he was certain this was something different. Not knowing the Wildcat [Felis silvestris] *had been extinct in England for many years, he assumed it was an 'English Wildcat'. The animal was twice the size of a big tom cat and very muscular. It was dark brown or grey, with darker stripes. Its short (9") tail was blunt-ended, its ears were tufted, and long fangs hung over the bottom lip. The description closely matches that of a Jungle Cat* [Felis chaus].

It is difficult to believe that a gamekeeper shooting a puma or leopard would assume it was a feral cat. But what of a smaller species? At close quarters, a jungle cat could be mistaken for a large, muscular, and long-legged domestic cat. Are some – if not all – our 'giant ferals' actually jungle cats?

..For a species to sustain itself over a long period would normally require a reasonably

large breeding population. However, as Dr Karl Shuker has recently pointed out, the jungle cat can hybridize with the domestic cat, and produce fertile offspring. This is highlighted by reports of apparent hybrids received by Dr Shuker from the West Midlands area.

In zoos, such hybrids tend to have the long legs of the jungle cat, and are often black in colour. There have been many sightings of slender, long-legged, black 'panthers' throughout the country.

Jan Williams – 'A Shadow And A Sigh', *SCAN News*, Spring 1993

This chapter was originally a two-part Lost Ark article of mine published by *FT* during 1993. Because it constitutes an investigation that I conducted during a particular time period a long while ago now, and which I have not returned to since, I am presenting it here in largely non-updated form, in order to preserve its historical context. For reasons of space, in the original published version of Part 1 the section on urbanised jungle cats in Britain was greatly reduced, but it is included here in its hitherto-unpublished unabridged form.

PART 1:
A NEW IDENTITY FOR BRITISH MYSTERY CATS?

For many months prior to 1989, sightings of a strange lynx-like creature were reported from the Ludlow area of Shropshire; but as so often happens, officialdom dismissed them as misidentifications of domestic cats, dogs, or foxes - until 3 February 1989. This was when farmer Norman Evans found near his grounds at Richards Castle, just outside Ludlow, the dead body of a very large, tall lynx-like cat with tawny pelage, tufted ears, a fairly short black-ringed tail, striking 1-in-long fangs, and stripes upon its upper limbs. It was an adult male specimen of the Asian jungle (swamp) cat *Felis chaus*, a species not native to Britain or anywhere else in Europe in historical times. Investigations revealed that it was a captive-bred escapee that had been living in the Ludlow countryside for a considerable time prior to being killed by a car.[1,2]

This was not the first jungle cat obtained in Britain. On 26 July 1988, an adult specimen was hit and killed by a car in Hampshire's Hayling Island. And as in the Ludlow case, its procurement followed months of discounted sightings locally of a large, odd-looking lynx-like beast.[3]

In autumn 1992, I succeeded in uncovering the current whereabouts of the Ludlow jungle cat, which had been preserved as a magnificent taxiderm specimen. After confirming that the relevant ministries and departments had no objection, I purchased it - because it not only provides unequivocal proof that non-native cats are indeed escaping from captivity and surviving in our countryside, but also (together with the Hayling Island cat) adds a hitherto-unrecognised yet potentially dramatic dimension to Great Britain's overall mystery cat situation.

There are certainly some much larger non-native cats roaming Britain, including pumas *Puma concolor*, European lynxes *F. (Lynx) lynx*, and black panthers *Panthera pardus*.[4] These could even breed in the wild here, but only if a male and female of the same species meet up - because interbreeding between the larger cat species rarely occurs; and even when it does, the offspring are almost invariably sterile. The jungle cat, conversely, constitutes a very significant exception to this, i.e. it does *not* need to

Alien Zoo

encounter others of its own species in order to mate and reproduce. This is because although it is three times larger than the domestic cat *F. catus*, the jungle cat is so closely related to it genetically that successful matings between these two species in captivity have yielded fertile hybrid offspring. Furthermore, these - after mating with other such hybrids or with either of their parental species - have themselves yielded offspring, and so on.[5] Interbreeding may also have occurred in the jungle cat's native Indian homeland.[6]

What has particularly intrigued me regarding the Hayling Island and Ludlow jungle cat incidents is that in both areas the sightings have continued long *after* these two specimens were obtained.[7,8] In some cases, the descriptions suggest the presence of other, still-uncaptured jungle cats (e.g. in the Hill Top

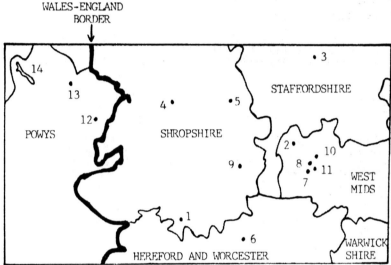

1 = Ludlow; February 1989 (procurement of dead jungle cat).
2 = Penn; summer 1991.
3 = Weston; summer 1991.
4 = Shrewsbury; August 1991.
5 = Telford; January 1992.
6 = Great Witley; May 1992.
7 = Tipton; late May/early June 1992.
8 = Great Bridge; June 1992.
9 = Bridgnorth; June 1992 (and spasmodically onwards).
10 = Wednesbury; November-December 1992.
11 = West Bromwich; December 1992.
12 = Welshpool; January 1993.
13 = Llanfyllin; January-February 1993.
14 = Lake Vyrnwy environs; February 1993.

Sightings map of possible jungle cats or jungle cat hybrids in mid-Britain as documented in this chapter
(Dr Karl Shuker)

district of West Bromwich, in the West Midlands;[9] in the Bridgnorth area, not far from Ludlow;[10] close to Lake Vyrnwy, just inside Wales's border with Shropshire[11]).

In others, however, although the cats described seem like jungle cats in general size and shape, their coat colours are not consistent with this species, as they include jet black (near Hayling Island[8]), almost white (Penn district of Wolverhampton in the West Midlands[12]), and pale brown with swirl-like

With the Ludlow jungle cat (Dr Karl Shuker)

Alien Zoo

blotches (Tipton, West Midlands;[13] jungle cat cubs have spots, which vanish as they mature). As I subsequently revealed to the media,[14,15] from my investigations of cat hybridisation and my field researches in the West Midlands I consider it plausible that these strange-coloured cats are hybrids between jungle cats and feral domestic cats.

This is because first-generation hybrids of jungle cat and domestic cat tend to resemble their jungle cat parent both in size and general outline (including their distinctive tufted ears and relatively short tail), but possess the coat colouration of their domestic cat parent.[5] This could therefore explain the odd cats that I have learnt about – with jungle cat shape and size but domestic cat colouration – in locations within or adjacent to those areas in Britain known to have bee frequented for a considerable time by adult jungle cats.

To obtain an independent, authoritative opinion, however, I communicated with internationally-renowned felid geneticist Roy Robinson regarding the likelihood of escapee jungle cats mating, and yielding hybrid offspring, with feral domestic cats in Britain. He agreed that this was certainly possible, given the close genetic relationship between the two species, and the general docility of the jungle cat despite its large size.[16]

Another noteworthy aspect to this scenario is the apparent presence of one or more jungle cats and hybrids in the industrialised regions of the West Midlands - where one might not expect such cats to survive. Yet in reality, in its native India the jungle cat associates closely with humans, living in deserted buildings and prowling in the immediate vicinity of villages.[17]

Indeed, it seems to constitute the feline equivalent there of our own urban foxes. Moreover, it may well be invading the same ecological niche here in the West Midlands and neighbouring regions, as demonstrated by the following examples.

Midlands-based car mechanic Ron Mills has a 24-hour breakdown service, and was called out one morning in June 1992 at about 3.30-4.00 am. Driving into Catherton Close, Tipton, to collect some tools from a workmate who lives here, Ron saw in his car's headlights a strange cat-like beast lurking by some dustbins. The cat froze, and Ron braked, sitting in his car about 30 ft away and watching the animal for several minutes. Convinced that it was not a domestic cat, Ron initially likened it to a small lynx, but after later seeing the Ludlow jungle cat he felt that it more closely resembled this.[13]

Ron informed me that its body was tall and thin, about 6 in taller than a domestic cat, but lithe and muscular. It was pale in colour and patterned with round swirl-like markings, but its most catching feature were its ears, which were noticeably pointed, triangular, and erect. After observing it in detail, Ron drove on to his workmate's house, causing the cat to flee. A couple of days later, a short report appeared in the local *Express and Star* newspaper concerning recent sightings by other eyewitnesses of a lynx-like cat encountered in this area. As pure-bred jungle cats do not possess body markings, could Ron's mystery cat have been a hybrid?[13]

At around the same time as Ron's sighting in Tipton, an unusual cat was spied at Quatford Caravan Park in Bridgnorth, Shropshire, by Robert Baker. He had just stepped out of his caravan one evening, at twilight, when the animal walked by, only a few yards away, apparently looking for food. Lighting conditions made it difficult for him to discern its colour, but he thought that it was tawny. The cat was about three times the size of a domestic cat, and noticeably tall, with conspicuously tufted ears. A few days later, another caravan owner dashed into Baker's caravan and told him that he had been extremely shocked to see a very large, strange cat close by, and several other sightings were made here during this same period.[10]

Back in the West Midlands: on 19 February 1993, I learnt from Wednesbury resident Margaret Cockbill that one dark evening two to three months earlier, one of her friends, living in St Luke's Road, Wednesbury, had seen an urban mystery cat that she later likened to a jungle cat after seeing a photo of the Ludlow jungle cat in various local newspaper reports from January 1993. At around 8 pm, she had opened her kitchen door to take some rubbish out to the dustbin when she spied, standing by the dustbin, a very large cat-like animal that she felt sure was neither a domestic cat nor a dog or fox. Its eyes were glowing at her in the light from the kitchen, and she was so shocked at seeing this creature that she rushed back inside the kitchen and slammed the door behind her.[18]

One evening at 8.30-9.00 pm in the week before Christmas 1992, forklift truck driver Fred Rackham was walking in West Bromwich's Hill Top area, not very far from his home in Tipton, when he saw beneath the street lights a feline creature running towards him. When it was about 20 yards away, however, it turned aside, fleeing down the road leading to a vicarage where it vanished from sight amid wasteland and undergrowth. Fred later saw a photo of the Ludlow jungle cat in the *Express and Star*, and in his opinion the animal that he had sighted was identical to it. He had taken particular note of its prominent head, big pointed ears, very powerful back legs, and the way in which it "ran tall" with head held aloft. It was over twice the size of a domestic cat, and seemed fairly dark in colour. He also mentioned that foxes, once commonly spied here, had not been seen in this area for the past 12 months - an intriguing detail that often occurs in reports of mystery cats.[9]

Similarly, I learnt from Harry Wilson, a horticultural market trader from northwestern Shrewsbury, that early one morning during the second week of February 1993, a postman had seen an *F. chaus*-like cat on the drive of a rather remote bungalow in the vicinity of Lake Vyrnwy, in Powys, Wales, not too far from the Shropshire border. Resembling a tall fox but definitely feline, it had a rufous coat, pointed ears, long limbs, and a tail that was neither thin nor bushy. Another local inhabitant had made three separate sightings of this animal.[11]

On 9 August 1993, former bookseller Bob Simmons spied a mystifying cat while walking his dog, Sam, near their home in Walsall, West Midlands. It was dusk, at around 9.30 pm, and while strolling across a grassy expanse between Helston Road and Truro Road, not too far to the right of Rushall Canal, they saw a large, fawn-coloured cat walking up towards Gillity Bridge. They watched it for about 30 seconds, at a distance of 40-50 yards, before it jumped over a fence and disappeared. Bob was certain that it was not a domestic cat, dog, or fox. Sam ignored it, as he does with cats - but he would have chased it if it had been a fox.[19]

The cat was about as tall as Sam, i.e. 18 in at the shoulder, but had a greater combined head and body length, i.e. exceeding 2 ft. Its tail was quite long, curving downwards, but not touching the ground. Its face was flat - "like a leopard's", in Bob's opinion. He could not recall whether or not its ears were tufted. Its fawn fur had paler patches, and resembled a ginger cat's rather than a tabby's. As there are plenty of foxes in this area, the cat may have taken up residence in a fox earth, and the direction in which it departed is very rural. Bob is positive that it was not a puma (and its tail was, in any event, far too short for a puma's), but was nonetheless a very large cat. Some pure-bred jungle cats do have fairly long tails; but if it was a jungle cat x domestic hybrid, this could also explain its fur's pale markings.[19]

Summing up: there would seem to be a realistic chance that if interbreeding between escapee jungle cats and feral domestic cats has indeed begun in Britain's countryside, and if the jungle cat genes can somehow avoid being substantially diluted by the preponderance of domestic cat genes (perhaps within a small, isolated area?), eventually a self-perpetuating strain of notably large hybrids could become established here - hybrids, moreover, that possess genes from a wild, non-native species (unlike hybrids produced by interbreeding between feral domestic cats and the native Scottish wildcat *F. silvestris*). The result would be a startling feline parallel to the situation already prevalent with native red deer *Cervus elaphus* and naturalised Asian sika deer *C. nippon*...not to mention an unpredictable addition to the British ecosystem.

I wish to express my especial thanks to Jan Williams for kindly contributing sightings from her own files and to everyone who submitted details to me of their mystery cat encounters.

Face to face with the Ludlow jungle cat (Dr Karl Shuker)

REFERENCES

1. ANON. (1989). 'Killer' found dead [Ludlow jungle cat]. *South Shropshire Journal* (Ludlow), 10 February.
2. ANON. (1989). Wild cat sightings. *South Shropshire Journal* (Ludlow), 17 February.

3. ANON. (1988). Hit-and-run driver kills mystery wildcat. *The News* (Portsmouth), 28 July.
4. SHUKER, Karl P.N. (1989). *Mystery Cats of the World: From Blue Tigers to Exmoor Beasts.* Robert Hale (London).
5. JACKSON, J.M. & JACKSON, J. (1970). The hybrid jungle cat. *Newsletter of the Long Island Ocelot Club*, 1 (no. 1): 45.
6. POCOCK, Reginald I. (1907). On English domestic cats. *Proceedings of the Zoological Society of London* (for 1907): 143-68.
7. WILLIAMS, Jan (1993). A confusion of cats. *Scan News*, no. 1 (January): 5-8.
8. MALORET, Nick (1990). Swamp cat fever. *Fortean Times*, no. 55 (autumn): 44-6.
9. RACKHAM, Fred (1993). Pers. comms, 2 & 4 February.
10. BAKER, Robert (1993). Pers. comm., 27 January.
11. WILSON, Harry (1993). Pers. comms, 29 January & 14 February.
12. WILLIAMS, Jan (1993). Pers. comm., 27 January.
13. MILLS, Ron (1993). Pers. comm., 27 January.
14. JONES, Alison (1993). The missing lynx! *Birmingham Evening Mail*, 29 January.
15. ANON. (1993). Danger cats 'roaming region'. *Wolverhampton Express and Star*, 30 January; plus various subsequent newspaper and radio interviews in 1993.
16. ROBINSON, Roy (1993). Pers. comm., 1 February.
17. GUGGISBERG, C.A.W. (1975). *Wild Cats of the World.* David & Charles (Newton Abbot).
18. COCKBILL, Margaret (1993). Pers. comm., 19 February.
19. SIMMONS, Bob (1993). Pers. comm., 21 August.

PART 2:
LOVECATS II – THE NEXT GENERATIONS?

Shortly after preparing Part 1 of my 'Lovecats' article for *FT* (which at that time was only planned as a single, self-contained article), I made contact with Gareth B. Thomas - the vet who had examined the corpse of the Ludlow jungle cat shortly after it had been discovered dead close to a farm at Richards Castle, Ludlow - and learnt of a very thought-provoking sequel, which in turn led me to prepare an all-new, second part to my original 'Lovecats' article.

Just a few days after he had examined the jungle cat, Gareth had received a visit from Jeanette Powell, who was living nearby at the time, in the Leintwardine area. She had brought in a male kitten, Jasper, to be castrated, and Gareth was startled to discover that he was more than a little reminiscent of the recently-deceased jungle cat! Indeed, he was so striking that Gareth took the opportunity of photographing him at various angles before the anaesthetic wore off after his operation.

Considerably larger than other kittens of comparable age, Jasper had grey fur that lacked markings on his flanks, a very prominent black dorsal stripe running along his back and down the entire length of his tail, stripes upon his limbs, a heavily-ringed tail with a black tip, black tufts at the ends of his noticeably broad-based ears, a white ring encircling each eye, white jaws, and notably large canine teeth. Although jungle cats are usually brown or fawn, grey specimens have been recorded, and the other features listed here for Jasper are typical for jungle cats, and were all exhibited by the Ludlow specimen.

Alien Zoo

Jasper, side view (Dr Karl Shuker)

Gareth learned from Jeanette that Jasper's mother was a grey feral domestic cat who had eventually been tamed by Dorothy Williams of Whitton Farm (also in Leintwardine) - near to where the cat had first appeared during the mid-1980s. Known as Mother, she had given birth to several litters since then - including the litter that contained Jasper, born in August 1988. I duly spoke to Dorothy, who stated that Mother had tufted, lynx-like ears, stripes on her limbs and tail, and ginger patches interspersed amid her otherwise grey-coloured fur, but unlike Jasper she had only an average-sized body.

Apparently, Mother was one of a number of unusual feral cats in the area that bore a degree of similarity to the jungle cat, and which were colloquially referred to by the locals as 'lynxes' or 'wild cats'. As the Ludlow jungle cat is believed to have been on the loose for up to 5 years, could these have been hybrid descendants of this non-native escapee? Certainly, Gareth considered it possible that Jasper was a second-generation hybrid descendant, i.e. a 'grandson', of the Ludlow jungle cat. If this is true, then at least one of Jasper's parents must have been a first-generation hybrid. His father was unknown, but the *chaus* characteristics of his mother, Mother, lend support to the possibility that she was one such crossbreed.

(Moreover, in summer 1997 I was contacted by Chris Holloway, who enclosed photos and descriptions of her cats Tiggy1 and Tiggy2, who were also born to Mother but in a different litter from that of Jasper. Significantly, Tiggy2 closely resembles Jasper.)

Although Gareth's photos of Jasper were very informative regarding his morphology, I was naturally anxious to observe him directly, and after receiving a kind invitation from Jeanette and her husband Robert to visit, in May 1993 I came face to face at last with her enigmatic pet. A sturdy, powerfully

Alien Zoo

Jasper, front view (Dr Karl Shuker)

muscular cat, Jasper was certainly very large, weighing 14 lb, but with tantalising ambiguity he recalled three different types of cat in appearance. What made this particularly confusing, moreover, was that two of these (both of which are domestic breeds) share features of the third, the jungle cat.

Jasper's face, chest, and banded limbs reminded me of the Egyptian mau (supposedly derived from an undetermined species of wild cat inhabiting ancient Egypt; this region is home to the African wildcat *F. silvestris lybica* and also to the jungle cat), which shares the jungle cat's white 'spectacles', its limb striping, white 'moustache' across its jaws. tufted ears with wide bases, and heavily-ringed black-tipped tail.

In contrast, his dorsal stripe and unmarked flanks, and once again his tufted ears, compared with those of the Abyssinian cat - a very distinctive domestic breed markedly similar in external appearance to the jungle cat. Two features that differentiated Jasper from the latter species, however, were his relatively short limbs, and his relatively long tail - the converse features characterise the jungle cat.

Nevertheless, Jasper did have one very unexpected surprise still in store for me - a brother, George. From the same litter as Jasper, George shared his distinct dorsal stripe and wide-based tufted ears, but was ginger and white in colour, and was more gracile in build, equipped with longer limbs and weighing 12 lb.

If Jasper had possessed George's build, his resemblance to a jungle cat would have increased further; yet even without this, it was easy to see why Gareth had been so impressed by his correspondence to this species. In addition, I learnt from Jeanette that over successive moults, Jasper was gradually acquiring brown fur; if his grey pelage were to be eventually replaced entirely with this new shade of fur, this would of course enhance even more extensively his resemblance to a typical jungle cat.

Could Jasper and George really be hybrid grandsons of the Ludlow jungle cat? Spurred by my encounter with this intriguing duo, I contacted a number of mammalian geneticists in Britain and overseas – including Dr Michael W. Bruford from the UK's Institute of Zoology and felid specialist Prof. Stephen J. O'Brien from the USA's National Cancer Institute - to determine whether blood samples taken from

Alien Zoo

George and Jasper (Dr Karl Shuker)

Jasper and George (not to mention Mother) would enable their taxonomic identity to be conclusively resolved - via chromosomal analyses, or via DNA hybridisation. Sadly, the information available was not encouraging.

At that time, no features unequivocally distinguishing the karyotype (chromosomal complement) of the jungle cat from that of the domestic cat had been documented, and DNA hybridisation techniques were insufficiently rigorous to facilitate the conclusive identification and differentiation of pure-bred jungle cats, domestic cats, and hybrids. As for anatomical analyses, the jungle cat is not sufficiently well-studied for these to offer a great deal of scope either.

Consequently, the lineage of Jasper and George remained a mystery. Were they, perhaps, nothing more than unusual pure-bred domestics whose jungle cat characteristics were due merely to an ancestry containing domestic breeds such as the Abyssinian and mau that themselves resemble this wild species? Or are they genuine, second-generation hybrid descendants of an escapee jungle cat? Only time - and technology - will tell.

(In 1994, Gareth Thomas, Jeanette Powell, myself, Jasper, and the taxiderm Ludlow jungle cat all featured in 'On the Trail of the Big Cats', an episode forming part of the then-forthcoming British television series 'Arthur C. Clarke's Mysterious Universe', filmed by Granite Productions; the big cat episode was first screened in the UK on 4 April 1995 by Discovery Channel.)

SELECTED REFERENCES

ANON. (1993). Zoologist who has a furry simple explanation. *Birmingham Post*, 30 January.
BAKER, Simon (1993). Pers. comm., 7 April.

BRUFORD, Michael W. (1993). Pers. comm., 28 April.

GEE, E.P. (1937). The size of the jungle cat (*Felis chaus affinis*). *Journal of the Bombay Natural History Society*, 39: 850-1.

GRAY, Annie P. (1971). *Mammalian Hybrids* (2nd edit.). Commonwealth Agricultural Bureaux (Farnham Royal).

HOLLOWAY, Chris (1997). Pers. comms, mid-June, 29 September & 24 October.

LEAKE, Jonathan (1994). The wild truth of one cat's legacy? *Wolverhampton Express and Star*, 27 April.

O'BRIEN, Stephen J. (1995). Pers. comm., 11 July.

SHUKER, Karl P.N. (1995). The coming of supercat! *Wild About Animals*, (January): 20-2.

THOMAS, Gareth (1993). Pers. comms, 15, 16 & 29 March.

THOMAS, Lars (1993). Pers. comm., 26 May.

WILLIAMS, Dorothy (1995). The wild one. *In:* MOORE, Joan (ed.) (1995). *Cat World Annual 1996*. Cat World Ltd (Shoreham-On-Sea), pp. 32-3.

WILLIAMS, Jan (1993). A shadow and a sigh. *SCAN News*, no. 2 (spring): 6-10.

I wish to thank in particular Gareth Thomas, Jeanette Powell, and Dorothy Williams most sincerely for their kind assistance in making this investigation possible.

SOME SIGNIFICANT POST-1993 PAW-NOTES.

Sadly, although DNA analysis techniques have advanced very considerably since 1993, they came too late to be of assistance in relation to this particular study. Tragically, George was killed by a car not long after my visit to the Powell family's home to see him and Jasper. And although Jasper lived for many years, attaining the ripe of age of 18, he too is now dead – and with him, therefore, has died any possibility of directly unlocking the riddle of his genetic identity.

A jungle cat (as formally identified when its body was subsequently taken to Paignton Zoo for examination) was killed at Black Dog Crossroads, near Warminster, Wiltshire, during summer 1996. Its origin remains a mystery.

In 1997, it was revealed that the bones of a jungle cat *F. chaus* dating back 20,000 years had been unearthed on the banks of the Thames at Aveley, Essex, by archaeologists working for Essex County Council. This is the first evidence that this species had ever reached Britain naturally, in prehistoric times (as opposed to escaping or being released from captivity by humans in modern times).

Finally: confirmation that viable, fertile hybrid offspring between the jungle cat and the domestic cat can indeed occur came from revelations during the mid-1990s that a dramatic new breed of 'domestic' cat had been formally registered in the USA. It is known as the chausie, for good reason - because it is the direct result of matings between jungle cats and domestic cats. Not only are chausies fertile, and therefore capable of mating and producing young with other chausies or with cats belonging to either of their parental species, they can also be very large. Some males weigh up to 30 lb, and stand 16 in at the shoulder - just like a fair few ABCs (alien big cats) on record from Shropshire and elsewhere in Britain. Of course, I won't say I told you so (but I did!).

REFERENCES

http://www.askwhy.co.uk/awfrome/15/marcusmatthews.html accessed 13 May 1999.
BROOKE, Chris (1997). The Essex sphinx. *Daily Mail* (London), 18 June.
HAWKES, Nigel (1997). Did kitty come from the Essex marshes? *Times* (London), 18 June.
ROBSON, Claire (1996). It pays to be chausie. *All About Cats*, 3 (December): 21-3.

A jungle cat (Keith & Jan Williams)

Pink-headed ducks (see p.66) ABOVE: ducks at Foxwarren Park, Surrey, in 1926 BELOW: *Journal of the Bombay Natural History Society* 1908

1998

THE GREAT *THYLACOLEO* HUNT. This is the title of yet another of the numerous cryptozoologically intriguing websites currently accessible on the internet. Set up by Elektromart Pty. Ltd in 1998, it offers extensive information concerning the officially extinct Australian marsupial lion *Thylacoleo carnifex*, favoured by many cryptozoologists as the identity of the elusive cat-headed yarri or Queensland tiger, but little in the way of precise details as to the afore-mentioned hunt for it.

Also intriguing is that the feline creatures described in reports offered up here as evidence of supposed living *Thylacoleo* specimens are from Victoria, and uniformly brown or black - a far cry from the principally Queensland-based, black-and-white banded yarris documented in previous cryptozoological works, but very reminiscent of ABCs (alien big cats) in the form of supposed oop (out-of-place) pumas and black panthers reported in Victoria, and also elsewhere in Australia. http://www.netstra.com.au/~elek/Thylacoleo/thylo1.html accessed 3 January 1998 [now 'The Quest For Thylacoleo' at http://www.thylacoleo.com].

Reconstruction of *Thylacoleo carnifex* (Markus Bühler)

Alien Zoo

A DEVIL OF A MYSTERY. Last year, in 1997, a Tasmanian devil *Sarcophilus harrisii* was captured alive and well at Balga, Western Australia, and was duly sent to Perth Zoo, where it still resides. What is so remarkable about this event is that in the wild state this famously ornery creature - second only to the thylacine *Thylacinus cynocephalus* as the largest species of modern-day carnivorous marsupial - is nowadays confined wholly to Tasmania, having officially died out long ago on the Australian mainland. So where did the Balga specimen originate? Several other oop (out-of-place) Tasmanian devils are also on record, the two best known examples turning up in Victoria in 1912 and 1971 respectively. Although assumed to have escaped from captivity, this has never been confirmed with any such specimen, and it is not a commonly-exhibited species in zoos anyway. Some of the bolder Antipodean cryptozoologists have speculated that perhaps the Tasmanian devil did not become extinct on the mainland after all - a theory often applied to sightings here of thylacine-like beasts too. *West Australian*, **5 January 1998.**

IN THE PINK AFTER HALF A CENTURY? A truly 'classic' extinct bird of Asia is the pink-headed duck *Rhodonessa caryophyllacea*, deemed demised since 1942; but its days among the deceased may (we hope) soon be numbered. Zoologist Peter Gladstone, great-grandson of British prime minister William Gladstone, hopes to achieve a measure of fame in his own right by rediscovering this species during a planned two-man expedition to a series of remote marshes in the mountains of southeastern Tibet, following an invitation by the Chinese government to survey Tibet's wildlife. All that he needs now are some sponsors willing to assist in defraying the expedition's cost - around £13,000. There have been several reported post-1940s sightings of this distinctive species of waterfowl, but none has been confirmed...so far. *Daily Mail*, **19 January 1998.**

A CHUPACABRAS CARCASE? On 25 September 1996, a small twice-weekly Louisiana newspaper, the *Dequincy News*, published a short report concerning the carcase of a strange beast, apparently killed by a car, which had been discovered on the Pearce road east of Temple-inland by Mrs Barbara Mullins. The size of a very large dog, it was covered in thick woolly fur, and had a baboon-like face. One of Mrs Mullins's photos of the beast was published in b/w with the report. Shortly afterwards, local investigator Michael White and some friends successfully followed up this account by seeking, discovering, and salvaging the mystery carcase, which was still lying on the roadside. On 20 August 1997, White posted some details on the internet concerning his investigations, in which he drew comparisons between the mystery beast's morphology and certain accounts of chupacabras (goatsucker) sightings in which this entity's face had been likened to that of a baboon. White's data has since been incorporated into a more detailed account, accessed by me during February 1998 and containing one of Mrs Mullins's pictures in colour, at **http://www.mindspring.com/~crypto/lachupa.htm accessed February 1998 [no longer online].**

SALLYING FORTH TO SELJORD. Swedish cryptozoologist Jan-Ove Sundberg is planning to launch an international 17-day expedition in August 1998 to Norway's Lake Seljord, in search of Selma, a monster claimed by a 250-year-old legend to inhabit its depths. Interested in the creature for 25 years, Sundberg led an unsuccessful expedition here back in 1977, but hopes that this latest effort, financed by the Seljord municipality in the Telemark region of southern Norway, will benefit from the great advances in relevant technology that have occurred in recent years. He believes that the monster may prove to be a new species of large eel, but Dr Torfinn Oermen from Oslo's Zoological Museum doubts that any large undiscovered eel exists in the cold lakes of the Nordic region. **http://www.nando.net, 25 February 1998 [no longer online].**

IS VIETNAM'S CRYPTOFAUNA TURNING TURTLE? Vietnam is justly celebrated among cryptozoologists for unfurling a veritable herd of new hoofed mammals in recent years, but its latest debutante is something very different. During the past five centuries, Hoan Kiem Lake, situated in modern-

Alien Zoo

day central Hanoi, has been associated with legends and stories of giant turtles (i.e. freshwater tortoises), and for many years three such creatures have been regularly reported here. On 24 March 1998, however, events escalated when a passing cameraman, who - wonder of wonders for cryptozoology - actually had a videocamera with him, filmed them while they surfaced to gulp air. After the film had been shown on television, Hanoi National University biologist Professor Ha Dinh Duc announced that he has been studying this outsized form of chelonian for the past decade, and believes that continued studies will reveal them to be a new species. A stuffed specimen is currently on exhibition at a small temple on an island in Hoan Kiem Lake, and according to Professor Duc its species is the world's largest freshwater tortoise, attaining lengths of up to 6.5 ft and weights of up to 440 lb. However, it has already been claimed that the narrow-headed soft-shelled turtle *Chitra indica* (another Asian freshwater

Porcelain ornament of a traditional Oriental dragon-headed turtle (Dr Karl Shuker)

tortoise) can attain lengths of 6 ft. In any event, the existence of what could be a new species of very sizeable reptile remaining 'undetected' by science while existing in full view of downtown urban Hanoi's teeming populace is sufficiently noteworthy in itself to encourage further investigation of these very remarkable animals. **http://cnn.com/EARTH/9804/13/vietnam.turtles.ap/, posted 13 April 1998 [no longer online].**

CRYING OUT LOUD - IT'S THE OZARK HOWLER! Veteran American cryptozoologist Ron Schaffner, editor of *Creature Chronicles*, has recently mentioned online a 'new' mystery beast called the Ozark howler, deemed by some to be a feline cryptid, which even has its own website, produced by the Howler Research Group, at http://www.geocities.com/Area51/Shadowlands/4479/index.html [no longer online]. One supposed howler eyewitness is Fred Sprout, who informed Ron that he had encountered a specimen of this cryptid one evening at dusk in mid-April 1998 at an unspecified location south of Branson in the Ozarks. According to Fred's description, the howler was very large, possibly 10-12 ft long and 4 ft at the shoulder, quadrupedal, and seemed extremely robust and powerful, especially at the front of its body. Its fur seemed black (but Fred did view it at dusk), and it possessed two big, curving horn-like structures that pointed to the front of its head. Although these may have been large ears, according to Fred "they weren't floppy", and in his opinion they certainly looked like horns. Notwithstanding its name, the howler made little noise. If we assume that the poor viewing conditions and Fred's own surprise at seeing the creature combined to exaggerate its apparent size, and that its 'horns' really were ears after all, it is possible that this mysterious animal was simply a very large feral pig, comparable to the notorious razorbacks of Australia. If, conversely, Fred's description is accurate, then cryptozoologists may perhaps need to seek a more exotic solution to this mystery (always assuming it to be a genuine one). *Creature Chronicles*, **at http://home.fuse.net/rschaffner 28 April 1998 [no longer online].**

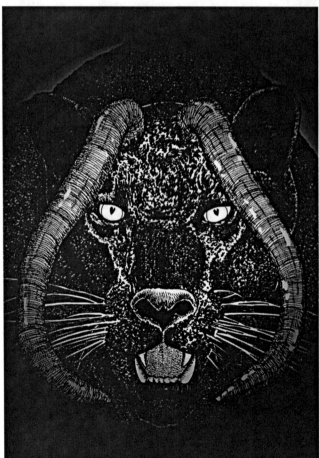

A HOWLER OF A HOAX. Not unexpectedly, a bizarre horned mystery creature called the Ozark howler and sometimes said to be panther-like in body form has proven to be a felid of the fraudulent kind. Its exposure and welcome removal from the list of genuine mystery beasts was achieved by the painstaking and very laudable detective work of veteran American cryptozoologist Loren Coleman, as revealed via the cz onelist discussion group established by fellow U.S. cryptozoologist Chad Arment. **cz@onelist.com, 10 May 1998.**

The Ozark howler (Andy Paciorek)

Alien Zoo

LAST OF THE MARSUPIAL MOHICANS. Marking the end of an era in English (un)natural history, the last known specimen of the famous Peak District wallabies has died. Since a number escaped from a private collection and in the late 1930s established a population in the upland area of Derbyshire and Staffordshire, the red-necked (Bennett's) wallaby *Macropus rufogriseus rufogriseus* has become one of Britain's most charming species of non-native, naturalised mammal. They thrived for many years, some even surviving the harsh winter of 1962-63, but their increasing fame brought danger in many different guises. Trophy hunters came, eager to bag an exotic species; stray dogs killed them too, as did speeding cars; and noisy inquisitive tourists constantly disturbed these reclusive mammals, all too often causing specimens to flee in panic, only to fall headlong to their deaths over cliffs. Now, they are no more. True, colonies of escapee wallabies have more recently become established elsewhere in the UK, but it was the Peak District population that put marsupials firmly on the map in Britain. RIP *rufogriseus*. ***Sunday Telegraph*, 24 May 1998.**

TENNESSEE'S CAT OF MANY COLOURS. On 17 October 1997, Bryan Long posted an online note revealing that about a year earlier, a group of hunters in Tennessee supposedly shot a big cat of such unusual appearance that they never formally reported it in case they were sent to jail. Nonetheless, details of its appearance eventually filtered out, and according to Bryan Long, who saw a photograph of its carcase, it was cheetah-like in form, but had a blood-red head and paws, a red line running from the back of its head to its tail (which was also red), and a golden-brown body patterned with black stripes and spots. Apparently, its carcase was skinned and the pelt later hidden in a basement. Curious to learn more, I emailed Bryan Long, and on 30 June 1998 I finally received a short email from him. In it, he revealed that the photo he had seen had belonged to a co-worker who had since sold it to a local university biology student (name unknown), the cat had been shot near Tennessee's Jackson County, and some local farmers claimed to have seen similar cats in the past but not for years now. Identities such as jaguar *Panthera onca* and jaguarundi *Felis* (*Puma*) *yagouaroundi* have been subsequently proposed by various people, but neither species fits the above-noted description. The jaguar does not have a red head, paws, dorsal line, or tail. And although the jaguarundi does have a red colour phase, called the eyra, this is red all over with no fur patterning, and is certainly not cheetah-like in form. Could the red colouration simply have been dried blood? **Bryan Long, The Cryptozoology Zone, on AOL's Parascope site, 17 October 1997 [no longer online]; Bryan Long, pers.comm., 30 June 1998.**

The Congolese mokele-mbembe
(William Rebsamen)

RETURN TO THE CONGO. A new expedition in search of the Congo's dinosaurian mystery beast, the mokele-mbembe, is planned for July-August 1998. The two-man team comprises 29-year-old Adam Davies, a graduate of Kings College London who currently works for Cable & Wireless in Wythenshawe, Manchester; and 26-year-old Andy Sanderson, a Novocastrian and fellow Cable & Wireless worker, this time at Warrington, Cheshire. According to their website, they plan to trek in the Likouala-aux-Herbes region, travelling down the Sangha River, and utilising the local pygmies as guides in their search for the elusive m-m. We at *FT* wish them well, and urge our readers to check out the Davies-Sanderson expedition website for updates on their progress, as posted directly from the Congo. **http://www.congo.exploration.mcmail.com/ accessed 13 June 1998 [no longer online].**

Alien Zoo

RETURN OF THE PARADISE PARROT? Named after its extraordinarily beautiful plumage, Australia's paradise parrot *Psephotus pulcherrimus* was last reported in the wild during 1926 in Queensland's Goondiwindi region, and the last captive specimen died in 1931. Since then, this spectacular species has been written off as extinct, although unconfirmed sightings have been reported spasmodically. The latest sighting was aired during a recent programme on ABC Radio, hosted by National Parks and Wildlife Service ranger Rick Nattrass, when a caller gave an accurate description of a paradise parrot that he claimed to have seen flying from a termite nest in the Goondiwindi region. Although he concedes that it may be a hoax, Nattrass is eager to investigate the report further, just in case specimens of this red-shouldered recluse have indeed survived undetected until now by science. *Illawarra Mercury*, **27 June 1998.**

CRYPTIC CHAMELEONS IN GREECE. A previously-overlooked population of unusual chameleons that may be of considerable zoological significance has been discovered living around the Pylos Lagoon in Greece by biologist-photographer Andrea Bonetti, who has been secretly studying them for the past two years. Not only are they larger than the normal, known species of European chameleon, *Chamaeleo chamaeleon*, they more closely resemble the African chameleon *C. africanus*. So how did they come to be living in Greece? As for their taxonomic status, they may prove to be conspecific with the African chameleon, or even a distinct, hitherto-unrecognised species in their own right. *BBC Wildlife*, **July 1998.**

MOVE OVER NESSIE, HERE COMES ESKIE. First it was Nessie, then Morag, Teggie, Lizzie - and now Eskie. The latest body of freshwater in the British Isles to claim a resident monster is Ireland's Lough Eske, sited three miles from Donegal Town. On 28 June 1998, several guests at Harvey's Point Hotel hurried down the pier to this nearby lake, claiming to have seen something in the water. When one of the hotel's staff, Seamus Caldwell, looked, he spied "something moving up and down at about 300 m out". *The Star*, **3 July 1998.**

Sketch of Oliver (Dr Karl Shuker)

UNMASKING OLIVER. After more than two decades of controversy, the identity of the mysterious bald-headed bipedal ape Oliver, now residing at a Texas animal sanctuary called Primarily Primates, has finally been resolved. Over the past 25 years, many identities have been proposed, including such dramatic suggestions as a totally new species of chimpanzee, and even a human-chimp hybrid. In 1997, however, Trinity University geneticist Dr John J. Ely and Texas University biologist Dr Charleen M. Moore announced that they would be conducting chromosomal and DNA analyses of Oliver in a bid to establish conclusively his taxonomic identity. Their findings have now been published, which reveal that Oliver has 24 pairs of chromosomes (normal for chimpanzees) and chromosome banding patterns distinguishing him from humans and from the bonobo (pygmy chimpanzee) *Pan paniscus* but consistent with those of the common chimpanzee *Pan troglodytes*. The results obtained by sequencing a 312 bp region of his mtDNA (mitochondrial DNA) D-loop region yielded a very close correspondence with central/western African chimps *P. t. troglodytes* - and especially with a specimen known to

Alien Zoo

have originated in Gabon. In short, Oliver is merely a common chimp most probably of western African provenance (an identity and origin that I predicted, incidentally, in *Man and Beast*, a Reader's Digest cryptozoology volume published in 1993). **John J. Ely, *et al.* (1998). Chromosomal and mtDNA analysis of Oliver.** *American Journal of Physical Anthropology*, **105: 395-403.**

SNAKES ALIVE! Nigel Dempster's famous celebrity-based gossip column in London's *Daily Mail* newspaper may not seem on first sight to be a likely source of cryptozoological information – which only goes to show how wrong you can be. Recently, he included details of a very curious encounter made by intrepid explorer Colonel John Blashford-Snell, in wildest Dorset. Walking through undergrowth near his home, Blashers was startled to see what he momentarily took to be a python, gliding through some grass nearby. On reflection, however, he decided that it was just an extremely large grass snake *Natrix natrix*. But was it? Blashers estimated it to be 6 ft long, and "grey with a pattern of black diamonds". Although some big grass snakes do occur, and can exhibit a range of colours, I have never read of a conclusively-identified British grass snake matching Blashers's description. Perhaps, therefore, it was an adder *Vipera berus* of over-estimated length, or even an escapee specimen of some non-native species? *Daily Mail*, **24 July 1998.**

THE STRANGE CASE OF THE CRYPTIC CRYPTO-MAGS! Longstanding colleague Alan Pringle recently brought to my attention a very intriguing mystery of the cryptozoologically-related kind. Back in 1970 or 1971, at a friend's home, he saw a complete set of a magazine partwork series containing at least 30-40 issues, on an encyclopaedic or scientific theme (possibly aimed at children), but whose title he cannot recall. However, it was not their contents, but their covers, that attracted his particular interest and attention, because each issue's back cover featured an animal or entity from the myths and legends of the world, including a number that have cryptozoological relevance. A very dramatic, full-colour illustration of the creature occupied one half of the cover's page, with accompanying text occupying the other half. The creatures that Alan can definitely remember appearing in this set of covers included the western dragon, eastern dragon, siren, tokoloshe, leshy, thunderbird, Midgard serpent, Assyrian winged bull, minotaur, centaur, bunyip, Egyptian ammut (soul eater), sphinx, and harpy. Others that he thinks may have been present include the werewolf, vampire, unicorn, cyclops, zombie, and banshee. Alan has never seen this partwork again (and has long since lost contact with the friend who owned it), but he can still vividly recall some of the back covers due to their very eyecatching nature. I have certainly never seen them, and despite their dramatic appeal they do not seem to have been reproduced in any other publication. So what was this mysterious partwork? If there is any *FT* reader who has seen it (or, better still, owns an edition of it!), or can offer any extra details, I'd love to hear from you. **Alan Pringle, several pers. comms, August 1998.**

PERUSING IN PERU. The latest volume of *Cryptozoology*, published by the International Society of Cryptozoology, has now appeared, and contains an interesting paper by Peruvian zoologist Dr Peter Hocking, reporting several mysterious Peruvian beasts seemingly unknown to science (his second *Cryptozoology* paper on this subject). These include: a reddish-brown mystery cat dubbed the 'jungle lion', said to be as large as an African lion and possessing long hair around its neck; a diminutive form of brown bear; and an all-black, wattle-lacking form of guan (gallinaceous bird). *Cryptozoology*, **vol. 12 (for 1993-1996 [but published in 1998]).**

I SPY AN IBIS. The blue-headed heron-like birds spied flying off from the Mahakam River in Kutai, Indonesian Borneo, by a team of Bornean and Dutch conservation researchers in early 1998 were distinctive and unfamiliar enough to warrant checking in a bird guide. Remarkably, they proved to be Davison's ibis *Pseudibis davisoni* - an endangered species previously thought to be confined to Vietnam. **Asia Intelligence Wire, Antara, 27 August 1998.**

Alien Zoo

A MEXICAN SABRE-TOOTH? A recent magazine article contained a frustratingly brief snippet claiming that in 1994, a Mr Roberto Guitierrez observed a creature resembling a living sabre-tooth in northern Mexico. Although I am aware of such reports from South America, this is the first Mexican report known to me. More details, anyone? *Science Illustrée*, **September 1998;** also *Illustreret Videnskab*, **September 1998.**

Do sabre-tooths still thrive in Mexico? (Dr Karl Shuker)

THE GIANT RAT OF CHELTENHAM. Reports of giant rats stalking the suburbs of Cheltenham, Gloucestershire, are not known for their frequency. But those emanating from a certain upmarket housing estate during August 1998 were finally backed by substantial proof - a bona fide 3-ft-long rodent with a shrill squeak, captured alive and conclusively identified by local exotic animals expert Terry Hooper as a South American coypu *Myocastor coypus*. The coypu was formerly a pest in the fens of East Anglia where it became established after escaping from fur farms. Although supposedly eradicated several years ago, a few specimens have turned up spasmodically since then across southern England. ***Sunday People*, 6 September 1998.**

Engraving of coypus

Alien Zoo

LASCAUX WAS NEVER LIKE THIS! For generations, South African anthropologists have marvelled at rock paintings in the Karoo, produced for untold centuries by the San hunter-gatherer people and depicting some very extraordinary entities - including mermaids. Recently, however, more have been discovered in Ezeljachtpoort, again in the Karoo, whose mermaid paintings are the most distinctive currently known. But how can they be explained? Opinion remains divided, with one school of academic thought pursuing the possibility that they represent the hallucinogenic experience of San shamans, and another proposing that they portray departed spirits, which feature extensively in San society. According to an elderly San man, however, they genuinely depict water maidens. Even today, mermaid sightings are often reported from this region. *Sun Times*, **13 September 1998.**

HOGGING THE HOG DEER. In October 1998, Sri Lanka's Department of Wildlife Conservation revealed that the hog deer *Axis porcinus*, hitherto believed to have become extinct on this island five decades ago, still survives here - thanks to a number of native inhabitants with an exotic taste in pet keeping. Unbeknownst to conservationists, who had been unsuccessfully seeking Sri Lanka's supposedly long-lost mammal for years, several villagers here have been keeping hog deer as pets! *South China Morning Post*, **2 October 1998.**

RETURN OF THE NATIVES. Although known still to thrive in Scottish and Welsh waters, the herring-related sea fish known as the allis shad *Alosa alosa*, which comes inland to spawn, recently took experts by surprise when it was found to be alive, well, and breeding in the rivers of Cornwall. Its discovery was due to the investigations of Alex Brown from the Environment Agency's Bodmin office, who revealed that a mystifying fish locally dubbed the scad or horse mackerel was the allis shad. *Wolverhampton Express and Star*, **18 September 1998.** Equally encouraging was the rediscovery in large numbers of the glutinous snail *Myxas glutinosa*, in Wales's Bala Lake. Except for a single live specimen found in Oxfordshire in 1991, this tiny species was last recorded from Britain in 1953. *Times, Daily Telegraph, Independent*, **23 October 1998.**

CHINESE WHISPERS ABOUT CHINESE PANTHERS? ABCs (alien big cats) of the black pantheresque variety have long been reported from Britain, continental Europe, North America, and even Australia and New Zealand, but now they seem to be invading more exotic realms too. According to a recent Xinhua News Agency report, Chinese scientists claim to have evidence that "American panthers" may be residing deep in the Qinling Mountains of northwest China's Shaanxi Province. Zheng Fengsong, a senior engineer of the Changqing Nature Reserve, spied one such beast here on 19 February 1998, which left behind "clear claw prints in the shape of a plum flower" that were duly photographed. Biologist Prof. Liu Shifeng, from Northwest China University, added that back in 1984, a hunter had informed him that he had shot just such an animal on the hillside near his home in Taibai County, but did not retain the carcase.

According to Liu: "[If] the rare American panthers are really proved to be there [in the Qinling Mountains], it will be a major discovery of wild animals". However, this is an odd report for various reasons. Firstly, black panthers (i.e. melanistic leopards) are not native to America (unless they are actually melanistic pumas?). Secondly, normal spotted leopards are known to exist in the Qinling Mountains, so what is so special about melanistic specimens also occurring there? Thirdly, just like the vast majority of cats, leopards do not typically leave behind clawed footprints (though it is not unknown for them to do so). **http://info.xinhua.org/ 7 November 1998.**

TAKING HEED OF THE TAKAHE. In 1948, 50 years after its last confirmed sighting, the large, flightless moorhen-related takahe *Porphyrio mantelli* was sensationally rediscovered near Lake Te Anau on New Zealand's South Island. Fifty years further on again, a reunion was held at Te Anau on

Alien Zoo

the weekend of 21-22 November 1998 by this famous bird's four co-rediscoverers - Dr Geoffrey Orbell, Neil McCrostie, and Rex and Joan Watson - a fitting celebration of one of the 20th Century's most momentous cryptozoological events. *New Zealand Evening Post*, **23 November 1998**.

Celebrating the takahe's anniversary! (Dr Karl Shuker)

PONIES OF A DIFFERENT COLOUR? Webmaster of the online cz@onelist cryptozoology discussion group, Chad Arment recently heard of a hitherto-obscure early report concerning "little ponies, striped with yellow and white" in New Guinea. Also noted by Gavin Souter in *New Guinea: The Last Unknown* (Taplinger: New York, 1963), this very brief report is from an 1888 book entitled *Adventures In New Guinea*. It was described by Souter as a "fanciful tale" of the adventures of Louis Trégance, a French sailor captured by an inland tribe called the Orangwoks (the owners of the ponies) who escort him to their city where he remains captive for the next nine years. Always assuming that this tale is indeed genuine (I have yet to find any mention of the Orangwoks outside of Trégance's book), more data regarding these decidedly odd-sounding animals, which according to Trégance were also extremely fleet-footed, would be very welcome. **Chad Arment, cz@onelist.com December 1998.**

FINDING OUT MORE ABOUT *FINDING OUT*. Since my Alien Zoo item appeared in *FT*116

(November 1998) concerning the mysterious magazine partwork from the early 1970s sporting lavish cryptozoologically-themed illustrations on the back covers, my source, Caledonian cryptozoologist Alan Pringle has succeeded in locating his friend from the early 1970s who owned a complete set. Sadly, he no longer does so, but he did remember the partwork's title, which was *Finding Out*. So if anyone does have any of its issues, please let us know, just in case we can trace the artist(s) responsible for their cryptozoological illustrations, and learn whether these may be available for republication. **Alan Pringle, pers. comm., December 1998.**

A caged puma in Cannich vindicated Ted Noble's claims of a mystery livestock predator (Dr Karl Shuker)

TED NOBLE AND THE CANNICH PUMA. In October 1980, farmer Ted Noble created a sensation when a cage, set on his estate at Cannich in the Scottish Highlands to catch the mysterious beast preying upon his livestock, caught a live female puma. Due to her elderly, tame nature, some sceptics doubted that this specimen, later dubbed Felicity, was the creature responsible, but Noble remained convinced that he had captured the true culprit.

Blissfully unaware of the controversy surrounding her, Felicity spent the remainder of her days at the Highland Wildlife Park, dying peacefully there in 1985. Sadly, Ted Noble too has now passed away, on 11 December 1998, aged 76, but his name will live on in the annals of Britain's long-running ABC

Alien Zoo

(alien big cat) saga. *Aberdeen Press and Journal*, **15 December 1998**.

COUGAR COMEBACK? Staying with pumas: The last confirmed specimen of a native wild cougar (puma) in Missouri, i.e. belonging to the eastern cougar subspecies *Puma concolor couguar*, was killed in 1927, but since then many sightings have been made, and some cougars have even been videoed in the wild here. Yet their behaviour suggested to experts that they were merely escapee cougars from captivity, and hence not genuine eastern cougars. On 13 November 1998, however, a cougar pelt was found lying beside a road near Houston in south-central Missouri's Texas County by a deer hunter, and it still had its head and feet attached. Researchers from Missouri University are studying the pelt itself for clues as to its former owner's taxonomic identity.

Meanwhile, DNA researchers from the University of California-Davis and the National Cancer Institute at Bethesda are analysing DNA in muscle and hair samples from the pelt, in a parallel bid to uncover whether this individual was a bona fide eastern cougar, rather than a non-native specimen. *St Louis Post-Dispatch*, **30 December 1998**.

CUMULATIVELY CRYPTOZOOLOGICAL. An interesting scientific paper of cryptozoological relevance was published during late 1998 in the *Journal of the Marine Biological Association of the U.K.* Authored by Dr Charles Paxton and entitled 'A Cumulative Species Description Curve For Large Open Water Marine Animals', the open water marine creatures that it is concerned with are those in excess of 2 m long and described by science from 1830 to 1995. From the resulting curve, it is estimated that a maximum of some 47 such species still await formal description, at a current rate of approximately one species per 5.3 years, and Paxton suggests that any imminent species descriptions are likely to be of cetaceans. *Journal of the Marine Biological Association of the U.K.*, **78: 1389-91**.

FACING PAGE: Felicity the puma (CFZ)

The Lost Ark

ARCHANGEL FEATHERS
- MORE THAN A DIVINE FLIGHT OF FANCY?

"All feathers relate to the human spirit and its innate connection to the Divine. Through feathers, we can learn to make our wishes reality. They empower our thoughts, and through them we can invoke Nature to confirm this link for us."

Ted Andrews – *Animal Speak: The Spiritual and Magical Powers of Creatures Great and Small*

From little acorns mighty oaks grow, or, as in this particular instance, from the briefest, 59-word footnote a 3800-word chapter flourishes – not to mention more than ten years of personal investigation into a hitherto-obscure but thoroughly fascinating (if highly frustrating!) subject.

THE TALE OF A TAIL – HOW IT ALL BEGAN

My knowledge of, and continuing interest in, those mystifying sacred relics known as archangel feathers began over a decade ago, shortly after purchasing an ornithological encyclopedia from 1896 entitled A *Dictionary of Birds*, edited by Alfred Newton and Hans Gadow.

Perusing its entry for a spectacular Guatemalan species of trogon called the quetzal *Pharomachrus mocinno* - which referred to the male's extraordinarily long tail plumes (indeed, as seen in flight the male quetzal inspired the Aztecs' belief in the green feathered serpent incarnation of their sky god, Quetzalcoatl) - I noticed the following tantalisingly-short footnote:

FACING PAGE: Quetzalcoatl (Andy Paciorek)

Alien Zoo

> He [18th/19th-Century French zoologist Baron Georges Cuvier] possibly had in mind the celebrated feather treasured in the Escurial as having come from the wing of the Archangel Gabriel. This might be thought to have been a Quetzal's, but the author of *Vathek* who saw it in 1787, says (*Italy with Sketches of Spain*, ii. p. 325) it was rose-coloured.

19th-Century engraving of male quetzal

My imagination was instantly captured, and captivated, by the romance of a divine archangel plume ensconced within the earthly abode of Madrid's palatial complex known as El Escorial (aka the Escurial), housing the Monastery of San Lorenzo - even though the rather more prosaic zoological voice intoning within my brain insisted that it evidently derived from some mortal albeit exotic avian form. Consequently, I was determined to uncover whatever I could about El Escorial's wondrous object of veneration.

BECKFORD BECKONS

As my starting point, I decided to seek out the identity of the unnamed author of the two tomes mentioned in the footnote, who had personally observed this fabled plume. That was soon accomplished – the person in question was William T. Beckford (1760-1844), an exceedingly wealthy and somewhat eccentric Englishman, who had travelled widely through Europe, visiting as many locations and establishments of note as possible. *Vathek* was a well-received Gothic fantasy novel that he wrote during his mid-20s, but of greater interest to me was the second volume of his other footnote-cited work, whose full title turned out to be *Letters From Italy, With Sketches of Spain and Portugal*, published in 1834. Obtaining a copy of this book, however, proved to be rather more difficult than I had expected, with even the normally indefatigable inter-library loans service stumbling – curiously, they could obtain the first but not the second volume for me.

The library angel, however, was evidently – and appropriately, given the subject of my research – taking an interest in my endeavours, because very shortly after I'd failed to elicit the elusive second volume via the loans service, I happened by sheer chance to spy a very pertinent book review in the *Daily Mail* newspaper for 6 June 1998. The review was of a newly-published biography of William Beckford!

Entitled *William Beckford: Composing For Mozart* (alluding to the fascinating fact that, as a child, Beckford was actually tutored in music by Mozart), its author was Dr Timothy Mowl, a Research Fellow in Bristol University's History of Art department. I lost no time in accessing this book, which proved to be an extensively-researched, exceedingly-detailed work, and after carefully perusing its contents I discovered on p. 195 the following mention of what I had by now come to refer to as the Gabriel

Alien Zoo

feather:

> A feather from the wing of the Archangel Gabriel, three feet long, rose-coloured and preserved in ambergris, may have gone some way to subduing his religiosity.

To have produced such a comprehensive account of Beckford's life, Dr Mowl had evidently succeeded in accessing all of his publications, so I wrote to him, explaining my interest in the Gabriel feather and enquiring whether he could send me the section dealing with it that had appeared in Vol. 2 of Beckford's *Letters From Italy...* Happily, Dr Mowl kindly obliged, forwarding to me the version reprinted in a compilation volume of Beckford's writings entitled *Vathek and European Travels*, published in 1891. And so, finally, I was able to read for myself Beckford's full account of his sighting (made on 19 December 1787) of this extraordinary icon at El Escorial:

> A spacious vault was now disclosed to me – one noble arch, richly panelled: had the pavement of this strange-looking chamber been strewn with saffron, I should have thought myself transported to the enchanted courser's forbidden stable we read of in the tale of the Three Calendars.
>
> The Prior, who is not easily pleased, seemed to have suspicions that the seriousness of my demeanour was not entirely orthodox; I overheard him saying to Roxas, "Shall I show him the Angel's feather? you know we do not display this our most valued, incomparable relic to everybody, nor unless upon special occasions." – "The occasion is sufficiently special," answered my partial friend; "the letters I brought to you are your warrant, and I beseech your reverence to let us look at this gift of Heaven, which I am extremely anxious myself to adore and venerate."
>
> Forth stalked the Prior, and drawing out from a remarkably large cabinet an equally capacious sliding shelf – (the source, I conjecture, of the potent odour [ambergris] I complained of) – displayed, lying stretched out upon a quilted silken mattress, the most glorious specimen of plumage ever beheld in terrestrial regions – a feather from the wing of the Archangel Gabriel, full three feet long, and of a blushing hue more soft and delicate than that of the loveliest rose. I longed to ask at what precise moment this treasure beyond price had been dropped – whether from the air – on the open ground, or within the walls of the humble tenement at Nazareth; but I repressed all questions of an indiscreet tendency – the why and wherefore, the when and how, for what and to whom such a palpable manifestation of archangelic beauty and wingedness had been vouchsafed.
>
> We all knelt in silence, and when we rose up, after the holy feather had been again deposited in its perfumed lurking-place, I fancied the Prior looking doubly suspicious, and uttered a sort of *humph* very doggedly; nor did his ill-humour evaporate upon my desiring to be conducted to the library.

Alien Zoo

'Annunciation', painted by an anonymous French (or Dutch?) artist, c.1370s, depicting the archangel Gabriel and the Virgin Mary

Irrespective of the monk's reluctance to reveal their most precious relic, one thing seems certain – the Gabriel feather did indeed exist, or, at least, it existed at the time of Beckford's visit in 1787. But what of now, over two centuries later? Was this sacred relic still to be seen at El Escorial, and were there any photographs of it?

Even before I had obtained the above passage from Dr Mowl, I had already formulated some ideas as to the possible ornithological origins of the Gabriel feather, but I needed some visual material to examine, as well as any information available as to its history. When and how had it come to be housed at El Escorial, and where had it been until then?

THE CURIOUS CASE OF THE MISSING FEATHER

Paying a visit to the monastery of San Lorenzo in El Escorial seemed the most direct approach, but on reflection, rather than simply arriving unannounced I decided to make first contact by letter. So I wrote to the monastery's director, Dr Pedro Criado Juarez, explaining my interest in the Gabriel feather and requesting details and sight of any illustrations of it that they may be able to send or loan to me. However, this is where the saga of my researches concerning this revered relic took a decidedly fortean turn.

On 21 January 1998, I received the following letter from Carmen Garcia-Frias Checa, the monastery's Keeper of National Heritage, which was kindly translated from Spanish into English for me by Hispanic cryptozoologist Scott Corrales:

> Dear Sir,
>
> I am sorry to inform you that there is no such thing as an archangel feather in the reliquaries of this monastery. Neither is there any bird feather or holy object which might be of ornithological interest.
>
> I do not know where did [sic] you get the information about the existence

of the said feather but I can assure you it is not mentioned by the most important chroniclers of El Escorial (among them Fray José de Siguenza who made an account of the monastery in his 1605 book "The Foundation of the Monastery of El Escorial").

Perhaps you could find some information about the Gabriel feather in the "Books of Donations" which contain a detailed account of all the objects donated to this monastery during the reign of King Phillip II [1527-98]. They consist of 8 handwritten volumes which are housed in the Archives of the Royal Palace in Madrid. The section of "the Books of Donations" dealing with relics has never been transcribed, the only published section being one dealing with objects of art, by J. Zarco, "Inventory of Jewels,...", *Boletín de la Real Academia de la Historia* [*Bulletin of the Royal Academy of History*], 1930.

Hoping to have been of service.

Yours sincerely

Curiouser and curiouser, as Alice might have said. In 1787, the Gabriel feather was famous across Europe and so devoutly venerated at El Escorial that it was shown only to those deemed deserving of the honour of gazing upon its radiant form. A mere two centuries later, conversely, its very existence, present and past, was being rigorously denied there. Why? Further letters from me to El Escorial seeking explanations for this anomaly failed to elicit any response. Clearly, therefore, this channel of communication and investigation was closed. True, I could have pursued the possibility that the Gabriel feather was indeed mentioned somewhere within the untranscribed sections of the Books of Donations, but this may well have led nowhere - especially if, as the monastery's Keeper had claimed, it was not alluded to in Siguenza's book of 1605 (in fact, my own researches suggest that this book was originally published even earlier, in 1598). This suggested that the feather had reached the monastery sometime after that date, and also, therefore, after the period of time covered by the Books of Donations.

Of course, it may still be held in El Escorial, but not in the monastery. This enormous complex also houses the Basilica of San Lorenzo el Real, the Royal Palaces of the Bourbons and Austrians, the college, the sacristy, a library, a picture gallery, a throne room, a museum of architecture, and several other major edifices – collectively offering a myriad of possible hideaways for a relatively small relic.

Alternatively, was it conceivable, I wondered, that sometime after Beckford's visit, the Gabriel feather had been exposed as a hoax, and so was quietly removed and disposed of, with all documented traces of this now-embarrassing artefact expunged from the monastery's records (or at least from its public records)? Or had some enterprising visitor with ornithological knowledge simply exposed its avian identity?

IN SEARCH OF AN (AVIAN) IDENTITY

There are a number of contenders on offer. Returning to that fateful footnote that set me off in hot pursuit of archangel feathers, a lengthy plume from the tail of the male quetzal had been one identity con-

Alien Zoo

A male Count Raggi's bird of paradise

sidered in the past, and there is no doubt that such plumes are very spectacular, and are of the same length as that quoted by Beckford for the Gabriel feather. Unfortunately, however, there is one glaring discrepancy with this explanation – the Gabriel feather was rosy in colour, whereas those of the quetzal's tail are emerald green.

Another possibility, at least on first sight, is a flamingo feather. Some of these can indeed exhibit a rosaceous hue, but they are neither lengthy enough nor exotic enough to warrant serious consideration.

In 1998, I published a couple of short articles on the Gabriel feather, concentrating upon its putative ornithological origin, and concluded that a much more satisfactory identity for it would be one of the elongate and absolutely gorgeous rose-suffused courtship plumes from a male Count Raggi's bird of paradise *Paradisea raggiana*, native to New Guinea. Expanding this notion in two of my subsequent books – *Mysteries of Planet Earth* (1999) and *Extraordinary Animals Revisited* (2007) – I revealed that bird of paradise skins (with those of Count Raggi's a particular favourite) were first brought back to Europe at least as long ago as 1522, by the survivors of the once-mighty expeditionary fleet of renowned Portugese explorer Ferdinand Magellan, and subsequent expeditions to New Guinea returned with many more. How fitting it would be if a feather once thought to be from an archangel of Heaven proved to have originated from a bird of paradise! Certainly, it would not be difficult to conceive how a feather from the skin of one of these heavenly-looking birds might have been taken to a European monastery and passed off as a genuine angel plume – the monks there would never before have seen anything remotely as exquisite as a bird of paradise feather. And the rest, as they say, would have been history...until, that is, in later ages, when someone with zoological knowledge came along and exposed its true identity.

FRAUDULENT FEATHERS?

Even worse, however, is the prospect that the feather may not even have been an authentic plume, but rather a composite, purposefully crafted from the skilful joining together of several smaller ones to engender a resplendent feather of inordinately eyecatching appearance and dimensions. Indeed, when first shown quetzal tail feathers, Cuvier had opined that they must be fakes, composites designed to fool unwary or gullible scientists. If the Gabriel feather has been unmasked as a hoax, it would not be unique. In his book *Roman Catholicism* (1962), Presbyterian Minister Loraine Boettner noted that a very showy plume previously displayed in one of Spain's many cathedrals and claimed to have fallen from one of Gabriel's wings when he visited Mary at the Annunciation was subsequently revealed upon investigation to be a magnificent ostrich *Struthio camelus* feather. (Could a rose-dyed ostrich plume explain the Gabriel feather?) Moreover, a fraudulent Gabriel feather (actually from a parrot) is even a central theme in a major work of classic medieval fiction – 'The Tenth Tale [told on the sixth day] – Friar Cipolla and a Feather of the Angel Gabriel', from *The Decameron* (c.1350 AD) by Giovanni Boccaccio.

OTHER GABRIEL FEATHERS

During my more recent investigations of El Escorial's elusive Gabriel feather, I have learnt of several other supposed archangel plumes allegedly held in other religious establishments, which I briefly summarised in *Dr Shuker's Casebook* (2008). For example, one of my Spanish correspondents, Isabela Herranz, mentioned that she had read in a Spanish guide book that there was said to be an angel feather in a small Navarra convent, but she was unable to discover anything more about this, because when she contacted the convent it claimed to have no knowledge of any such feather. Veteran *FT* correspondent Ted Williams recalled reading about angel feathers too, but didn't offer any specific details. In his privately-published book *A Truth Seeker Around the World* (1882), D.M. Bennett included the following noteworthy snippet about angel feathers held at a monastery on Mount Athos, Greece, referring to them as:

> ...identical feathers which came from the angel Gabriel's wings at the time he visited the Virgin and told her about her being overshadowed by Yahweh, or the Holy Ghost.

In early 2008, I discovered that a Gabriel feather is also reputedly preserved at the Basilica di Santa Croce in Gerusalemme, Florence, Italy. I have since sent a number of emails of enquiry to the Basilica via its official website regarding its archangel feather, but as is so often the case when dealing with these enigmatic objects, I have not received any reply. Consequently, if there is any reader out there who has relevant information, I would be delighted to receive it, as my investigation is currently at a standstill, with no new leads to pursue.

A GABRIEL FEATHER IN ENGLAND!

The article that forms the basis of this chapter was published by *FT* in its January 2009 issue, and on 14 January 1999 I received the following fascinating email from Chris Woodyard of Dayton, Ohio:

> As a relic-fancier, I was very interested in your article on the relic of the Archangel Gabriel's feather at El Escorial. I'm afraid I can't help you with the location of that particular relic, but I have found a reference to more feathers from the wings of the Archangel Gabriel a little closer to home: in the church of St John the Baptist, Pewsey, Wiltshire. They may still be displayed behind glass in a recess in one of the pillars where they were discovered during restoration work in 1888. The story is that Zacharias, John the Baptist's father, clutched at the Archangel after receiving the news of his son's forthcoming birth and was left with a handful of feathers. A Crusader from Pewsey found them in the Holy Land and brought them home to the

Alien Zoo

local church. The story (along with a photo of a feather faintly seen in the glassed-in recess) is found in *Timpson's Country Churches*, John Timpson p. 125.

And sure enough, when I consulted that book, there on p. 125 was a colour photo by acclaimed church photographer Christopher Dalton depicting what looked rather like a white goose feather sealed inside a dusty, wood-framed, glass-fronted recess. Continuing Chris's email:

> Although I cannot recall the reference, I have also read that some feather relics may have actually been dried plant fronds. Featherwork embroidery imported from New Spain [i.e. Mexico] enjoyed a vogue in Spain and Portugal during the reign of Philip II and many fine examples of religious featherwork still survive. (See http://www.ladap.org/library/article_detail.php?idf5, although the artisan [who, by a wonderful coincidence, just happens to be named Don Gabriel!] doesn't mention some significant collections. He does mention the types of feathers used.). It would not surprise me to hear that the sacristies of El Escorial contained feathers or featherwork from New Spain or even further afield, even if they were not the product of Archangelic moulting.

The possibility of at least some Gabriel feathers being the product of skilful ornamental featherwork is another interesting option well worth further investigation. So too is the dried frond notion, especially as the extremely long leaves of the raffia palm tree *Raphia regalis* (and related species) were sometimes passed off to naïve crusaders during the Middle Ages as feathers from the gigantic mythical bird known as the roc. I actually own some green-dyed specimens of these leaves and can confirm that to the casual, zoologically-untrained eye they do look deceptively plume-like, as can be seen opposite. Needless to say, I was also very thrilled to learn from Chris that there was apparently a Gabriel feather here in England, so I lost no time in emailing the church of St John the Baptist in Pewsey, and finally on 9 April 2010 I received the following informative email from Pewsey's local historian, Roger Pope:

> I cannot find any reference to the feather in Canon Bouverie's book on Pewsey which was written in 1890 and is the earliest book on Pewsey Church. I will however look in the Church wardens Register if I can find a copy. They do go back to c. 1600. There is a reference in [Arthur] Mee's *Wiltshire*. It is somewhat dismissive but this is what Mee says: "One of the massive square piers of the nave arcades has a small recess to protect a queer trifle for which all travellers look, for it is a relic of the days of superstition when people believed they could pick up on Earth a feather from an angel's wing. The traveller who climbs on a chair and looks in this recess will find in it long white feathers from a swan or a goose. They are dusky with age and it is said that they have been here since the Church was dedicated to St. John the Baptist, an old legend saying that the Angel Gabriel dropped them from his wing when he announced the good news to the Madonna". Although that was written in 1939 it is as good a description as would be written today.

One of my roc feathers (Dr Karl Shuker)

Alien Zoo

I emailed back to Roger to thank him for this much-valued information, and to ask whether these feathers are still retained in the church, and on 14 April I received the following reply:

> They are exactly the same today as they were in Mee's description. Nothing has changed.

So even though they do seem to be the feathers of mere waterfowl rather than anything of divine origin, I am no less delighted, because at least I have finally tracked down some contemporary claimants from this most ethereal category of holy relics. Having said that, on 28 July 2010, I received another extremely illuminating communication by email regarding the Pewsey feathers, offering a very different take on their status, this time from acclaimed paranormal-occult author Andrew Collins. Andrew first wrote about them in his book *From the Ashes of Angels* (1996), but had not observed them himself at that time. Nowadays, however, he lives quite near Pewsey and has viewed them through their glass panel on several occasions:

> With regards the Pewsey angel feathers. The idea that they are angel feathers was merely the suggestion of those who found them during restoration work in the nineteenth century. It was assumed that they must have been brought back from the Holy Land at the time of the crusades, and were thus most probably a holy relic consisting of angel feathers. There is absolutely no basis for this identification. Plus holy relics are more likely to have been placed somewhere in the vicinity of the high altar. More likely is that the feathers' placement in a concealed niche within a pillar nearest to the north door, well away from the altar, is an act of archaic folk magic. If they are goose feathers, and this was the identification when they were first uncovered, then this might well be the clue.
>
> A graffito image of a grey lag goose [*Anser anser*] is carved in the same place on one of the two pillars closest to the north door in the church of Waltham St Lawrence church in the neighbouring county of Berkshire. See Michael Bayley, *Caer Sidhe: vol. 1 - The Night Sky* (Capall Bann, 1997). The author interprets the graffito as a symbol of the Cygnus constellation, which in the past was a goose in parts of Britain. Swans and geese were, until well into the twentieth century, strongly associated in Britain with the transmigration of the soul back to a northern placed heaven, which was probably associated with both the Milky Way and the Cygnus stars (see my book *The Cygnus Mystery*, 2006). I went to Waltham St Lawrence church, but could not see the carving, but it was certainly there as a drawing of it is found in the book quoted here.
>
> Obviously north doors in churches were associated with the devil, simply because this was the old pagan direction of heaven, particularly in Germanic and Nordic mythology. I have a distinct feeling that in medieval times the north door of churches was still seem as the access point to a northerly placed heaven in the minds of some rural communities, and that the goose and swan, and of course their feathers, were symbols of a safe transition to the next life. It was perhaps for this reason that the feathers were purposely concealed in a niche by the north door in Pewsey church.

Heartened by this long-hoped-for success, I have now sent off some enquiries once again concerning the medieval European examples documented above, but, as before, I have yet to receive any response. Perhaps, however, I should not be too surprised at the steadfast silence confronting my attempts to track down these elusive – and illusive? – items. After all, as I mentioned when writing about them just over a decade ago in *Mysteries of Planet Earth* (1999):

> The cynic may say that our mundane mortal world is incapable of retaining for any length of time such fragrant, divine traces as the plumes of angels. In reality, of course, we must acknowledge that the concept of angels bearing feathery avian wings is a relatively recent one, nurtured principally by Renaissance artists and hence dating back only a few centuries, with no foundation in early theological lore. Even so, it felt somewhat comforting to know that a few relics with claims to angelic fame (justified or not) existed on earth – which is surely now a little sadder for their absence.

REFERENCES

BECKFORD, William T. (1834). *Letters From Italy, With Sketches of Spain and Portugal, Vol. 2.* Baudry's European Library (Paris).

BECKFORD, William T. (1891). *Vathek and European Travels.* Ward Lock (London).

BENNETT, D.M. (1882). *A Truth Seeker Around the World.* Privately published (New York).

BOETTNER, Loraine (1962). *Roman Catholicism.* Presbyterian & Reformed Publishing House (Phillipsburg - New Jersey).

COLLINS, Andrew (1996).*From the Ashes of Angels: The Forbidden Legacy of a Fallen Race.* MichaelJoseph (London).

COLLINS, Andrew (2010).Pers. comms, 28 July.

LEWIS, Peter (1998). The eccentric, his wife and a dangerous liaison [review of: MOWL, Timothy (1998). *William Beckford: Composing For Mozart.* John Murray (London)]. *Daily Mail* (London), 6 June, p. 39.

MEE, Arthur (1950). *Wiltshire: The King's England.* Hodder & Stoughton (London).

MOWL, Timothy (1998). *William Beckford: Composing For Mozart.* John Murray (London).

NEWTON, Alfred & GADOW, Hans (eds) (1896). *A Dictionary of Birds.* Adam & Charles Black (London).

NICKELL, Joe. (2007). *Relics of the Christ.* University of Kentucky Press (Lexington).

POPE, Roger (2010). Pers. comms, 9 and 14 April.

SHUKER, Karl P.N. (1998). Angel feathers and feathered snakes. *Strange Magazine*, no. 19 (spring): 24-5.

SHUKER, Karl P.N. (1998). In search of the Gabriel feather. *Wild About Animals*, 10 (May-June): 38.

SHUKER, Karl P.N. (1999). *Mysteries of Planet Earth: An Encyclopedia of the Inexplicable.* Carlton (London).

SHUKER, Karl P.N. (2007). *Extraordinary Animals Revisited.* CFZ Press (Bideford).

SHUKER, Karl P.N. (2008). *Dr Shuker's Casebook: In Pursuit of Marvels and Mysteries.* CFZ Press (Bideford).

SHUKER, Karl P.N. (2010). Two of a fortean kind. *Fortean Times*, no. 265 (August): 51.

TIMPSON, John (1998). *Timpson's Country Churches.* Weidenfeld & Nicolson (London).

WOODYARD, Chris (2009). Pers. comm., 14 January.

Illustration of pool frog from 1908

1999

REBORN IN THE SOUTH. South Australia has recently seen the discovery of a bird species never previously reported within its borders, and also the rediscovery of a very mysterious lizard last seen here back in 1934. A colony of spinifex birds *Eremiornis carteri*, a small brown warbler-like species, was spotted last year in the far north of the state by scientists participating in the Biological Survey of South Australia, a major wildlife project funded by this state's government. And the lizard, *Egernia kintorei*, a species of skink known locally as the tjakura, was refound following a diligent search for it in their lands by the Pitjantjatjara aborigines, within whose mythology this reptile features extensively, despite being 'lost' for over 60 years. *Adelaide Advertiser*, **9 January 1999.**

NOT SO LUCKY, AFTER ALL. In January 1999, a small frog named Lucky died of old age, and with its death Britain lost a whole species of native animal. For Lucky, captured in a small Norfolk pool back in 1993, was the last-known specimen of pool frog *Pelophylax* (*Rana*) *lessonae* known to exist in Britain. Having said that, a number of persons have since claimed to know where other specimens can still be found. Ironically, it had traditionally been assumed that the pool frog had been introduced to Britain from the continent by Victorian collectors less than two centuries ago. However, in early 1999, just prior to Lucky's death, Coventry University zoologist Dr Chris Gleed-Owen had discovered the thigh bone of a pool frog at Chopdike Grove, Lincolnshire, which is a Saxon settlement dating from 600 to 950 AD. *Times, Eastern Daily Press*, **14 January 1999.**

NOTHING TO CROW ABOUT. British twitchers may soon be adding a new but rather unwelcome addition to their ornithological checklist. The Asian house crow *Corvus splendens* is an aggressive bird that raids other birds' nests, killing fledglings and devouring eggs, and flocks will attack ducks, cats, and even humans. To quote UK ornithologist Chris Mead: "They're like something out of the horror film *The Birds*". Now this menacing species, which has endangered the survival of several smaller

Alien Zoo

birds, is infiltrating Europe, having already begun to breed in the Netherlands, and experts fear that it may reach Britain. Yet its western invasion owes more to stowaway subterfuge than to typical migratory means, because the crows are expanding their distribution range by taking refuge on maritime vessels; they first reached Europe aboard a container vessel. **Sunday People, 31 January 1999.**

A GREYER SHADE OF BLACK. Despite countless reports of such felids, not a single specimen of a black puma has ever been obtained and conclusively identified in North America. Hence I was very interested to learn a while ago from correspondent Keith Foster that he planned to investigate reports that a black puma had been killed in Oklahoma back in the 1970s. True to his word, Keith did indeed investigate, but hopes of a cryptozoological scoop were foiled when he discovered that the specimen in question had actually been grey, not black. Nevertheless, this is at least one such case that has been thoroughly investigated and can now be eliminated from any further consideration - a worthwhile achievement in itself. **Keith Foster, pers. comm., 1 February 1999.**

KEEPING UP WITH THE KAPRE. On 12 March 1999, Californian bigfoot researcher Bobbie Short plans to fly to Manila, to seek evidence for the existence of one of the bigfoot's less familiar relatives, a Filipino bipedal man-beast known as the kapre. Reported from the islands of Luzon and Samar, and said to stand up to 8 ft tall, the kapre is covered with hair but has a human-like face, hands, and feet. Seemingly omnivorous, it is particularly fond of fruit such as mangoes, pineapples, and papayas, as well as fish, land crabs, and even the local rats. The Philippines have a history of man-beast legends and lore, as previously documented by Mark A. Hall (*Wonders*, September 1993). **NewsFlash Service, 24 February 1999.**

THINK PINK. The cz@onelist cryptozoology discussion group has yielded some fascinating snippets of previously-overlooked or little-publicised data since its establishment in 1998. One of the latest stems from the online revelation by cryptozoological researcher Richard Muirhead that Edward Schafer's book *The Vermilion Bird* (1967), concerning life in T'ang Dynasty China (618-907 AD), refers to a race of black elephants with small pink tusks in Hsun and Lei, corresponding to the Leizhou Peninsula and southeastern Guangxi Province. It appears that this peculiar form of pachyderm has been formally dubbed *Elephas maximus rubridens* by zoologist Dr P.E.P. Deraniyagala, who used as his type specimen a depiction published in 1925 of an antique Chinese bronze statuette, held at Chicago's Field Museum of Natural History. **Assorted cz@onelist communications, February-March 1999.**

RETURN OF CUBA'S IVORYBILL? After an official rediscovery in 1986, the spectacular Cuban

Engraving of ivory-billed woodpecker

Alien Zoo

ivory-billed woodpecker *Campephilus principalis bairdii* seemed to have slipped back into official extinction. Recently, however, a new flurry of reports has emerged, which seem sufficiently convincing to have inspired a team of ornithologists to plan a week-long search for this elusive bird. **Chad Arment, cz@onelist.com, 5 March 1999.**

SUMATRAN RHINOS IN INDIA TOO. The Sumatran rhinoceros *Dicerorhinus sumatrensis* has been widely believed extinct in the Indian subcontinent since the early 1920s. In March 1999, however, hopes for its survival here were raised, following an official announcement by Anwarudding Choudhury, chief executive of the Rhino Foundation, that several specimens had been sighted close to India's border with Myanmar (formerly Burma). ***Vancouver Sun*, 13 March 1999.**

QUEST FOR A QUAIL. For much of the 20th Century's second half, India was famous ornithologically for four 'classic' extinct birds - Jerdon's courser *Rhinoptilus* (*Cursorius*) *bitorquatus*, the forest spotted owlet *Athene blewitti*, pink-headed duck *Rhodonessa caryophyllacea*, and Himalayan mountain quail *Ophrysia superciliosa*.

More recently, however, the first two have been rediscovered, and there have been quests for and claimed sightings of the third. Now, WWF-India is planning to launch a search by local people during November 1999-January 2000 for the fourth. Ever since it was formally described back in 1846, the Himalayan mountain quail has been the most elusive member of this quartet - so much so that there is still debate as to the colour of its legs and beak.

Painting of Himalayan mountain quails from 1846

Since February 1997, however, WWF-India has been collecting, compiling, and reviewing all available information regarding this shy species, which it will now be utilising in order to present local searchers with as much data as possible concerning their cryptic quail quarry. ***The Hindu*, 23 March 1999.**

FINDING OUT EVEN MORE ABOUT *FINDING OUT*. Time for another update regarding my search for the elusive weekly British partwork magazine *Finding Out* (which began in 1962) and its issues' cryptozoologically-illustrated back covers (see *FT*116, *FT*119). In one recent letter, Phil Hide of Aylesbury informed me that he actually owns most issues of this partwork up to volume 11, and that it was published by Purnell & Sons. Unfortunately, none of those particular issues' covers has any cryptozoology painting!

Alien Zoo

Happily, however, I also received a letter from Alex Lamprey of Cardiff, who owns seven issues from volumes 16 and 17, and each of these issues' covers does depict a mystery beast or legendary entity. These are: sea serpents (vol. 16, #3), bunyips (v.16, #10), the Scottish cailleach-bheur or blue-faced hag of the Highlands (v.17, #2), the morrigan (v.17, #4), gnomes (v.17, #8), the Little People (v.17, #9), and the lamassu (v.17, #10).

Obviously, therefore, this series of illustrations did not begin until sometime after volume 11. The artist responsible for them was Alex McBride, and cryptozoologist Alan Pringle, who first informed me about *Finding Out*, is now attempting to trace McBride to discover if he still owns the originals. My thanks to Phil Hide and Alex Lamprey for kindly contacting me with their information. **Alex Lamprey, pers. comm., 14 January 1999; Phil Hide, pers. comm., 25 March 1999.**

In the absence of Alex McBride's version from *Finding Out*, here is the cailleach-bheur as envisaged by acclaimed fantasy artist Andy Paciorek (Andy Paciorek)

TIGERS IN BORNEO? In a short but fascinating recent paper, zoological researcher Erik Meijaard surveys a wide range of hitherto little-publicised reports and publications dealing with the remarkable possibility that tigers exist, still unrecognised by science, in the large, primarily Indonesian island of Borneo. Tiger skins and skulls have also been recorded here (though whether these are from elusive indigenous specimens or have been imported from elsewhere is unknown), and alleged sightings made as recently as the 1990s are noted. If the tiger, wholly exterminated or greatly reduced in number throughout its known modern-day distribution range, were ultimately shown to thrive in Borneo (where it had officially died out by the beginning of the Holocene Epoch, roughly 12,000 years ago), it would be one of the most outstanding cryptozoological revelations of modern times. *Cat News*, **spring 1999.**

CRYPTIDS ON DISPLAY. Two notable cryptozoological attractions of 1999 can be found at opposite ends of the globe. Since early this year, 'Tasmanian Tiger: The Mystery of the Thylacine', a magnificent exhibition celebrating, and mourning, Tasmania's most enigmatic creature, has been touring Australia. It surveys many aspects of thylacine *Thylacinus cynocephalus* history, biology, mythology, marketing, and, of course, the tantalising possibility of its continuing survival. Nearer home, 'Loch Ness 2000' is a major new exhibition based at Drumnadrochit, which deals with the 'official' natural history of this famous lake, but inevitably features Nessie too. Designed by veteran Ness naturalist Adrian Shine, 'Loch Ness 2000' was formally opened in **June 1999** by renowned explorer Sir Ranulph Fiennes.

A WHALE OF A TALE. As revealed online by Norwegian cryptozoological researcher Erik Knatterud, Norway's NRK1 television station broadcast a brief but fascinating news item on **2 June 1999** featuring footage filmed recently by a journalist while taking an evening coastal walk at Aalesund. Af-

Alien Zoo

ter spotting a dead whale floating on the sea surface, he saw through his binoculars a narrow, unidentified aquatic beast approach it and proceed to feed off it. The creature measured 25-30 m long, 1.5 m across, and sported a huge square fin protruding above the water surface. No known species of fish this size occurs along this coast, it did not resemble a shark, and the director of Aalesund Aquarium was unable to identify it.

CRYPTO-CUBA! And as revealed in 'Forbidden Depths', a TV documentary first screened on **6 June 1999**, by Discovery Channel, several new species of marine fishes have lately been discovered in the sea around Cuba. These include the orangeblotch gaper *Chaunax suttkusi* (related to cusk eels), a new spiny ray, a duckbill flathead, and a tiny *Philomenus* goby.

CZ BIBLIOGRAPHY. In response to popular demand, I have compiled and placed online in my website for ready access by fellow researchers a comprehensive listing of cryptozoological and zoomythological books, which currently provides full bibliographical details for 555 separate volumes. Details of further books, especially foreign-language and newly-published tomes, would be greatly appreciated. Check it out at: **http://members.aol.com/karlshuker/ [transferred in September 2008 to: http://www.karlshuker.com].**

THOSE DRATTED SQRATS AGAIN! One of the media's favourite areas of animal anomalies is the exotic, if not downright impossible, zoological hybrid. A popular perennial among these is the sqrat - claimed to be a bona fide crossbreed of squirrel and rat. This unlikely mammalian mongrel recently reappeared in a brief newspaper report, claiming that New Yorkers are reporting sightings of such creatures, whereas city officials are discounting them as urban legends. Certainly, no verified cases are on record, and as squirrels and rats belong to entirely separate taxonomic families of rodent, there is little chance that, even if such hybridisation did occur, any viable offspring would result. Moreover, there seems to be no plausible scenario to explain why rats and squirrels would attempt to interbreed anyway. *Denver Post*, **28 June 1999**.

MOA OR LESS A MYSTERY. July 1999 marked the beginning of a proposed five-year hunt for a living moa, and also any surviving oop (out-of-place) moose *Alces alces*, in Fjordland, on New Zealand's South Island. The hunt is led by Keith Armstrong, formerly from Canada, now a master brewer in New Zealand's West Coast.

Whereas the giant *Dinornis* moas are certainly long extinct, some cryptozoologists hold out hope that at least one of the smaller moa species may still lurk undetected by science amid the forests of Fjordland. Based upon various recent reports, there is also talk that the moose, a non-native species introduced from Canada to South Island at the 20th Century's onset but widely assumed to have died out here since, may also be lingering reclusively. ***Christchurch Press*, 1 July 1999.**

With statue of giant *Dinornis* moa in Auckland, New Zealand
(Dr Karl Shuker)

Alien Zoo

ABC IN NZ? Staying in New Zealand: this dual-island country is not well known for reports of alien big cats. Nevertheless, on 13 or 14 July 1999 (different accounts give different dates) two British tourists saw, and photographed, what they claim was a black, alsatian-sized but longer-bodied, cat-like beast, walking across a Lindis Pass paddock in Mackenzie Country, about 59 km from Omarama, towards Queenstown on South Island. After photographing it (sadly, the photo, like all classic ABC shots, merely portrays an unidentifiable black 'something'), Mark and Deborah Greening reported their encounter to the Queenstown police, but sceptics have dismissed their sighting as simply an over-sized black feral domestic cat. ***Southland Times*, 16 July 1999; *Otago Daily Times*, 17 July 1999.**

SEND IN THE CLONES. The New Zealand huia *Heteralocha acutirostris*, famous among birds due to the female and male having different-shaped beaks, officially became extinct in 1907; and although several sightings have been reported since (including one claimed in 1992 by Danish cryptozoologist Lars Thomas), none has been confirmed.

A pair of huias, showing their dimorphic beaks (the female's was long and curved, the male's short and straight)

In July 1999, however, a conference attended by biologists, bioethicists, and Maori representatives was held in New Zealand to discuss the exciting possibility of resurrecting the huia by cloning, utilising preserved DNA samples. Cyberuni, a firm based in New Zealand and California, is offering to help fund the project if suitable DNA samples can be found. Some researchers have objected to the plan, claiming that the huia's extinction was a natural process demonstrating its non-viability as a species.

In reality, however, part of its decline was due to over-hunting, because its tail plumes were prized for ceremonial head-dresses and fashion accessories, and stuffed specimens were eagerly sought by museums and private collectors. Consequently, if the huia could be restored, humanity would merely be redressing the balance. ***TIME Digital*, 26 July 1999.**

SERPENT SEEKING. Science has yet to confirm the existence of an anaconda specimen measuring 30 ft or more, but two planned expeditions to the Amazon jungle swamps, as announced in late July 1999, hope to change all of that. One of these, led by Johnny Arnett from Cincinnati Zoo, boldly aims to snare a record-breaking anaconda and bring it back to the zoo for exhibition. The other expedition, conversely, led by New World Expeditions official Bill Cacciolfi, is content merely to obtain videos and photos of any such serpent. ***Wolverhampton Express and Star*, 29 July 1999.**

LION AROUND IN CHILE. Even more exotic than black panthers stalking Britain or pumas in the Australian bush are reports of African lions *Panthera leo* roaming the semi-desert highland region of northern Chile. However, these ABCs, six in number, are fully confirmed, having escaped in February 1999 when the truck transporting them crashed while driving through Chile to Bolivia. At present, they are posing problems for the local herders, reports claiming that they have already killed around 600 domestic llamas (not lamas, as at least one media account soberly announced!). Hence the Chilean au-

Encountering a giant anaconda (William Rebsamen)

Alien Zoo

thorities have proposed capturing these leonine llama-slayers alive, with the aid of big game hunters and specially-constructed cages. **Steve Newman's EarthWeek, 15 August 1999; Loren Coleman, cz@onelist.com 15 August 1999.**

A NOSE FOR MYSTERY. While visiting Macas in southern Ecuador's Morona-Santiago province in July 1999, Spanish cryptozoologist Angel Morant Forés noticed a very mysterious stuffed mammal in a small shop. Although he did not purchase it, Angel did take colour photos, which reveal a superficially mole-like creature, 35-40 cm long, with white fur, broad brown markings on its back, webbing on all four feet, short tail, and also a short but distinctive nasal trunk. Back home in Spain, Angel showed his pictures to five mammalogists, four of whom could not identify the creature. The fifth suggested a South American yapok (water opossum) *Chironectes minimus*, but as Angel has noted, there are various problems with this identification. Unless it has faded, the stuffed mystery beast's fur colour does not match the yapok's; only the hind feet are webbed in the yapok; both sexes of yapok have pouches whereas the mystery beast did not seem to be pouched; and the yapok certainly does not possess the mystery beast's trunk. Looking at the pictures, I was reminded of those Old World aquatic insectivores known as desmans - related to the moles, but of amphibious propensity, with webbed feet and a prominent nasal trunk. Could Angel's mystery beast be an undiscovered New World equivalent? The shop owner has since refused to sell Angel the specimen, but the Shuar Indians of Macas claim that this cryptid is common in local rivers, so perhaps a specimen can be captured and examined. **http://perso.wanadoo.fr/cryptozoo/expeditions/ecuador_eng.htm accessed 5 September 1999 [now http://pagesperso-orange.fr/cryptozoo/expeditions/ecuador_eng.htm].**

Reconstruction of the Ecuadorian mystery beast's possible form in life, based upon the stuffed specimen's appearance as seen in Angel's photos (Tim Morris)

ANGEL'S AMAZING TECHNICOLOR DREAMCAT. During his field trip to southern Ecuador, Angel Morant Forés also learnt of several other mystery animals said to inhabit this country's Amazonian jungles. They include: a small reddish-brown peccary called the esakar-paki; a white-coated cat with solid black spots known as the shiashia-yawá; a huge dark-grey cat with massive paws called the tapir tiger or pamá-yawá because it hunts tapirs; the water tiger or entzaeia-yawá, bushy-tailed and bigger than a jaguar; a pack-hunting semi-aquatic cat termed the tsere-yawá; and, most fascinating of all,

Alien Zoo

the aptly-named rainbow tiger or tshenkutshen. According to the Shuar Indians in the Macas region, this last-mentioned mystery cat is reputed to be the size of a jaguar, and black in colour, but ornately decorated with several stripes of different colours – black, white, red, and yellow – on its chest, "just like a rainbow", in the words of one native hunter interviewed by Angel. Said to inhabit the Trans-Cutucú region, Sierra de Cutucú, and the Sangay volcano area near Chiguaza, Ecuador's mystifying rainbow tiger is described by the Shuar as having monkey-like forepaws and being an exceptionally good tree-climber, leaping from tree-trunk to tree-trunk at great speed, and greatly feared as an extremely dangerous animal. One such cat may well have been killed in 1959 by Policarpio Rivadeneira, a Macas settler, while walking through the rainforest of Cerro Kilamo, a low mountain near the Abanico River. He had seen the creature leaping from tree to tree and, scared that it would attack him, shot it. When he examined it, he discovered that it was a jaguar-sized cat, but instantly distinguishable from all cats that he had ever seen by virtue of the series of multicoloured stripes running across its chest, and also by its simian forepaws. Sadly, Rivadeneira does not appear to have retained the creature's carcase, or even its pelt, so as yet there is no physical evidence available to verify this extraordinary felid's existence. **http://perso.wanadoo.fr/cryptozoo/expeditions/ecuador_eng.htm accessed 5 September 1999 [now http://pagesperso-orange.fr/cryptozoo/expeditions/ecuador_eng.htm].**

Yowie (Richard Svensson)

A STUNNING PLAN! In an attempt to prove once and for all that Australia's alleged man-beast, the yowie, really does exist, Larry Lesh, a bigfoot hunter from the USA, has proposed launching an expedition to Australia's Great Dividing Range in the hope of capturing a living specimen - with the assistance of stun guns. In media accounts, Lesh has been quoted as saying: "The idea was to close within 21 feet of a yowie and shoot. Twin barbs would penetrate the yowie's skin - and then 50,000 volts would bring him down. A couple of plastic riot cuffs for the wrists and feet and you've just captured a yowie". The Australian authorities, however, clearly see things differently - having lost no time in refusing Lesh permission to import the stun guns. *Brisbane Sunday Mail*, **5 September 1999.**

DEM BONES, DEM BONES... Scientists at Simon Fraser University in British Columbia, Canada, are presently examining a series of bones that have so far not been assigned to any known species of animal. As these remains have sparked interest among bigfoot investigators, it is to be hoped that DNA analyses will be carried out, just in case they are indeed of cryptozoological significance. **John Kirk, cz@onelist.com, 6 September 1999.**

NOSING OUT THE TRUTH? In Alien Zoo for *FT*128, I reported a stuffed mystery beast superfi-

Alien Zoo

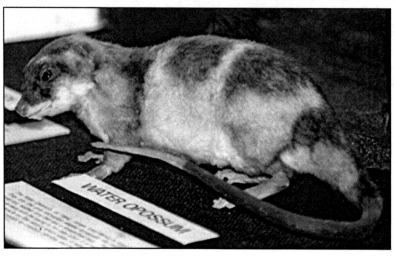

cially resembling a yapok or water opossum *Chironectes minimus* but readily distinguished by its short nasal trunk, no pouch, and interdigital webbing on all four feet, which had been spotted by Spanish cryptozoologist Angel Morant Forés in a shop at Macas while visiting Ecuador in July 1999. I have since learnt from Angel that this odd specimen was recently examined by Ecuadorian biologist Dr Didier Sanchez, who claims that it had been considerably manipulated by its taxidermist, accounting for its absence of a pouch, and also its trunk, which Sanchez believes to be artificial. Sanchez has now purchased the specimen, and will be examining it thoroughly at Quito to determine its identity conclusively. **Angel Morant Forés, pers. comm., 12 September 1999.**

A taxiderm yapok (Dr Karl Shuker)

NEW YORK'S MONSTROUS EXHIBITION! On 12 October 1999, the world's best-preserved specimen of a giant squid, measuring 25 ft long and weighing 250 lb, went on public display at the American Museum of Natural History in New York. A male *Architeuthis kirkii*, netted by fishermen in 1997 off the coast of New Zealand in 1997 and donated to the museum by that country's National Institute of Water and Atmospheric Research, this remarkable specimen is so big that its two longest tentacles had to be folded in order for its colossal bulk to fit inside its specially-designed fibre-glass, windowed display case! It will be on show for the next 2 years, so if you're in the vicinity of New York City's Upper West Side, be sure to drop in and see for yourself this spectacular representative of what is unquestionably one of the world's most incredible yet least-known creatures. The museum already has two life-size models of giant squid on view. One of these hangs from the ceiling opposite the real specimen in the Hall of Biodiversity; a 105-year-old papier-mâché model, it measures 42 ft long. The oldest model on display at the museum, it was purchased in 1895. The other model battles a sperm whale - its best-known predator - in a diorama in the adjacent Hall of Ocean Life. **http://www.amnh.org/museum/press/feature/squid.html 12 October 1999.**

Engraving of giant squid (this particular specimen was unsuccessfully captured off Tenerife by the French gunboat 'Alecton' on 30 November 1861)

Alien Zoo

HERD ABOUT THIS? A small herd of feral cattle whose ancestors were abandoned 25 years ago on the small, isolated Orkney island of Swona has recently been classified as the world's newest breed of cattle. Left behind on Swona when the isle's inhabitants left for Mainland, the original herd has gone through several generations, and although derived from Aberdeen Angus and Short Horn crossbreeding it has established a breeding pattern and morphology unique to itself. Discovered by zoologist Professor Stephen Hall from de Montford University in Leicester, it is the first new entry in the *World Dictionary of Livestock Breeds* for more than a hundred years. ***Wolverhampton Express and Star*, 21 October 1999.**

THE END OF A CRYPTOZOOLOGICAL ERA. After a lifetime of pioneering cryptozoological research, Dr Bernard Heuvelmans, popularly referred to as the Father of Cryptozoology, has donated all of his vast archives of material amassed during the past 50 years to the Musée Cantonal de Zoologie in Lausanne, Switzerland. In recognition of this unique and extremely sizeable bequest, on 12 October 1999 the museum formally opened its 'Bernard Heuvelmans Department of Cryptozoology'. Heuvelmans's archives include 1000 books, 100 archive boxes (containing approximately 25,000 pages of documents), 12,000 photo-slides, 25,000 photographs, and numerous cryptozoological magazines, as well as cryptozoological specimens such as plaster casts of bigfoot prints and tusks from pygmy elephants. **http://www.lausanne.ch/musees/zool/cryptozoologie/default.htm [no longer online] and http://perso.wanadoo.fr/cryptozoo/ [now http://pagesperso-orange.fr/cryptozoo/] 23 October 1999.**

WHEN EVEN IS ODD. All of the countless recorded specimens belonging to the world's 3000 or so species of centipede have an odd number of legs - all but one, that is. The 'even one out' is a remarkable specimen lately uncovered in Whitburn, near Sunderland, by postgraduate student Chris Kettle from Sunderland University, which has 48 pairs of legs, instead of the 47 pairs normally sported by its particular species. News of this extra-limbed oddity attracted great interest at a recent zoological conference attended by Kettle in Poland, and his specimen is now being studied in Italy. ***Electronic Telegraph*, 28 October 1999.**

GLOWING WITH PRIDE? A longstanding seeker of mysterious reptilian creatures is British herpetologist Mark O'Shea. Later this year, at least in the United States but possibly also in Britain, he and Trinidad-based snake expert Hans E.A. Boos will be seen on television not only seeking but also successfully finding a highly controversial creature first brought to attention by Ivan T. Sanderson in his book *Caribbean Treasure* (1937). Largely forgotten afterwards until documented in my book *Mysteries of Planet Earth* (1999), the animal in question is the glowing lizard of Trinidad. According to Sanderson, the male of the rare tejid *Proctoporus* (=*Oreosaurus*) *shrevei* can emit light from a series of porthole-like markings running down its flanks, but attempts by other scientists to induce this behaviour, which if genuine is unique among reptiles, have not been successful. Mark has promised me that the glowing lizard's secret will finally be revealed on his programme, but he will not release any details beforehand, so keep a look out for his programme in the 'Animal Planet' documentary series if you're hoping to see the light, literally! [The programme filmed a momentary, glow-and-fade effect of only very modest proportions, subsequently dismissed by some herpetologists as mere reflectivity, and certainly far removed from the dramatic pulses of light alleged by Sanderson.] **Mark O'Shea, pers. comm., November 1999.**

SEEING THROUGH SANDERSON'S TRANSPARENT CATFISH? While on the subject of Sanderson's book *Caribbean Treasure*: Another controversial claim made by him in it is that, while visiting Aripo Cave (where he found the glowing lizard) in Trinidad's Northern Range, he also encountered a completely transparent catfish, in a pool at the foot of this cave's first vertical drop, whose presence

Alien Zoo

he was only able to discern by virtue of its shadow. Surely, however, if the fish had indeed been completely transparent, it would not have cast a shadow, because all light would have passed directly through it. Very strange.

A Nepal dragon, visualised with crocodilian head and jaws (William Rebsamen)

LEGLESS IN NEPAL! I've recently learned some fascinating crypto-news from veteran field cryptozoologist Bill Gibbons. Bill is currently involved in a new planned TV venture, in which a production company is hoping to fly him out to various remote localities around the world to film him in search of some particularly exotic mystery beasts. The four presently under consideration for documentary treatment are the Congolese mokele-mbembe, which Bill has sought on previous occasions; the orang bati or flying man of Seram, Indonesia [which has its own separate chapter elsewhere in this present book of mine]; and two cryptids that are new to me.

One is a putative pterosaur known as the ropen, hailing from New Guinea [again, this has its own separate chapter elsewhere in the present book]. The other is a monstrous reptile known as the Nepal dragon, which sounds so extraordinary that I am quoting Bill's own description of it in full:

NEPAL: We will explore the jungle valleys to find the legless, 40 ft dragons which prey on cattle (sounds like a giant snake). They are well camouflaged like a tree log, and use their luminous green eyes to attract their prey, which they will devour with their enormous jaws. Missionary Reverend Resham Poudal, who has worked in Nepal for 25 years, camped overnight right next to one of these sluggish creatures, and made a close examination of the animal before continuing on his way with his entourage.

If such an astonishing animal is more than imaginative native folklore, and is indeed ophidian in identity, it would be of stupendous weight, explaining why it is so sluggish (in contrast, giant aquatic anacondas reported from South America would be buoyed by their watery domain, rendering them much more mobile). Even so, the specimen allegedly encountered by the missionary named above by Bill must have been exceptionally inactive, having resisted the temptation to gobble up its inquisitive human observer camping well within range of its gaping maw! **Bill Gibbons, pers. comm., 8 November 1999.**

HEADING FOR DISAPPOINTMENT. In late 1999, zoologist Dr Jack Giles of Sydney's Taronga Zoo returned to Australia from a visit to Vietnam, with some tantalising news. While in Vietnam he had been shown an old photo of a dead, complete specimen of the holy goat or linh duong *Pseudonovibos spiralis*.

Formally described and named as a new species of ungulate (hoofed mammal) in 1994, it was - and still is - known to science only from its spiralled horns (and even some of these are nowadays deemed to be fakes). Hence a photo of an entire specimen would have been of great zoological value.

Tragically, however, the photo was of such poor quality that little could be discerned. And to make matters even worse, its head, which might have afforded important clues regarding its possible relationship to other ungulates, was being sat upon by the hunter who had killed it and was therefore totally hidden from view!

This picture presents another conundrum. Is this what the dead beast in the above-noted photo looked like in life? Depicted in the *San Cai Tu Hui*, a Chinese encyclopedia by Wang Chi and Wang Si Yi dating from 1607, could this still-unidentified spiral-horned ungulate, referred to by them as the ling, be one and the same as the equally mystifying holy goat *Pseudonovibos spiralis*? **Prof. Colin P. Groves, pers. comm., 15 December 1999.**

The Lost Ark

KICKING UP A STINK
- BARBARA WOODHOUSE'S POUCHED SKUNK.

> *Experts cannot account for a skunk with a pouch, and try to persuade me that she was a 'possum. But she was no 'possum: she had the bushy tail of a skunk and was identical with the skunk picture in Cassell's* Book of Knowledge. *She did have a pouch: I examined her closely.*
>
> Barbara Woodhouse – *Talking To Animals*

I've said it before, but it's well worth repeating - mystifying creatures can turn up in the most unexpected locations, and the following example is certainly no exception.

To those of us of a certain age, the name Barbara Woodhouse is fondly associated with the staccato cry "Walkies!", uttered by a Joyce Grenfellesque lady of the genteel English schoolma'am variety that, sadly, seems to have quietly expired in these much coarser and more belligerent modern times. She acquired national - indeed, international - fame rather late in life, aged 70, when in 1980 her idiosyncratic show 'Training Dogs the Woodhouse Way' was first screened on British television and soon attained cult status, as a result of which she became one of the most recognisable, and parodied, personalities of the '80s.

Nevertheless, there seemed to be no connection between the redoubtable Ms Woodhouse and cryptozo-

---

FACING PAGE: Engravings of hog-nosed skunks

Alien Zoo

ology - at least, that is, until I was reading through her autobiography *Talking To Animals* (1954) a while ago. While perusing a section concerning her life as a young woman training horses in Argentina, I stumbled upon the following fascinating, but very perplexing, paragraph:

> Shortly after the storm [she had been describing the aftermath of a very violent storm that had hit their estate the previous evening], the foreman's little son came rushing up to say that all his pet rabbits had gone and that in the cage instead was a baby skunk. The mother had perished in the storm and lay dead by the cage. How that living little skunk had got into the undamaged cage, and the rabbits out of it, was beyond our understanding. In the mother's pouch were two dead babies. Experts cannot account for a skunk with a pouch, and try to persuade me that she was a 'possum. But she was no 'possum: she had the bushy tail of a skunk and was identical with the skunk picture in Cassell's *Book of Knowledge*. She did have a pouch: I examined her closely.

Woodhouse then went on to describe how she attempted to care for the alleged baby skunk by rearing it and feeding it in a cotton wool-lined pocket of her riding skirt, noting that it successfully fed and survived in this makeshift pouch for a week before ultimately dying after escaping from the pouch one night and becoming severely chilled.

Skunks, of which there are at least ten recognised species, were traditionally classed as mustelids (members of the weasel family), but more recently, based upon genetic studies, these infamously malodorous mammals have been allocated a taxonomic family of their own. However, although they do exhibit quite a diversity of morphologies, none of them has a pouch – a taxonomically-significant anatomical feature specific to marsupials. Moreover, only the hog-nosed skunks (genus *Conepatus*, comprising 4-5 species, depending upon opinion) are native to South America, and only two of these species are known to occur in Argentina.

Consequently, I find myself in agreement with the unnamed experts who claimed that Woodhouse's 'pouched skunk' was a 'possum - or, to be accurate, an American opossum, of which many species in several genera have been described. Having said that, the fundamental problem with this identity is that none of the known species of American opossum bears any real degree of similarity to a skunk.

True, the black-shouldered opossum *Caluromysiops irrupta* has distinctive black shoulders, a black dorsal stripe, and dark feet and tail that contrast markedly with the much paler fur on the rest of its body, but it is hardly skunk-like. And the distal portion of its tail is unfurred and rat-like, thereby bearing no resemblance to the uniformly furred tail of a skunk. Conversely, there is another species, *Glironia venusta*, which is actually known as the bushy-tailed opossum because of its unusually thick, densely-furred tail; however, it lacks any black-and-white fur colouration reminiscent of a skunk's, and as with the previous species its tail's distal portion is unfurred.

Even the yapok or water opossum *Chironectes minimus*, whose distinctive black and pale grey fur may conceivably invite comparisons with skunks by observers poorly acquainted with these latter mammals, can be readily eliminated from further consideration by virtue of its very slender, wholly unfurred tail. As for the thick-tailed opossum *Lutreolina crassicaudata*, its pelage (especially in females) also has dark and light markings, though these are far less prominent than those of the yapok; however, it has a thicker tail than the yapok, but this is still far less bushy than that of a skunk. Moreover, of the species

Alien Zoo

Engraving of yapok

noted here, only the yapok and thick-tailed opossum are native to Argentina anyway.

So what could Woodhouse's pouched skunk have been? I have even considered briefly the possibility that the adult female animal found dead was a genuine skunk that was heavily pregnant, and that the shock of the storm had caused one of her babies to be born prematurely, with Woodhouse mistaking this mother skunk's vagina and uterus for a pouch! However, this all seems highly improbable, especially as Woodhouse was someone with considerable experience from a very early age at caring for and handling animals.

I would have dearly loved the opportunity to contact Barbara Woodhouse in order to elicit more details concerning her baffling little beastie, but, sadly, she died in 1988, long before I discovered her account of it in her book. There is still one way, however, of shedding, perhaps, just a little more light on this mystery.

Does anyone out there have a copy of Cassell's *Book of Knowledge*, which I am assuming must date from around the late 1920s or the 1930s, bearing in mind that Barbara Woodhouse was born in 1910 and lived in Argentina for more than three years during her 20s? If so, I'd love to see its picture of a skunk, because this would give some idea of what her supposed pouched variety looked like (bearing in mind that there are several very different skunk morphologies, depending upon the species in question). That in turn may provide clues as to its real identity – unless, of course, by any remote chance it really was a pouched skunk, and thereby constituted a still-undescribed and dramatically different species?

REFERENCE

WOODHOUSE, Barbara (1954). *Talking To Animals.* Faber & Faber (London).

2000

AN ABC IN EE. Reports of ABCs (alien big cats) rarely emerge from East Europe, so it is interesting to learn of a recent case from the Czech Republic. Since Boxing Day 1999, several people have claimed sightings of a puma-like beast in fields near the village of Jinacovice, which happens to be just a few miles from Brno Zoo. Villagers believe that the creature must have escaped from there, but the zoo has denied this. A number of dead roe deer that had clearly been killed by some carnivorous animal have also been found in this vicinity, and their remains have been examined by the zoo's director, Martin Hovorka. Although villagers cite these deaths as proof that the Brno Beast is real, Hovorka deems it more likely that the deer were killed by martens, foxes, or dogs. This is because the deer were attacked and killed from the front, whereas in Hovorka's opinion: "Pumas and other cats don't kill that way. They attack from the rear and they also feed on the rear portions of the body". Nevertheless, a group of hunters is hoping to seek out and catch the elusive deer-killer, to reveal whether an ABC really is prowling out there. ***Blesk*** **(Prague), 7 January 2000.**

BRAZILIAN MONKEY BUSINESS. Those cryptozoologists who believe that Loys's 'ape' *Ameranthropoides loysi* (the controversial ape-like entity reputedly encountered and famously photographed on the Venezuela-Colombia border in c.1920 by Dr François de Loys while leading a geological expedition, and pictured opposite) was an unknown species of giant, bipedal spider monkey have hitherto been scoffed at by others, because there has been no physical evidence that any such primate ever existed. However, during the very first scientific survey of the southern hemisphere's longest cave - eastern Brazil's 60-mile-long Toca da Boa Vista in Bahia - a team co-led by Californian palaeontologist Dr Walter Hartwig has uncovered thousands of fossil mammal bones...including the skull of a 55-lb spider monkey. Such a creature would have been more than twice the size of any living species of spider monkey. A worthy ancestor, perhaps, of Loys's 'ape'? ***National Geographic*, January 2000.**

Alien Zoo

THE *OTHER* MILLENNIUM BUG. Just when you thought it was safe to go back to the keyboard - I'm here to tell you that the millennium bug is alive and well and living in Australia! Happily, however, it is not the computer-nobbling version. Instead, as announced by Dr Ebbe Nielsen, head of the Australian National Insect Collection, it is a hemipteran - the taxonomic order of insects known scientifically as bugs. A new species of water strider, related to the familiar pond skaters that race across the water surface of pools and lakes, it will be formally dubbed the millennium bug when it is scientifically described within the next year, and gobbles up nothing more significant than flies, not files (computer files, that is!). [In 2001, the millennium bug was formally christened *Drepanovelia millennium*.] **Matthew Bille, cz@onelist.com 8 January 2000.**

MORE THAN A MYTH? The most famous mythical entity of Australian aboriginal Dream-Time belief is the giant rainbow serpent. Although zoologically deemed to be fictitious, it may have been based upon a real creature after all. In a fascinating paper, Queensland University zoologist Dr Michael Lee, Dr John Scanlon from New South Wales University, and colleagues have proposed that the rainbow serpent might have been inspired by preserved memories of bygone encounters between early aboriginal settlers Down Under and a now-extinct genus of Aussie reptile comprising two species of huge python-like snake - the wonambi. Measuring 10-16 ft long, with a cross-section the size of a large dinner plate, it would have been big enough to swallow very large creatures, and thus could have readily given rise to legends of immense, all-consuming rainbow serpents. So perhaps it did survive long enough to have coexisted for a time with the first humans to reach Australia. ***Nature*, 27 January 2000.**

CRYPTO-CAMBODIA! During a Fauna and Flora International expedition to Cambodia from 25 February to 7 March 2000, three remarkable crocodiles were observed in marshes and rivers within a valley at the base of western Cambodia's Cardamom mountain range, near to its border with Thailand. What made these animals so remarkable is that they proved to be specimens of the Siamese crocodile *Crocodylus siamensis* - a highly-endangered species hitherto claimed by some zoologists to exist nowadays only in captivity. ***Daily Mail*, 18 March 2000.**

Looking out over Loch Ness (Dr Karl Shuker)

GUST AT LOCH NESS. During late March 2000, Swedish explorer Jan-Ove Sundberg led a week-long GUST (Global Underwater Search Team) survey at Loch Ness, using hydrophones day and night in the hope of recording sounds of the cryptozoological kind within its deep waters. The hydrophones have a range of 5 miles in ideal conditions, and unlike sonar equipment, which might disturb any creatures that may exist here, they do not create any disturbance when recording. It is expected that any findings obtained by the GUST survey, dubbed 'Nessie 2000', will be made public in due course. ***The Scotsman*, 20 March 2000.**

WHISKERED WONDERS. As lately pointed out online by palaeontologist Darren Naish in the cz@onelist discussion group, ornithologist Philip Chantler's recently-published book *Swifts: A Guide to the Swifts and Treeswifts of the World* (Pica Press: Sussex, 2000) contains a short but informative section on unidentified swifts, principally dark in colour and reported variously in Africa and China. It also contains intriguing reports of mysterious swifts sighted in several widely-spaced localities around the world but all loosely termed 'whiskered swifts', as they sport pale grey, whisker-like markings running backwards from their mouth. In the past, these odd birds have been claimed by some to be one or more new species, but Chantler suggests that they may either be individuals of known species that were spied carrying feathers in their mouths, or had their mouths full of collected feathers that therefore stretched the feather tracts on their cheeks, creating the striped effect. **Darren Naish, cz@onelist.com 28 March 2000.**

MAKING A FUSS ABOUT FOSSAS. The fossa *Cryptoprocta ferox* is a large superficially cat-like yet civet-related carnivore indigenous to Madagascar. There was formerly a much bigger version here too, the giant fossa *C. spelea*, measuring 6 ft long, weighing around 37 lb, and as strong as a leopard. However, this formidable beast is supposed to have died out many centuries ago - but did it? There have long been rumours and unconfirmed native reports of a giant-sized fossa existing in the more remote regions of Madagascar. Consequently, in November 1999, biologist Luke Dollar trekked to one

Engraving of the fossa *Cryptoprocta ferox*

such region, the so-called Impenetrable Forest in the northeast of this island, officially known as Zahamena National Park, in the hope of spying one of these cryptic mammals. Sadly, he did not do so. Nevertheless, in view of the fact that this locality is indeed virtually impenetrable, with no detailed maps of its interior ever having been made and much of its forest expanse still unexplored, hope must surely remain that this uninviting area may yet unveil a major surprise or two for cryptozoologists. ***Discover*, April 2000.**

Alien Zoo

THE BEAST OF BONAVISTA. A very curious cryptid was spotted in the early part of April 2000 (but not 1 April, thankfully!), when Bob Crewe was driving his truck along Cape Bonavista, Newfoundland, one morning. Spying it lying in the water, he described it as snake-like, measuring about 30 ft, and when he blew his truck's horn the creature raised its head out of the water on a 4-5-ft-long neck, revealing a snout-like projection. As he watched, it swam away towards the lighthouse at the tip of the cape, moving very rapidly with its head still raised above the surface, tilted slightly forward. Despite being a Bonavista resident, Crewe had never seen anything like it before. *Telegram* **(St John's), 6 April 2000.**

FROM MONSTROUS FOSSIL TO MYTHICAL MONSTER? The Museum of Fine Arts in Boston contains a Greek vase dated at c.550 BC, depicting a scene from Homer's Trojan epic *The Iliad*, in which the hero Heracles is rescuing a maiden called Hesione from a formidable sea monster. The monster is represented by a disembodied head, which had hitherto been dismissed as a wholly imaginary form. Recently, however, classical folklorist Dr Adrienne Mayor from Princeton University studied this vase, and is convinced that the portrayed head is actually an accurate depiction of a fossil skull, possibly belonging to a long-extinct antlered giraffe called *Samotherium* - named after the Greek island of Samos where this huge mammal's fossils were first unearthed. Mayor has since suggested: "Now the painting takes on a new meaning: a sudden and surprising exposure of a huge fossil skull on the coast of Troy may well have been the inspiration for the Homeric tale". **Discovery.com News, 12 April 2000.**

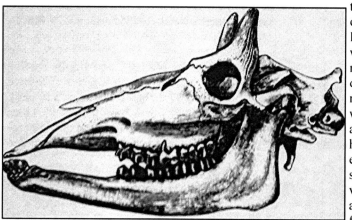

Samotherium skull (Dr Adrienne Mayor)

BEWARE OF THE OWL! Over the years, a fair number of reports have appeared in the British media concerning escapes into the countryside of European eagle owls *Bubo bubo* - one of the world's largest owl species, but not native to Britain. Yet surprisingly few follow-up reports, documenting the recapture of these aerial abscondees, have subsequently surfaced. So what has happened to all of these mega-owls? Have they managed to survive here? As far as the villagers in Stebbing, Essex, are concerned, there is good reason to support this possibility, as they - or, more specifically, their pets - have lately been subjected to a reign of terror by one such escapee. Nicknamed Eddie and of unknown origin, this adult male eagle owl has swooped down after Jennifer Nichol's mongrel dog Muffin, as well as the pet cat and ducks owned by neighbour Mavis Butson, and has devoured some ducks from the village pond. More recently, however, Eddie has turned his attention to a captive female eagle

A European eagle owl (Dr Karl Shuker)

Alien Zoo

owl called Charlie owned by villager Robert Spalding, regularly giving voice to loud and lusty mating cries. Spalding has attempted to capture him, using chicks placed inside an empty cage nearby as bait, but so far Eddie has proved too cunning. ***Daily Mail*, 19 April 2000.**

AN EZITAPILE EXOTIC. A decidedly weird wotzit has lately been reported by the inhabitants of Ezitapile in South Africa's Eastern Cape. Sighted in the surrounding forest area with its long tail coiled around a tree, it has been described as being yellow in colour, and serpentine in overall form, but with a horse-like head, a mane running down its back, and a body the shape of a 20-quart barrel! According to Captain Mpofana Skwatsha of the Aliwal North police, the local livestock become agitated whenever this bizarre-sounding beast is close by - which is hardly surprising! **Discovery Channel Earth Alert, 26 April 2000.**

SEEKING MONSTERS OF THE SEA. His late father would have been very proud of Jean-Michel Cousteau, for in October 2000 this son of the world-famous maritime explorer Jacques Cousteau plans to set sail in search of two of the oceans' most mysterious monsters. One is the giant squid *Architeuthis*, known from various stranded carcases to exist, but never spied alive in its natural marine environment by a scientist, which Cousteau Jnr will search for in New Zealand's deepwater Kaikoura Canyon. Even more controversial is the megalodon shark *Carcharocles megalodon*, a huge relative of the great white

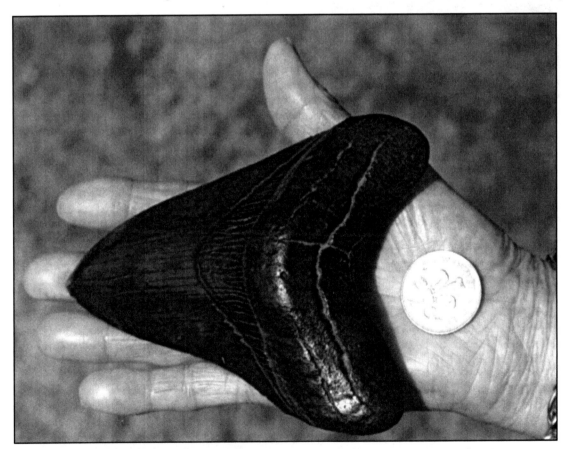

A huge tooth from the jaws of a megalodon shark (Dr Karl Shuker)

shark but supposedly long extinct. However, many reports of sharks resembling extra-large great whites have been filed from the Pacific Ocean over the years that have inspired hope among some cryptozoologists and ichthyologists that the megalodon may still linger today, and Cousteau Jnr plans to seek this monstrous species in the waters off New Caledonia. *Adelaide Advertiser*, **29 April 2000.**

Photo of Ivan Mackerle
(Ivan Mackerle)

A CZ TRIP DOWN UNDER. In late May 2000, famous Czech explorer-cryptozoologist Ivan Mackerle sets off from Prague for Cape York, in Queensland, Australia, to begin a lengthy cryptozoological foray planned to continue until the end of June.

His major quarries are the elusive Queensland tiger, popularly believed by mystery beast investigators to be a surviving species of thylacoleonid or marsupial lion; and the equally evanescent giant lizards believed by some cryptozoological researchers to be extra-large varanids (monitors), possibly even surviving representatives of the officially extinct giant monitor *Megalania prisca*. **Ivan Mackerle, pers. comm., 19 May 2000.**

PHOTO FIRSTS. In June 2000, researchers from the New York-based Wildlife Conservation Society announced that the first photos ever taken of tigers in the jungles of Cambodia had been snapped during a year-long monitoring of this long-neglected expanse of wild habitat, previously off-limits to scientific study due to civil upset and war. These precious pictures confirm that this dangerously rare big cat does still survive here, along with more than 60 other mammalian species also photographed during the society's survey. And in neighbouring Thailand, confirmed sightings and photos of the hairy-nosed otter *Lutra sumatrana*, the world's rarest otter species, have been obtained during a biodiversity study in Phru Toa Dang Swamp Forest - the first time that this scarcely-known species had been seen for many years. **Associated Press, 6 June 2000.**

CRYPTOZOOLOGY MEETS MASTERMIND! On 19 June 2000, cryptozoology achieved a notable new claim to fame when Mike Meakin from London became the first contestant on the BBC's famously cerebral, long-running quiz 'Mastermind' (at that time transferred from BBC1 on television to BBC Radio 4) to nominate cryptozoology as his chosen specialist subject (with questions set by yours truly!). Although, sadly, he did not win his heat, Mike gave a creditable performance, and certainly brought to the attention of quizophiles everywhere a wide range of intriguing cryptids, from Mongolia's electrifying death worm to putative pterodactyls in Zambia. **'Mastermind', broadcast by BBC Radio 4 on 19 June 2000.**

IRISH ELK SURVIVED ICE AGES. In my book *In Search of Prehistoric Survivors* (1995), I presented a selection of circumstantial evidence suggesting that the massive-antlered Irish elk or giant deer *Megaloceros giganteus*, traditionally believed to have become extinct around 12,600 years ago, prior to the end of the Ice Ages and the Pleistocene epoch, may in fact have survived into the present (Holocene) epoch.

Alien Zoo

At that time, however, there was no firm palaeontological support for this possibility - but now there is. New radiocarbon dating research conducted on the bones from a near-complete skeleton of an adult male Irish elk found on the Isle of Man in 1819 reveals that they date from 9430 BC, i.e. well within the Holocene. *Nature*, **19 June 2000.**

Statue of Irish elk at Crystal Palace, London (Dr Karl Shuker)

SQUID'S IN! In April 2000, the biggest specimen ever recorded of *Kondakovia longimana*, a highly reclusive deepsea squid, was discovered washed up on an Antarctic beach by biologist Amanda Lynnes of the British Antarctic Survey. Measuring 7.5 ft from the hooks on its hunting tentacles to the tip of its tail fin, and weighing almost 66 lb, it was being attacked by a flock of giant petrels, but scientists were able to photograph it and retain its hard, parrot-like beak. *BBC News Online*, **3 July 2000.**

THE CAMERA NEVER LIES(?). Ichthyologists are currently in heated dispute about a certain highly controversial photo of an Indonesian coelacanth *Latimeria menadoensis*. This second species of coelacanth was officially discovered in 1997, when US biologist Mark Erdmann spotted a dead specimen on sale at a fish market in Sulawesi. The following year, one was captured alive off Sulawesi, and was filmed swimming in a tank by Erdmann. A photo of this specimen was published by Erdmann and

Alien Zoo

two colleagues in the journal *Nature* on 24 September 1998. Recently, however, a French team of coelacanth researchers submitted a photo to *Nature* depicting an Indonesian coelacanth that one of their members claimed had been caught off Java in 1995. Apparently, the specimen was later lost, but if the photo is genuine, this latter specimen's procurement would pre-date Erdmann's discovery of the Sulawesi specimen - and hence of the Indonesian species itself - by two years.

But is it genuine?

As revealed in a recent *Nature* report by Heather McCabe and Janet Wright, the fish in the photo submitted by the French team "bears a striking resemblance to one that appeared in *Nature*" (i.e. Erdmann's filmed fish, in 1998). This is despite the fact that whereas Erdmann's fish was alive and swimming when photographed, the French team's was supposedly dead and inanimate, lying alongside some other fishes on a slab. One of Erdmann's colleagues, Prof. Roy Caldwell, who co-authored the *Nature* paper of 1998, examined the French team's photo using a picture-editing computer programme, and concluded: "I am 100% certain the image is a fake". As for the French team: the member who originally claimed to have taken the photo, Georges Serre, maintains that it is authentic, though he now states that it was a friend who took it.

Conversely, a second member of the team, ichthyologist Dr Bernard Séret from the National Museum of Natural History in Paris, concedes that the two photos do seem to show the same coelacanth specimen, adding: "This is very embarrassing". ***Nature*, 13 July 2000.**

THYLACINE, OR JUST A SHAGGY DOG STORY? On 11 June 2000, during an expedition to Australia in search of cryptids, Czech cryptozoologist Ivan Mackerle was driving along a track through deep forest from Portland Road south to Archer River in Queensland's Cape York Peninsula when suddenly an odd-looking canine beast appeared on the road up ahead. Stopping the car, Mackerle's team was able to video it as it faced them, then turned right, before disappearing into the forest. Although they felt that it resembled an alsatian, it was biscuit-coloured and seemed to have lateral stripes - as does the reputedly extinct, superficially dog-like thylacine (Tasmanian wolf) *Thylacinus cynocephalus*. But is that what it was?

Their film was later shown on Czech television, and Ivan remains undecided as to the creature's identity. However, having seen stills from the film (two of which are reproduced on the facing page courtesy of Ivan Mackerle), I am concerned that the creature's legs are far too long for a thylacine's. Its lateral stripes are also too long and its ears seem too big. Moreover, according to Ivan its tail was brush-like, whereas a thylacine's is stiff and slender. Many supposed sightings of mainland thylacines have been reported in modern times, but none has yet been confirmed. ***Blesk*, 21 July 2000; Ivan Mackerle, pers. comm., 22 July 2000.**

NEW BEASTS FROM BRAZIL? Dr Marc van Roosmalen is famous for lately discovering several new species of marmoset and South American monkey. During his survey along both banks of Brazil's Rio Aripuanã earlier this year, however, he also spied, or collected reports of, several other seemingly-undescribed mammalian forms.

One of the most intriguing of these is a strange cat resembling a black jaguar but distinguished by its

FACING PAGE: A thylacine in Cape York Peninsula? (Ivan Mackerle)

Alien Zoo

white collar and tufted tail-tip. Known to the locals as the onça-canguçú, a skull and pelt of this mysterious felid have been promised to Roosmalen by Indian hunters. Equally elusive at present is an odd type of brocket deer referred to locally as the veado branco, which appears to be intermediate in form between the area's two known species - *Mazama americana* and *M. gouazoubira*. **Marc van Roosmalen, pers. comm., 3 August 2000.**

BIG BIRDS ON THE WING IN SCANDINAVIA. *FT* recently forwarded to me a most interesting e-mail from reader Malcolm Sewell, in which he refers to the apparent existence of two very sizeable (and mysterious!) birds in Sweden. According to a Swedish friend of his, the Scandinavian Arctic is home to a very rare but extremely large eagle known locally as the kungstorn or king's eagle, with a wingspan in excess of 2 m. Also present there is an even bigger owl, called the slaguggla or strike owl, said to boast a wingspan of approximately 3 m. His Swedish friend informed him that one of these mega-owls once lived close to his house in central Sweden, and he remembers hearing the "eerie, low frequency whooshing of it passing overhead". Furthermore, both of these 'big birds' are allegedly able to lift prey at least as large as a reindeer calf! What could they be? Of particular note is that the strike owl's claimed wingspan is far bigger than even those of the mighty European eagle owl *Bubo bubo* and great grey owl *Strix nebulosa* - both natives of Scandinavia and among the world's biggest owls. **Malcolm Sewell, pers. comm., 6 August 2000.**

STRANGERS ON (AND AROUND) THE SHORE. Britain has witnessed visits from some unexpected but very welcome travellers lately. In July, news emerged that a pair of European spoonbills *Platalea leucorodia* had been spotted building a nest at a nature reserve in Dumfriesshire, Scotland. Although this species is an annual non-breeding visitor to the UK, it last bred here on a regular basis during the 1600s, due to its preferred marshy habitat being drained. It is believed that this latest pair made their way here from the Netherlands. A month later, an even more surprising visitor was recorded - an 18-inch tuna *Thunnus* sp, netted off the coast of Rye, in East Sussex, by fisherman Jimper Sutton. Tunas normally inhabit the Mediterranean Sea, and Sutton's specimen is believed to be the first taken in British waters. ***Sunday People, Sunday Mirror*, 16 July 200; *Sunday People*, 20 August 2000.**

MICRO-CLOUD CREATURES AHOY? The possibility of unknown 'sky beasts' or 'cloud creatures' living exclusively in our planet's atmosphere has often been touted (as in Trevor James Constable's books dealing with 'critters'), but no evidence that any life forms can actively thrive here has been obtained - until now. Innsbruck University researcher Dr Birgitt Sattler and a team of fellow Austrians have lately revealed that many different kinds of bacteria were growing and even reproducing within a cloud that passed over a meteorological station at the peak of Mount Sonnblick, near Salzburg. Samples of water droplets extracted from the cloud contained approximately 1500 bacteria per millilitre of water, but their various species have yet to be identified. ***Wolverhampton Express and Star*, 24 August 2000.**

UNCOVERING A CZ TREASURE TROVE. Australian National University zoologist Prof. Colin P. Groves has discovered and described several new species of large mammal during the 1980s and 1990s, and may soon be adding even more to his tally. After 40 years of searching for a long-lost collection of zoological specimens procured during the late 1800s by Pierre Marie Heude, a French Jesuit missionary based in China but also travelling in Indochina, Prof. Groves and an international research team recently tracked it down. It proved to be a series of grimy, neglected crates in the Beijing Institute of Zoology and the Natural History Museum in Shanghai. These now-priceless boxes contain the skeletons of over a thousand large Asian mammal specimens, which in Groves's opinion may well include several previously unknown species, as well as providing the first morphological evidence for certain species that are now extinct. ***Canberra Times*, 5 September 2000.**

Dr Marc van Roosmalen with the black pelt of an onça-cangucú on the floor to his left (Marc van Roosmalen)

Alien Zoo

A HEAVENLY REDISCOVERY? The last verified sighting of Australia's exquisitely-beautiful paradise parrot *Psephotus pulcherrimus* was in 1927, in the Burnett River region of Queensland.

Since then, it has been written off as extinct, despite several later, unconfirmed reports. Following recent claimed sightings by two reputable sources at a publicly-undisclosed location, however, the Queensland Parks and Wildlife Service Nature Search plans to instigate a formal search for this elusive species in October 2000.

Rather than seeking the bird directly, however, the team plans to search for a moth that fed upon its droppings, because if they succeed in finding specimens of that insect in the area, this would lend weight to believing in the paradise parrot's own continuing survival there. *Courier Mail* **(Queensland), 8 September 2000.**

John Gould's 19th-Century painting of a pair of paradise parrots

A SECRET IN ST LOUIS. I've recently received some interesting information from American veterinary surgeon Dr George Stoecklin. According to William Pflieger's authoritative tome *The Fishes of Missouri* (1975), there are no species of cave fish inhabiting the caves of St Louis, Missouri - the nearest species living about 100 miles away. However, Dr Stoecklin informed me that he had read a book by Hubert and Charlotte Rother entitled *Lost Caves of St Louis* (2004), which referred to one called Cherokee Cave that had been sealed in 1964 but which had been visited by the book's authors a little while earlier, where they had seen blind cave fishes.

Intrigued by this, Stoecklin has been looking into the matter, and has learnt that the cave's sealed-up entrance is due to be excavated by a local high school teacher and his students. Stoecklin and I are now awaiting news concerning this dig, in the hope that once it is complete, the fishes, if they do indeed exist, will be located and formally examined, in case they constitute an unknown species. Having said that, it is known that the pool did, and may still, contain amphipods, which are small, pale-coloured aquatic crustaceans. So it is certainly possible that these were seen and mistaken for fishes, especially on account of the poor viewing conditions present inside the cave. **George Stoecklin, pers. comms, 24 & 30 August, 18 & 27 September 2000.**

KOKAKO 2000. Starting on 17 October 2000, ecologist Rhys Buckingham will lead the most extensive search ever mounted for one of New Zealand's most elusive birds, the South Island kokako *Callaeas cinerea cinerea*. Distinguished from its North Island counterpart by its orange-and-blue wattles (those of North Island kokakos are entirely blue), this slaty-plumaged endemic passerine has long been written off as extinct, but spasmodic reports of its very distinctive bell-like call have inspired Buckingham to seek it on many occasions - and with such diligence that in June 2000 he was awarded New Zealand's Order of Merit. This latest search, dubbed 'Kokako 2000' and planned to last 4 months, is sponsored by the environmental lobby group Ecologic Foundation, the Department of Conservation, several major New Zealand companies, and private donations. It begins in a region of the Kahurangi National Park, and its participants are armed with digital audio and photographic recording equipment. They will also be seeking feathers for DNA tests. *Nelson Mail*, **5 October 2000;** *Christchurch Press*, **6 October 2000.**

BRINGING BACK THE BUCARDO? The last-known specimen of the bucardo or Pyrenean wild goat *Capra pyrenaica pyrenaica* was found dead in Spain's Ordesa National Park by gamekeepers during January 2000, having been killed by a falling tree. Now, however, it may become the first extinct mammal to be resurrected by cloning. On 8 October 2000, Advanced Cell Technology, Inc. (ACT) announced that the Spanish government had accepted its offer to recreate the bucardo, in collaboration with other scientific associates, via the cloning technique of interspecies nuclear transfer, using preserved bucardo tissue, and domestic goats as surrogate mothers. [In January 2009, a single cloned bucardo was indeed born, but died just 7 minutes later, due to physical lung defects.] **http://www.noonanrusso.com/www/act2/act2.html/ 8 October 2000.**

MAKING FOR THE MEKONG. During the early part of October 2000, Exeter-based cryptozoological investigator Richard Freeman set out for Thailand, to begin a TV documentary-sponsored search for the Mekong River dragon - one of oriental Asia's lesser-known water monsters. We at *FT* wish him well, and await further news of his quest with interest. **Darren Naish, cz@egroups.com 12 October 2000.**

Richard Freeman with a naga (CFZ)

Alien Zoo

SCANDINAVIAN 'BIG BIRDS' UPDATE. In *FT*139, I requested further news from readers re two supposedly unidentified types of mystery 'big bird' - the *slaguggla* ('strike owl' - a giant owl) and the *kungstorn* ('king's eagle' - a giant eagle) - said to inhabit the Scandinavian Arctic. *FT* recently heard from John Kahila, who states that *slaguggla* is a Swedish name for the ural owl *Strix uralensis*. Although certainly a big owl, this species is not known to attain the dimensions claimed for the mystery *slaguggla* by the source that I cited in my earlier item. John Kahila also mentioned that the seabird species dubbed the royal tern *Thalasseus maximus* in English is called the *kungstärna* in Swedish. Yet despite the similarity between the names *kungstärna* and *kungstorn*, it is difficult to conceive how this very modest-sized gull-related seabird could be the basis of reports concerning a huge eagle. However, I also received word from Danish zoologist Lars Thomas, who suggested that *kungstorn* was actually a mis-spelling of *kungsörn*, which refers to the golden eagle *Aquila chrysaetos*; and stated that the ural owl is indeed called the strike owl, because of its savage tendency to strike out at anything it encounters. It would appear, therefore, that neither of these mystery 'big birds' is mysterious, but merely derived from exaggerated accounts of real birds. **John Kahila, pers. comm., 19 September 2000; Lars Thomas, pers. comm., 12 October 2000.**

OFF TO THE CONGO. I recently learnt from team member Adam Davies that 'Dino2000', the long-awaited search for the mokele-mbembe (see *FT*125), would finally be setting off in mid-November for a month-long foray around Lake Tele in the People's Republic of the Congo. Among the Congolese contingent is Dr Marcellin Agnagna from Brazzaville's Parc de Zoologie. The British team members hope to be back home in time for Christmas. **Adam Davies, pers. comm., 3 November 2000.**

SEEKING THE CYANIDE MOTH. The New Forest burnet *Zygaena viciae*, famed for its ability to produce a cyanide compound to deter birds from eating it, seemingly died out in Britain almost 80 years ago when it was last reported in its native Hampshire haunt in 1927. In 1963, however, a colony was unexpectedly discovered in Argyll, Scotland, but this subsequently vanished, and the species was once again thought to be lost to Britain until 1990, when 12 were rediscovered. Since then, its existence here has been closely guarded; but in 2000, Scottish researchers proudly announced that its numbers had now risen to 1000. Thus, although it is still one of Britain's rarest moths, its future here does seem more secure than at any time during the past eight decades. ***Daily Telegraph*, 20 November 2000.**

MAYBE A MANATEE? The cz@egroups.com discussion group has lately posted one or two brief messages concerning the supposed discovery of a new subspecies of manatee in Africa by an expedition from the Genoa Aquarium, led by a Dr Nessi. Needless to say, the thought of any large mysterious water beast being discovered by someone called Nessi is likely to raise the suspicions of any sceptic; and as I've been unable to uncover any further information concerning this alleged find, *FT* would welcome any additional news regarding it! **Matthew Bille, cz@onelist.com 14 November 2000; Zack Clothier, cz@onelist.com 15 November 2000.**

With a life-sized manatee statue at Sea World, San Diego (Dr Karl Shuker)

Alien Zoo

ALL LIT UP IN THAILAND. English cryptozoological investigator Richard Freeman has returned home from his search in northern Thailand's jungles for various mystery beasts, including the naga - a giant snake familiar in Hindu and Buddhist myths, but also said genuinely to exist in Indochina's more remote parts. Although he did not spy it himself, he amassed a sizeable collection of local eyewitness reports, including one from the abbot of a Buddhist monastery in the rural village of Phon Pisai, and another from Officer Suphat, Phon Pisai's chief of police. The reports consistently describe a gigantic, black, scaly snake bearing an erectile crest on its head. Richard suggests that the naga, if it truly exists, might be a surviving species of huge archaic snake known as a madtsoid. While in Thailand, he also penetrated some remote caves that may not have been explored before by westerners, but where a naga exceeding 18 m long had been reported. Again, he didn't see it himself, but what he did find was a small colony of midge larvae (occupying a patch about 30 cm square) living on the cave roof that attracted prey using beads of luminous saliva suspended on 15-cm-long threads of silk. A comparable species, *Arachnocampa luminosa*, is a famous sight in New Zealand's Waitomo cave system (though in much larger colonies), which I saw for myself when visiting these caves in November 2006, but as far as Richard is aware, there are no such insects recorded from Thailand. Not expecting to find anything like these, Richard did not take any collecting tubes when exploring the narrow cave passages, so was unable to obtain a specimen for examination. Hence the intriguing question of whether the Thai midges constitute an undiscovered species presently remains unanswered. *Animals & Men*, December 2000; Richard Freeman, pers. comm., 3 December 2000; cz@egroups.com 7 December 2000.

Statue of a legendary multi-headed naga in the Royal Palaces complex, Bangkok, Thailand (Dr Karl Shuker)

THAILAND'S MAMMOTH MYSTERY. Staying in Thailand: on 3 December 2000, Princess Rangsrinopadorn Yukol of Thailand released to the local press some hazy aerial video stills, taken by her while flying in a royal helicopter over Chiang Mai's Om Koi district, that she claimed were of a herd of about 20 very unusual elephants. Dubbing them 'Thai mammoths', she described them as being bigger than normal elephants, with long tusks, and long hair on their back. The princess, a noted forestry conservationist, had made the video back in 1984, but had publicly released the stills now because she stated that a USA conservation team had recently claimed to have seen these animals too, in the same area. Her sighting was also substantiated by Samarn Prangwacharakorn, the district chief of Doi Saket, who stated that there were 27 elephants, which were very wild. A search, comprising two expeditions led by the princess, was promptly launched, but on 7 December they were halted, following security concerns expressed by Plodprasop Suraswadi, director-general of Thailand's Royal Forestry Department (RFD), who invited them to join an RFD expedition instead. Meanwhile, Kasetsart University forestry lecturer Naris Poompakphan examined the video stills and opined that the animals were merely common Asian elephants. No further news regarding this intriguing episode has emerged to date. *The Nation* (Bangkok), 5 & 8 December 2000.

Alien Zoo

SLOTH-SEEKING IN PATAGONIA. News of a notable crypto-search planned for 2001 was announced in December 2000 by Charlie Jacoby. A century earlier, following its launch, London's *Daily Express* newspaper sent Jacoby's great-grandfather, Hesketh Prichard, to Patagonia to investigate reports of living ground sloths. Sadly, he failed to discover the real thing, but now, Jacoby hopes to do just that, by retracing his ancestor's footsteps in Patagonia, inspired by researches undertaken by other modern-day sloth-seekers elsewhere in South America. These include Dr David Oren of Brazil's Goeldi Museum, tenacious pursuer of a mysterious Amazonian creature called the mapinguary, which he believes to be a living ground sloth. ***Daily Express*, 15 December 2000.**

Statue of giant ground sloth at Crystal Palace, London (Dr Karl Shuker)

The Lost Ark

REVIEWING THE ROPEN
- A NEW CRYPTID FROM NEW GUINEA.

Yes, I did search for giant living pterosaurs in Papua New Guinea, and I criticize the philosophy that causes Westerners to thoughtlessly dismiss any eyewitness report that suggests a living pterosaur; I censure nobody, however, for simply doubting such creatures still fly, for the idea defies an entrenched Western belief. About the ropen, believe what you will; what do I think about this creature? Such is the power of the testimonies of the eyewitnesses I've encountered, that it's as real to me, almost, as if I had stared it in the face. How can I deny the credibility of the eyewitnesses I've interviewed? With no other reasonable interpretation of these testimonies, I believe the ropen is a living pterosaur.

Jonathan Whitcomb – *Searching For Ropens*

Several notable cryptids (mystery animals) have been reported from New Guinea and its offshore islands over the years, including the thylacine-like dobsegna, the tapir-like devil-pig or gazeka, the enigmatic freshwater shark of Lake Sentani, and what may be an undescribed bird of paradise species from Goodenough Island. However, the latest cryptozoological riddle from this enormous yet still poorly-explored island is also its least-publicised.

Although I have briefly mentioned it in two of my Alien Zoo columns, no detailed account of New Guinea's winged mystery beast, the ropen, has appeared in any fortean publication - until now ['now' being January 2002, when the article upon which this present chapter is based was published in *FT*154].

Alien Zoo

UNRAVELLING THE ROPEN – A *RHAMPHORHYNCHUS* DOPPELGÄNGER

I first heard about the ropen during a series of communications in late 1999 and early 2000 from veteran cryptid-seeker Bill Gibbons, who was hoping to search for it as part of a planned television documentary (as yet, sadly, unmade). According to Bill, who later provided data to the cz@egroups online cryptozoological discussion group too, two different kinds of unidentified flying mystery beast allegedly exist in or around New Guinea, but the name 'ropen' is often, confusingly, applied to both of them. The 'true' ropen supposedly frequents Rambutyo (=Rambunzo), a small island off the east coast of Papua New Guinea (PNG) in the Bismarck Archipelago, and Umboi, a slightly larger isle sited between eastern PNG and New Britain. With a wingspan of 3-4 ft, a long tail terminating in a diamond-shaped flange, and a long beak brimming with sharp teeth, its description is startlingly reminiscent of an early prehistoric pterosaur known as *Rhamphorhynchus*.

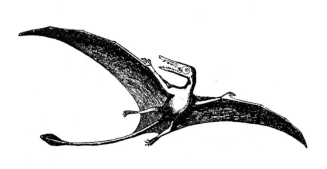

Early engraving of *Rhamphorhynchus*, a ropen lookalike?

Feared greatly by the PNG and Solomon Islands people, the ropen hides or sleeps during the day in caves on Rambutyo and Umboi, but is on the wing at night. Attracted to the stench of decaying flesh, it has reputedly been known to attack human funeral gatherings here - including one instance when western missionaries were present. Missionaries have also spied this creature on nocturnal coastal fishing trips, and it will attack native fishing vessels by snatching fishes from the nets as they are being hauled in by the fishermen. According to Bill, moreover, just six years ago the residents of an Umboi village called Gumalong watched as a ropen flew from nearby Mount Bel, down over the jungle valley, and directly over Gumalong as it headed out to sea.

DISTINGUISHING THE DUAH

An even more formidable winged cryptid supposedly exists on PNG's mainland, where it is referred to as the duah ('ropen' and 'duah' come from different native languages, but both apparently translate loosely as 'demon flyer'). With huge leathery wings spanning up to 20 ft, a fairly long neck, and bony crest, it is said to resemble *Pteranodon*, North America's mighty Mesozoic pterodactyl. Intriguingly, however, locals state that the duah has glowing underparts. Bill claims that in 1995, a missionary saw one as it flew by a lake, and that another missionary more recently claimed to have seen several in a mountain cavern.

The next important source of ropen news that I obtained was a letter by Robert F. Helfinstine, of Anoka, Minnesota, published in issue #33 of *Ancient American* during autumn 2000. In his letter, Helfinstine noted that he is acquainted with four of the eight members of an expedition (which in-

Alien Zoo

Is this what the duah looks like? (Dr Karl Shuker)

cluded Dr M.E. Clark, a retired professor from Illinois University) that in 1994 sought "a large flying creature...called 'Ropen'" on PNG (presumably the duah, therefore, not the smaller, true ropen of the offshore islands). Interestingly, Helfinstine revealed that the natives claim this creature's mystifying light derives from glowing patches on its underside, which can be actively turned on and off, like a firefly's bioluminescence. He also reported a bizarre-sounding story of how one local had encountered a sleeping ropen, and had tied a nearby log to one of its legs to prevent it from escaping - only to see, upon returning with some companions, the fully-awake ropen flying away with the log hanging from its leg! According to another story, one of these powerful creatures had lifted a woman up off the beach and had transported her aloft to a mountain about four miles away.

Helfinstine also noted that this mystery beast is so attracted to decaying flesh, digging up freshly-buried human corpses to carry them away, that graves here have special coverings to prevent it from desecrating them. Yet although the expedition spent some time seeking a ropen cave in the mountains, none was found. On a second expedition, several ropens were allegedly seen, but not photographed. Helfinstine claimed that missionaries and Peace Corp workers in New Guinea report that these animals fly from one offshore island to another. In January 2001, Helfinstine reiterated all of these details during a telephone conversation with Mark K. Bayless, a herpetological colleague of mine from Berkeley, California.

Alien Zoo

UGOS – UNIDENTIFIED GLOWING OBJECTS

In February 2001, while seeking ropen information online, I noticed a short item on a Special Creation website called 'Genesis Park' (www.genesispark.com). Various extra details were added some time later. It states that during two expeditions to PNG, Dr Carl E. Baugh of the Creation Evidence Museum learnt from the locals about a glowing flying beast termed the ropen. He also succeeded in spying such a creature himself at night, using a monocular night scope, and he photographed a strange print in the sand the following morning.

In September 2001, one of my many correspondents, Brian Irwin, revealed on the crypto-list@yahoogroups.com discussion group that while in New Guinea during July-August 2001 he had visited some offshore islands allegedly home to the true, smaller ropen. On Rambutyo, locals claimed that it was quite easy 30 years or so ago to see up to three of these flying at night, glowing (like the bigger PNG duah), on the island's uninhabited eastern side, but only one is seen at a time there today. According to an eyewitness called Ralph, interviewed by Brian, one night about twelve years earlier a ropen had dive-bombed his boat while he and a friend had been fishing on Rambutyo's east coast. After hitting the boat, the ropen had fallen into the water, where it splashed about for a time before flying off again. On nearby Manus Island, the local school's headmaster informed Brian that he had once seen one of these glowing beasts sitting on a branch of a tree on Goodenough Island, in the Milne Bay Province. And on Umboi, many locals claimed to have seen the ropen at night, including a policeman and a government employee during Irwin's own short stay there. It is thought to live near the top of Mount Bel, and is most frequently seen in the early morning on the island's western side. Sadly, Brian did not spot one of these entities himself.

Of course, winged mystery beasts from this region of the globe are nothing new - cryptids that may be undescribed species of exceptionally large bat have already been reported from the far-eastern Indonesian island of Seram (the orang bati or 'flying man' – see separate chapter in this present book) and Java (the ahool). Moreover, there are even precedents for luminescent flying cryptids on file - including the Namibian flying snake (see my books *The Unexplained*, 1996, and *Mysteries of Planet Earth*, 1999), and a *Rhamphorhynchus*-like creature reported in the vicinity of East Africa's Mount Kilimanjaro that glows in the dark. Highly-reflective scales or adhering phosphorescent fungi may be responsible. Several major cryptozoological discoveries have lately been made in New Guinea - most notably a panda-patterned whistling tree kangaroo called the dingiso *Dendrolagus mbaiso* in 1994. Perhaps the ropen and duah will be next.

UPDATE

In the years that have passed since the publication in January 2002 of my *FT* article that gave rise to the above chapter, cryptozoological and public interest in the ropen and duah have increased dramatically. Paul Nation of Granbury, Texas, has led no less than three expeditions to New Guinea seeking this creature, and in 2006 fellow field investigator Jonathan Whitcomb authored an entire book on the subject, entitled *Searching For Ropens: Living Pterosaurs in Papua New Guinea*, in which he provided extensive documentation of his own searches and those of others. As quoted from Wikipedia (accessed by me on 22 April 2010), some of the major highlights in recent ropen history are as follows:

> In late 2006, Paul Nation, of Texas, explored a remote mountainous area on the mainland of Papua New Guinea. He videotaped two lights that the local natives called "indava." Nation believed the lights were from the biolu-

minescence of creatures similar to the ropen of Umboi Island. The video was analyzed by a missile defense physicist [Clifford A. Paiva] who reported that the two lights on the video were not from any fires, meteors, airplanes or camera artifacts. He also reported that the image of the two lights was authentic and was not manipulated or hoaxed.

In 2007, cryptid investigator Joshua Gates went to Papua New Guinea in search of the Ropen for his TV show 'Destination Truth'. He and his team also witnessed strange lights at night and could not confirm what they were.

In 2009, the television show 'Monster Quest' conducted an expedition in search of the "demon flyer" but found no evidence of the creature. Later, they had a forensic video analyst examine the Paul Nation video. The analyst could not definitely conclude what was causing the lights, but ruled out vehicles and campfires believing the footage was of a pair of bioluminescent creatures perched in a tree that later take flight.

While they fail to provide firm evidence for the ropen's reality, it is clear from these findings that this winged mystery beast is associated with mysteries that have yet to receive a satisfactory explanation.

SELECTED REFERENCES

ANON. (2001). The "ropen" of New Guinea. http://www.genesispark.com/genpark/ropen/ropen.htm accessed 1 February & 3 October (by which time its details had been updated).

BAYLESS, Mark (2001). Pers. comm., 18 January.

CTRINGFARM@aol.com (2000). Ropen, flying cryptid of New Guinea. cryptolist@egroups.com 26 November.

GIBBONS, Bill (1999). Pers. comm., 8 November.

GIBBONS, Bill (2000). Ropen. cz@egroups.com 22 October.

HELFINSTINE, Robert F. (2000). Letter [re ropen]. *Ancient American*, 5 (no. 33; autumn).

IRWIN, Brian (2001). 'Ropen' of New Guinea. cryptolist@yahoogroups.com 19 September.

JURASEK, Todd (2002). Pers. comm., 16 January.

SHUKER, Karl P.N. (2000). Alien Zoo – Roping in the ropen. *Fortean Times*, no. 133 (April): 20.

SHUKER, Karl P.N. (2001). Alien Zoo - Roping in another ropen. *Fortean Times*, no. 142 (January): 21.

WHITCOMB, Jonathan (2006). *Searching for Ropens: Living Pterosaurs in Papua New Guinea.* BookShelf Press (Livermore); (2007). 2nd edit.

WIKIPEDIA (2010). Ropen. Accessed on 22 April.

The sheet music of the famous song from 1924 that gave the Lake Okanagan monster its modern-day Ogopogo nickname (Dr Karl Shuker)

2001

OGOPOGO NEWS. In October 2000, Prof. Roy P. Mackal was seeking Ogopogo of British Columbia's Lake Okanagan with a Japanese film crew. In a phone conversation with fellow mystery beast investigator Bill Gibbons in January 2001, details from which were posted by Bill to the cz@egroups.com discussion group, Roy allegedly noted that during the search he saw what may have been three juvenile lake animals with their embryonic sacs still attached, and the team made very strong sonar contact with two adult specimens. In December 2000, Roy underwent triple bypass heart surgery, and is now recovering at home. *FT* and our readers wish him a speedy return to full health. **Bill Gibbons, cz@egroups.com 2 January 2001.**

A CRABBY FIND DOWN UNDER. Oop (out of place) animals come in all shapes and sizes, and turn up in the most surprising places. One of the latest examples is the unexpected capture in Port River, in Adelaide, South Australia, of *Charybdis japonica* - a shallow-water species of crab normally found in Japan, China, and Malaysia. It was soon identified by way of its distinctive paddle-like back legs, but a search for other specimens was unsuccessful. Happily, moreover, it was a male, rather than a female with eggs, but as readily conceded by Will Zacharin, acting fisheries director, how it came to be living in Port River remains a mystery. *Adelaide Advertiser*, **20 January 2001.**

PITY ABOUT THE PELT. The enigmatic Andean wolf *Dasycyon hagenbecki* is known only from a single pelt housed at Munich's Zoological State Museum. Analyses of hair samples taken from it in the past have been controversial and inconclusive. Recently, however, Guillaume Chapron, a French researcher with a longstanding interest in this mysterious creature, took skin samples for genetic examination to a laboratory with experience in wolf DNA analysis - the first investigation of this pelt for some time. Sadly, however, the results were unsatisfactory, because the pelt samples were contaminated with human, dog, wolf, and pig(!) DNA, and the pelt had been treated chemically at some stage in the past, causing problems with the analysis. **Guillaume Chapron, pers. comm., 5 February 2001.**

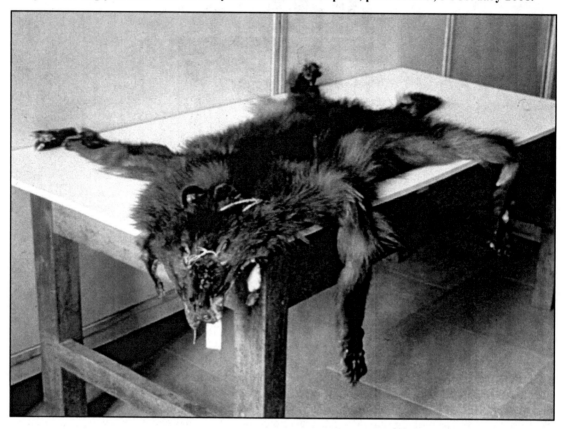

The unique Andean wolf pelt (Alan Pringle)

OGOPOGO UPDATE. Clarification of Bill Gibbons's claim on 2 January 2001 that Prof. Roy P. Mackal had allegedly spied three juvenile Ogopogos last autumn while filming at Lake Okanagan with a Japanese film crew (*FT144*) has recently emerged from Roy via veteran Ogopogo researcher Arlene Gaal. In fact, what happened was that while filming, Roy interviewed a security guard who had had a summer sighting of Ogopogo. Based upon the description given, Roy had suggested that what the guard had seen may have been a juvenile Ogopogo, recently birthed, with possible foetal membrane attached. Roy had no sightings himself, but sonar hits were recorded. **Arlene Gaal, cz@yahoogroups.com 8 February 2002.**

NEWS OF A NEST. In the search for the reputedly extinct South Island kokako *Callaeas cinerea cinerea* (see previous Alien Zoo items re this New Zealand wattlebird), gold-miner Des Gavin has lately reported not only hearing and seeing what he believes to be one of these birds, but also finding a possible kokako nest, in a remote area of South Otago. The exact location has not been disclosed, but longstanding kokako seeker Rhys Buckingham and Department of Conservation officials will be speaking to Gavin, and Buckingham hopes that analyses will be conducted on feathers and other samples from the nest. *Southland Times*, **14 February 2001.**

Alien Zoo

MOKELE-MBEMBE GOES WEST! To the west of the Congo's Likouala swamplands, allegedly home to the elusive mokele-mbembe, is Cameroon, a country hitherto little-associated with reports of putative sauropod dinosaurs. In November 2000, however, veteran m-m seeker Bill Gibbons and American creationist lecturer David Woetzel became the first Westerners to penetrate the forests and swamps fringing Cameroon's Boumba and Loponji Rivers, where they hoped to uncover evidence that the mokele-mbembe may also exist here. They were not disappointed. Unlike the Likouala natives, who have become well-accustomed to Westerners seeking long-necked mystery beasts, being interviewed by cryptozoologists was a totally new experience for the Cameroon pygmies acting as guides to Gibbons and Woetzel. Yet they too claimed that a long-necked swamp dweller existed here, which they termed the li'kela-bembe, and claimed to be as big as an elephant with a snake-like head, long neck, and long powerful tail. Moreover, when shown a range of animal pictures, they readily selected representations of sauropods as its identity. Heartened by this news, in February 2001 Gibbons returned here in the company of American cryptozoologist Scott Norman and a BBC television crew, hoping to film this extraordinary creature, We await their return and film with interest. **Bill Gibbons, pers. comm., 21 November 2000;** *Concord Monitor*, **2 March 2001.**

The late Scott Norman (William Rebsamen)

SPONGING SECRETS FROM THE SEA. Around 65 million years ago, sponges existed in massive underwater reefs, analogous to coral reefs. Today, however, sponge reefs are long gone, except for their ancient fossil remains - or so it was thought. In March 2001, Dr Kim Conway from the Geological Survey of Canada formally announced the discovery of four colonies of glass-like sponges comprising giant reefs covering more than 1000 square kilometres in the sea off British Columbia's Queen Charlotte Islands. Occurring at a depth of approximately 250 m, with each individual sponge standing 1.5-2 m high, and spanning at least 3 m across, their presence was first suspected following sonar surveys back in 1984. However, they were not actually viewed until 1999, during an 18-day submarine foray. So far, only the most northerly colony has been studied. Incorporating three predominant, siliceous species, they take a variety of forms, resembling massive domes 2 m high and 4-5 m across, or comprising a mass of delicate radiating tubes, each only 1-2 mm thick. Due to their extreme fragility, however, they are in danger of being severely damaged by bottom trawling. *Vancouver Sun*, **9 March 2001.**

A MONGOLIAN MYSTERY. An Asian anomaly is an unidentified species of hoofed mammal reported from the Delger Hangai somon territory, in Mongolia's Dund Gobi aimak. It is described as being the size of a goat and brown in colour, with the front part of its body shaped like a saiga *Saiga tatarica* (a trunked Asian antelope) bent forward, but the back part high like that of a musk deer *Moschus* sp. This Mongolian mystery ungulate also has a pair of small spreading horns, and a running ability exceeding that of sheep and goats but less than that of antelopes. Apparently, these animals were first sighted a few years ago in the Hulhairhan mount by local people, but were not reported as the locals wanted to protect them. Grazing in groups of 4-5 in rough, rocky terrain, their numbers increased and they can now be found in the Zoson Tegel mountains too. **http://www.mol.mn/news/eng_main.php3?myDir=March/eng_Mar20.html 20 March 2001 [no longer online].**

Alien Zoo

BEMUSED IN BILI. A century ago, a large ape skull was found in Bili, northern Congo, since when controversy has persisted as to whether this remote area may harbour an undiscovered population of gorillas. Much more recently, a plaster cast of a big ape footprint was also obtained here. Thus in March 2001, a National Geographic Radio Expedition that included Harvard University/Leakey Foundation primatologist Dr Richard Wrangham and Dr Christophe Boesch from Leipzig's Max Planck Institute for Evolutionary Anthropology visited the Bili forest in search of clues as to the mysterious Bili ape's identity. Pictures of gorillas did not seem familiar to the local people, but the ape's reality was confirmed by the discovery of a typically anthropoid ground nest. Ape faeces were also found, and were analysed to learn what it eats - from which Wrangham concluded that it was not a gorilla but a chimpanzee, and which team leader Karl Amman feels may be of undescribed taxonomic status. http://www.npr.org/programs/RE/archives/index_biliape.html 26 March 2001 [no longer online].

Will the yeti soon be formally recognised?
(Dr Karl Shuker)

JUST A HAIR'S BREADTH AWAY? That's how far the yeti may be from official zoological recognition, thanks to a mystifying sample of hair discovered in the hollow of a cedar tree in eastern Bhutan by a new British scientific expedition.

Led there by a local hunter who claimed that a yeti had recently visited the tree, zoologist Dr Rob McCall collected some hair from the hollow and sent it to the Oxford Institute of Molecular Medicine.

Here it was examined by geneticist Prof. Bryan Sykes, one of the world's leading experts in DNA analysis. Prof. Sykes later announced that although the hair did contain DNA, they had not been able to identify it. "It's not a human, not a bear nor anything else we have so far been able to identify. It's a mystery and I never thought this would end in a mystery.

We have never encountered DNA that we couldn't recognise before". *The Times*, 2 April 2001; 'To the Ends of the Earth' TV documentary, Channel 4, screened 2 April 2001.

Alien Zoo

'MAMI WATER' A MANATEE? One of the most curious creatures featured in traditional West African folk stories is a mermaid-like entity known as 'mami water', said to be half woman and half fish. However, Ghanaian research fellow Dr Mamaa Entsua-Mensah, from the Institute of Aquatic Biology of the Centre for Scientific and Industrial Research (CSIR), together with scientists from the Ghanaian Wildlife Department, have recently proposed that this enigmatic being may be based upon a real animal - the West African manatee *Trichechus senegalensis*. Manatees and the related dugong or sea-cow, collectively termed sirenians, have long been popular identities among zoologists for mermaids. However, many mermaid reports describe exotic, beautiful humanoid entities seemingly very different from the decidedly homely sirenians. Moreover, the mami water issue is particularly problematical, as this latter being is reputedly scaly, which sirenians are not. **http://allafrica.com/stories/200104050058.html 5 April 2001.**

A STEED THAT BLEEDS. According to the ancient Chinese 'Shiki' histories, there was once an amazing breed of horse living in a region along the Silk Road covering modern-day Tajikistan and Uzbekistan that could run at such high speed it was able to travel almost 4000 km in a day. Moreover, when it ran, it sweat blood. Scientists have traditionally dismissed these claims, and explain the supposed blood-sweating phenomenon as either the result of infection by a parasitic worm, or an illusion created by a horse's arteries standing out at the surface of the skin so that its skin appears red. In mid-April 2001, however, Japanese equine specialist Hayato Shimizu revealed that during August 2000 he had photographed a bona fide blood-sweating horse in Central Asia's western Tian Shan mountain range. He claims that the horse bled after running for a time at full speed, its shoulder bulging before blood mixed with sweat trickled down its flank, and that locals state that near Tian Shan such horses do indeed still exist. *Japan Times*, **15 April 2001.**

A MALAYSIAN MAN-BEAST. So too might a certain shy crypto-primate, said to inhabit the vast Endau-Rompin park in the East Coast of Peninsular Malaysia. Known locally as the hantu jarang gigi, it has been described by eyewitnesses as standing about 3 m tall, covered all over with dark brown hair, and with a footprint measuring approximately 45 cm (roughly twice the size of an adult human's shoeless foot). A number of sightings have been reported near a tributary of the Endau River known as Sungai Kencin, where fish bones have been discovered scattered on the ground, as if the creature had lately eaten a meal of fish. Reports of a comparable man-beast, termed the ensut-ensut, have emerged from Malacca, the most recent being a sighting by villagers of an adult fleeing for shelter out of a burning jungle with its youngster. *New Straits Times Press*, **25 April 2001.**

MADIDI MYSTERIES. Still with mystery apes: explorer Simon Chapman's latest book concerns his search in Bolivia's Madidi region for the mono rey, an elusive ape-like entity synonymised by Chapman with the (in)famous *Ameranthropoides loysi* photo. Although he failed to find the mono rey, his book did contain a couple of tantalising snippets that were new to me. One was his claim that until recently, a local Bolivian actually owned a pelt from a mono rey, which was then purchased by "a gringo" who took it home and sent it (or samples from it) off for DNA analysis, but the results (if any) were never revealed. No details were given in Chapman's book as to who the "gringo" was, where he came from, or where the pelt/samples was sent. The other snippet, which Chapman had apparently attempted unsuccessfully to substantiate, was that a living mono rey had allegedly once been exhibited for a time at Santa Cruz Zoo! *The Monster of the Madidi: Searching For the Giant Ape of the Bolivian Jungle* **(Aurum Press: London, 2001).**

HERE'S HESSIE, ONLINE! In *FT*125, I reported the airing on Norwegian TV during June 1999 of some recently-filmed video footage showing an alleged elongate sea serpent feeding off the carcase of a dead whale at Hessa, Aalesund. The distance between sea serpent and camera was 900 m. This intrigu-

Alien Zoo

ing footage of what has since been nicknamed Hessie can now be accessed online, at: http://www.mjoesormen.no/Bilder/hessievideo1999.asf **Erik Knatterud, cz@yahoogroups.com 8 May 2001.**

END OF THE LINE FOR THE THYLACINE? The last confirmed thylacine or Tasmanian wolf *Thylacinus cynocephalus* died in Hobart Zoo in 1936. Since then, numerous sightings of thylacines have been claimed, but never verified, resulting in this enigmatic creature being dubbed the world's most common extinct animal. In May 2001, however, its alleged extinction became official - at least as far as the State Government of Tasmania is concerned, which announced that its official verdict on the status of this island's most famous native species was "presumed extinct". Michael Aird, Leader for the Australian Government in the Upper House, stated that this official view was based upon the lack of any hard evidence of the thylacine's survival for 50 years - the accepted international criterion. He also revealed that although the State Government "supports responsible efforts in trying to establish the existence of the thylacine", it no longer had a detailed contingency plan for its possible future rediscovery, and would not be financing any searches. In contrast, the Tasmanian Museum and Art Gallery aims to conduct its own AU$250,000 thylacine search and hopes to attract corporate funding. **http://themercury.com.au/0,3546,2034295%5E3462,00.html 24 May 2001.**

Thylacines, painted by Henry Constantine Richter, 1845

ON THE LOOKOUT IN LOUGH REE. Nearer to home, in late June monster seeker Jan-Ove Sundberg and his Global Underwater Search team (GUST) are heading to Lough Ree in County Galway, Ireland, in search of its fabled horse-eel inhabitant. Claimed to have the head of a horse and the body of a huge eel, it is one of many such creatures reported in loughs all over Ireland down through the ages. This is in spite of the fact that these bodies of water often appear too small to sustain any form

Alien Zoo

of large aquatic animal, leading some cryptozoologists to suggest that if horse-eels do exist, perhaps they migrate overland from one lough to another, rather than remaining permanently in any one lough. Sundberg believes that reports of the Lough Ree horse-eel indicate that it fits the description of a Norwegian lake monster called Selma, previously sought by Sundberg and his team. http://www.ireland.com/newspaper/breaking/2001/0612/breaking69.htm 12 June 2001 [no longer online].

SERBIAN SWAMP SQUID? Carska bara is a big swamp in Serbia where reports of strange aquatic monsters periodically surface - as do the monsters themselves, allegedly! According to a recent investigation by Serbian cryptozoological enthusiast Marko Sciban, one night a Bosnian sailor called Slaven heard loud smashing sounds in the water, which he likened to the sound made by lots of tails hitting the water, accompanied by loud bubbling and croaking noises. Another man, Sima, who is responsible for this lake, was so frightened that he fired several shots over the lake's surface. Both have suggested that a squid-like tentacled monster exists here, which surfaces every few evenings. Marko hopes to uncover further information regarding this hitherto-obscure cryptid. **Marko Sciban, pers. comm., 24 June 2001.**

Wooden yowie statue at Kilcoy, Queensland (Seo75/ Wikipedia)

NINETY-FOUR NEW SPECIES. Proving that the age of zoological discovery is far from over, 94 new species of wildlife, including frogs, lizards, and fishes, as well as many insects, have been recorded from the Wapoga River Area of western Irian Jaya, on the island of New Guinea. The discoveries were made during a survey by Conservation International's Rapid Assessment Program (RAP). *Oryx*, **July 2001.**

ON THE TRAIL OF THE YOWIE. Australian cryptozoologist Rex Gilroy, who is also director of the Hominid Research Centre at Katoomba, New South Wales, plans to launch a major new expedition in late July, lasting six months, in search of Australia's supposed man-beast, the yowie. Financed by American backers and featuring American bigfoot researcher Todd Jurasek in the team, their itinerary includes investigating not only recent claimed sightings of yowies but also apparent freshly-abandoned camp sites where stone tools have been found in the Morton National Park. *Canberra Times*, **8 July 2001.**

LOUGH REE MONSTER... A RealVideo-format video of a television news segment aired lately by the national Irish news agency RTE concerning GUST's 'Operation Horseel' expedition in June to

Ireland's Lough Ree in search of its fabled horse-eel has now appeared online. Featuring veteran lake monster researcher Peter Costello, it can be accessed at: http://www.rte.ie/news/2001/0706/nationwide.html **Nick Sucik, cz@yahoogroups.com 17 July 2001.**

AND LAMPREYS. As for the Lough Ree monster itself: according to GUST leader Jan-Ove Sundberg, the team believes it possible that it is a giant form of lamprey - an eel-like but jawless sucker-mouthed vertebrate - and hopes to return next summer with a full-scale expedition. In the meantime, GUST made a reconnaissance expedition to Sweden's Lake Storsjon in July, Jan plans to visit another of Sweden's 23 'monster' lakes during early August, and GUST returns to Norway's Lake Seljordsvatnet later that same month. **http://www.cryptozoology.st/ accessed 17 July 2001.**

HAPPY ANNIVERSARY, STEVE! July 2001 marks the tenth anniversary of Nessie seeker Steve Feltham's vigil at the lochside of Loch Ness. For the past decade he has diligently albeit as yet unsuccessfully scoured the loch's waters for a sign of its legendary inhabitant, and has lived throughout those ten years in nothing more luxurious than an old library van. Now, however, he hopes to swap it for a double-decker bus, converting the bottom deck into a Loch Ness exhibition, and is planning his next decade here. *FT* wishes him well, and hopes that his patient persistence will ultimately pay off. ***Daily Record*, 18 July 2001.**

Steve Feltham at Loch Ness (Steve Feltham)

SILVER SURPRISE. Since their discovery in Spain in 1964 and 1965 respectively, two eycatching silver figurines portraying a coelacanth-like fish have attracted a great deal of cryptozoological curiosity. This is because it has been widely suggested that they may have derived from Mexico and date back as far as the 17th Century. If so, then they considerably precede the discovery in 1938 of the first-known species of living coelacanth, *Latimeria chalumnae*. Consequently, this has substantiated belief that an unknown species of modern-day coelacanth might conceivably thrive in the Mexican Gulf. Now, however, this tower of speculation has apparently been toppled like a cryptozoological house of cards, following the publication of a new paper by veteran coelacanth researcher Prof. Hans Fricke and Marseille colleague Dr Raphael Plante. For in their paper, they present a new study revealing that these enigmatic silver artefacts were modelled directly upon *L. chalumnae*, and are only of modern-day origin after all. ***Environmental Biology of Fishes*, 61 (August 2001).**

SEPTEMBER IN SUMATRA. An intrepid cryptozoological field researcher is Adam Davies, famous for his recent Congolese search for the mokele-mbembe. On 11 September 2001, Adam is broadening his mystery beast horizons by leading a 3-man, 3-week expedition to Sumatra, seeking its elusive man-beast, the orang pendek. While there, 'Expedition Orang Pendek' will be liaising with veteran orang pendek investigator Debbie Martyr, who has spent several years here - participating in an ongoing search sponsored by Fauna and Flora International and also by the WWF among others. Moreover, next year Adam is hoping to set forth on the trail of an even more dramatic mystery beast - the formidable Mongolian death worm of the Gobi. **Adam Davies, pers. comm., 6 August 2001.**

TAZ IN OZ? As previously discussed in Alien Zoo (***FT*109**), live specimens of the Tasmanian devil *Sarcophilus harrisii*, a species that officially became extinct long ago on the Australian mainland, have nonetheless occasionally and inexplicably been found here, especially in Victoria. This has led some scientists to speculate whether this burly marsupial may have indeed survived to the present day in parts of mainland Australia after all. Now, this intriguing possibility is finally being seriously investigated - by a scientific team at Victoria's La Trobe University. They are analysing the DNA of four Tasmanian devils discovered at different times and in different locations in Victoria during the past century, to de-

19th-Century painting of Tasmanian devil by John Gould

termine whether they are merely escapees from captivity here or bona fide representatives of a hitherto-undiscovered remnant native Victorian population. *The Mercury*, **Hobart, 19 August 2001.**

Mongolian death worm based upon eyewitness descriptions (Ivan Mackerle)

DOCUMENTING THE DEATH WORM. Ivan Mackerle is the Czech explorer famous for bringing the Mongolian death worm or allghoi khorkhoi to widespread cryptozoological attention via his expeditions to the Gobi desert in search of this reputedly lethal creature during the 1990s.

He has now published the most detailed documentation currently available concerning his investigations of this cryptid and other Mongolian mysteries, in a new book entitled *Mongolské Záhady*. It is divided into three sections, dealing respectively with mysterious but hitherto little-publicised UFO-like lights and alleged alien beings seen in and over Mongolia, the Mongolian death worm, and Mongolia's supposed man-beast termed the almas. *Mongolské Záhady* **(Ivo Ƶelezný: Prague, 2001).**

AND NOW THERE ARE THREE. After longstanding zoological speculation as to whether the African elephant *Loxodonta africana* actually comprises more than one species, DNA studies have now provided the clearest evidence yet that it should be split taxonomically into two distinct species. Kenyan researchers working with Maryland geneticists have published the results of their work in the leading American scientific journal *Science*, revealing that the West African forest elephant is significantly different genetically from the larger plains elephant of East Africa. This discovery underlines previously-recognised external differences between the two forms, the most notable of which is that the forest elephant's ears are smaller and rounder than the bigger, pointed ears of the plains elephant. Consequently, when the Asian elephant *Elephas maximus* is also added to the list, this means that there would now appear to be three valid species of elephant alive today - and perhaps even more if, as some cryptozoologists believe, the controversial African pygmy elephant is also a legitimately discrete species in its own right. *Science*, **24 August 2001.**

A STRANGE SHARK. More proof that the sea still holds many fishy surprises comes from Colombian biologist Dr Sandra Bessudo, who features in a recent article by François Sarano concerning her ongoing search for a still-unidentified but very sizeable shark in the eastern Pacific waters off Colombia. The shark has been observed and even photographed on several occasions by divers, who have variously dubbed it 'bongo' and 'the monster' - on account of its size, estimated by them to be a very impressive 6 m long. Judging from their photographs, this Colombian mystery shark behaves like an *Odontaspis* (small-tooth sand tiger) shark, but differs from the two known species of this genus not only by its greater size but also by the very advanced position of its dorsal fin. *30 Millions d'Amis*, **September 2001.**

CRYPTOZOOLOGICAL KYRGYZSTAN. The former Soviet Republic of Kyrgyzia, now an independent central Asian country renamed Kyrgyzstan, has not previously attracted much attention from cryptozoologists. But all of this may now change - thanks to the recent discovery in its mountainous Aktalinsky region of mysterious footprints reminiscent of those of man-beast entities previously reported further south in the neighbouring Pamir mountains of Tajikistan 20 years ago. Measuring 45 cm long and 30 cm wide, they were apparently found by a frontier guard, who claimed that he saw them outlined against the clay bank of a mountain river flowing through this remote region. **WENN/Nick Sucik, cz@yahoogroups.com, 6 September 2001.**

BATTY OVER A BARBASTELLE. In September 2001, news emerged that a male barbastelle *Barbastella barbastellus* had been found alive in Briddlesford Wood on the Isle of Wight - the first time that this species of bat had been discovered on the island for over a century.

Indeed, until as recently as five years ago, only a single barbastelle colony was known anywhere in the U.K. **http://www.ananova.com/news/story/sm_400704.html 17 September 2001 [no longer online].**

Barbastelle bats

NEW BIRDS ON THE WING. The British magazine *Birdwatch* recently published a most interesting article by acclaimed birder Jon Hornbuckle concerning newly-discovered species of bird, some of which have still to be formally described due to their discoverers having more pressing commitments - which may seem like heresy to some of the more fervent members of the cryptozoological community, but is a fact of life in mainstream zoology! Having said that, many of these new birds are relatively unspectacular passerines (perching birds). In addition, Hornbuckle refers to some still-uncollected, unconfirmed birds, including one that particularly intrigued me - a possible undiscovered species of paradigalla (a crow-like bird of paradise) sighted in 1992 by David Gibbs while exploring the uninhabited Fakfak Mountains in southwestern Irian Jaya, New Guinea. ***Birdwatch*, October 2001.**

NEW FOR THE NEW FOREST. In the New Forest's northern edge, near Landford, Southampton University entomologists Dr Dave Goulson and Ben Darvil recently found a species of carder bumblebee that, although widespread in continental Europe, had never been recorded from Britain before. Known scientifically as *Bombus hypnorum*, the tree bumblebee, it is characterised by its distinctive colour combination of orange thorax, black abdomen, and white tail. ***Daily Mail*, 5 October 2001.**

ENTOMOLOGICALLY ENDOWED. Moreover, a new butterfly for Ireland, distinguished from the otherwise-similar common wood white *Leptidea sinapis* by virtue of its sizeable genitalia, has recently been discovered. Known as Real's wood white *Leptidea reali*, and identified by scientists from the Ulster Museum and Butterfly Conservation, it occurs in both Northern Ireland and the Republic of Ireland. **http://www.ananova.com/news/story/sm_414094.html 10 October 2001 [no longer online].**

CZ IN OZ. Saturday 20 October 2001 saw the first cryptozoological conference to be held in Australia, which proved to be a great success. Entitled 'Myths and Monsters 2001', and organised by Ruby Lang (publisher of online magazine *Strange Nation*), it was held at History House in Sydney's Macquarie Street, and was well-attended throughout the day, lasting from 9 am to 6 pm.

Proceedings began with an hour-long paper by Paul Cropper, outlining the history of Australian cryptozoology. Other speakers included Australian Yowie Research founder Dean Harrison, Malcolm Smith on Aussie sea monsters, Peter Chapple on ABCs Down Under as well as the thylacine, Prof. Gary Opit

Alien Zoo

Bunyip by Richard Svensson

Did the Aussie elephant humanoids resemble this example from an early 17th-Century bestiary by Aldrovandi?

on the bunyip, Tony Healy on 'high strangeness' in yowie reports, and Bill Chalker on the Fernvale flap of 1927 (in which some truly bizarre entities were reported, such as 'big bird' creatures and 'elephant humanoids').

The conference ended with an hour-long debate by the speakers concerning the evidence for Australia's mystery beasts. **Australian Associated Press, 20 October 2001;**

http://www.strangenation.com.au/cryptoconf.htm October 2001 [now at http://www.strangeark.com/craig/MMPapers_Aust.pdf].

BEST FOOT FORWARD. In an earlier Alien Zoo (*FT*151), I noted that explorer Adam Davies and two colleagues would be conducting a 3-week Sumatran expedition in September, seeking its elusive man-beast, the orang pendek.

They have now returned, having obtained some putative orang pendek hair samples and also an excellent footprint cast, which exhibits the highly distinctive toe placing (the big toe set well back and at an angle from the other toes) characterising this cryptid.

Having been shown photos of the footprint by Adam, I recommended that he should submit them to Canberra-based primatologist

Alien Zoo

Adam Davies with cast of orang pendek footprint (Andy Sanderson)

Prof. Colin P. Groves for further examination, and that he should send the hair samples for DNA analysis to the Oxford Institute of Molecular Medicine, which quite recently examined some alleged Bhutan yeti hair (Alien Zoo, *FT*147). Adam followed both of my suggestions, and we now await with interest these authorities' findings - which may, if positive, provide the best evidence so far obtained for the orang pendek's reality as a still-undescribed species of primate. **Adam Davies, pers. comms, October 2001;** *The Times*, **27 October 2001.**

PISTOL-PACKING PRAWN. In November 2001, marine biologist Matt Slater, from Newquay's Blue Reef Aquarium, revealed that he had recently been given an unusual bright-orange shrimp with extra-large pincers that had lately been caught off Cornwall's Falmouth Bay by fisherman Timmy Bailey. But what attracted Slater's particular attention was the loud, incredible sound that the shrimp kept making with its claws - resembling a pistol being shot, or bubble-wrap being popped. This extraordinary activity, coupled with its distinctive appearance, enabled Slater to identify his strange captive as a pistol shrimp - a species of crustacean normally found in the Mediterranean, and last reported off the British coast way back in 1914. Moreover, at much the same time that Bailey had caught his pistol shrimp, fellow Cornish fisherman Micky Burt had captured another Mediterranean vagrant locally. It was a sponge crab, so-named because this odd crustacean grows a sponge on its back to enhance its camouflage talents. And in late November 2001, the first-ever barracuda recorded in British waters, measuring 90 cm and weighing around 5.5 kg, was captured off Cornwall's Lizard - even though this infamously savage species of fish normally does not journey closer to Britain than the Bay of Biscay. Experts have speculated that the exceptionally warm October of 2001 was responsible for enticing these marine creatures far away from their normal, balmier waters. *Daily Mail*, **2 & 27 November 2001.**

NEWS FROM NORWAY. Water monster researcher Erik Knatterud has launched a website comprising a national register of Norwegian sea serpents. It also contains details of Norwegian lake monster sightings. Among the most notable recent reports is one dating from the second half of November 2001, concerning a creature measuring approximately 45 ft long that was sighted resting at the surface of Lake Mjoesa's south end, then diving. This lake is situated north of Oslo and Gardermoen airport. **http://www.mjoesormen.no accessed 30 November 2001.**

Alien Zoo

MEETINGS WITH THE MAPINGUARY. Back in 1993, zoologist Dr David Oren, based at that time in Belem, Brazil, published a *Goeldiana* paper on his researches concerning the mapinguary - an Amazonian mystery beast that he believes may be a surviving species of ground sloth. At that time, he had not actually spoken to anyone alleging to have killed a mapinguary. However, since then he has conducted expeditions in search of the beast during which he spoke to seven different local hunters who claim to have shot specimens of this creature, as well as to another 80 people who have supposedly seen one. These and other details have now been published by Oren in a second paper, updating his mapinguary investigations. Whether he will be able to continue further with them, however, is another matter, as Oren has recently moved several thousand miles away from the Amazon, to take up a new post in Brasilia. ***World Conservation Union's Edentate Specialist Group Newsletter*, winter 2001; Reuters, 18 December 2001.**

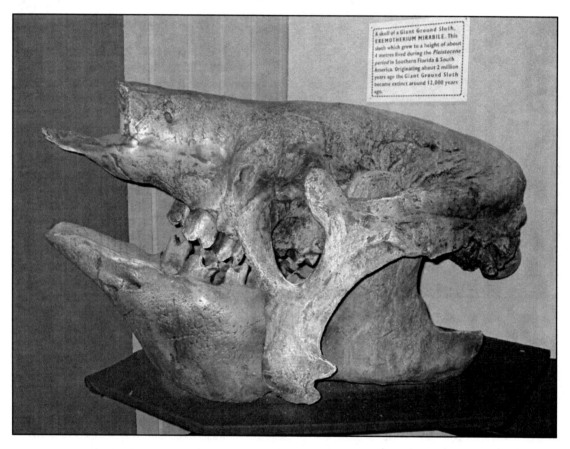

Skull of giant ground sloth at Dorchester Dinosaur Museum (Mark North)

The Lost Ark

HORSING AROUND IN TIBET - IN PURSUIT OF THE PIG-PONY.

We were in a very unexplored area, the primitive pre-Buddhist area of Tibet to the north of Lhasa, not far from the Chinese border. We weren't able to proceed on our intended route because the passes were blocked with snow. So we took another way, into Riwoche, which is where we found the little monster.

It is pony-size, about 4 ft, a little like a donkey but with small ears, hardly any nostrils and a rough coat. It has a black stripe down its back, stripes on its back legs and a black mane. I thought it looked like cave drawings of horses, although a friend of mine says it looks like a pig.

Dr Michel Peissel, quoted by Suzanne Lowry in online article at http://alpha.mic.dundee.ac.uk/ft/ft.html

It was October 1995, during a seven-week, six-member expedition to northeastern Tibet, and the team was retracing its steps to this once-independent country's lofty capital, Lhasa. Unfortunately, their planned route had been rendered impassable by impenetrable blockades of snow, and so they were forced to navigate along a scarcely-traversed alternative path, through a remote region called Riwoche, which led to a mysterious valley.

Anyone imagining that I am about to launch into a Shangri-La-inspired storyline would be totally wrong - actually, it was the Lost World that I was thinking of, for the creatures that the team discovered here seemed to have stepped out from the pages of prehistory. True, they were not living dinosaurs, but zoologically they were scarcely less special. Suddenly, a horse came into view, but it was unlike any that Western observers had seen before. At first, the explorers wondered if it could simply be a freak specimen, but then they saw a second, and a third, until eventually a sizeable herd was present - and

Alien Zoo

A Riwoche horse (Tim Morris)

they were all exactly the same.

In the words of the team's leader, French ethnologist Dr Michel Peissel: "They looked completely archaic, like the horses in prehistoric cave paintings," i.e. from Neolithic times.

In overall appearance, this remarkable creature, since dubbed the Riwoche horse, resembles a somewhat portly donkey-like pony with short legs - one of Peissel's colleagues unkindly likened it to a pig. Standing almost 4 ft high, with a wedge-shaped head, it has a rough beige-coloured coat, harsh black mane and long black tail, stripes on its lower hind legs, and a black dorsal stripe (all very primitive equine features). Like many montane mammals, its ears are short (to prevent heat loss), and its nostrils are extremely small.

Enquiries revealed that these horses run wild in the valley, but are captured with lassos by the pre-Buddhist Bon-po people dwelling here, when transportation is required. Once they have served their purpose, however, they are released to roam the valley again.

The valley itself is 70 miles long, and is sealed off on either side by sheer passes around 16,000 ft high. In a practical sense, therefore, it is an island, wholly separated from the outside world. Team member Dr Ignasi Casas, a veterinarian, believes that it is the valley's insularity that has preserved the Riwoche horse as a bona fide living fossil. Its many highly primitive characteristics have been retained simply because there has been no opportunity for interbreeding with other forms of horse, which would have diluted these characteristics.

Plans were made for the Riwoche horse's taxonomic affinities with other types of horse to be investigated via comparative DNA analyses undertaken at the Royal Agricultural College in Cirencester by geneticist Dr Steven Harrison, using blood samples taken from some specimens. Particular attention would be paid to comparisons with DNA samples from Przewalski's horse *Equus ferus przewalskii*, a western Mongolian subspecies of wild horse that remained unknown to Western science until formally documented in 1881, following its discovery by Russian explorer General N.M. Przewalski. Although larger than Tibet's newly-revealed 'pig-pony', it shares many of the latter beast's primitive features, including its donkey-like build and short legs. Even so, as Peissel later revealed in his book *Tibet: The Secret Continent* (2002), those tests that were conducted did not reveal notable genetic differences from other domestic breeds of horse. So he does not consider the Riwoche horse and Przewalski's horse to be directly related.

Alien Zoo

Przewalski's horses

Irrespective of this, the Riwoche horse's discovery is doubly noteworthy, because the reason for the expedition's presence in Tibet was to learn more about another equine enigma, the Nangchen horse. Revealed to Western science by the self-same Dr Peissel in 1993, this is a large, powerful creature, but of seemingly independent origin from Arabian and Mongolian breeds. Its lungs and heart are abnormally large - adaptations, presumably, to high-altitude life.

In his book *Living Treasure* (1941), cryptozoologist Ivan T. Sanderson recalled encountering a herd of so-called wild horses some years earlier while animal collecting on Haiti. In contrast to herds of recently run-wild horses on the island that sported a wide range of shapes, sizes, and colours, these horses were all precisely the same as one another. They shared a pinkish-brown coat interspersed with silvery-grey hairs, a silvery mane becoming dark brown terminally, a long dark-tipped tail, a white dagger-like blaze running down the brow and muzzle, and short dirty-white socks on all four feet. They were also said to attack with great aggression the polymorphic feral pretenders sharing their island. Sanderson sent the skeleton of one of Haiti's wild horses to the United States for study, where it was found to be of modern-day descent, not a prehistoric relic, but with unusually slender bones - not previously realised, because in life these horses are noticeably fat.

One final living fossil of the equine variety can apparently be found very much closer to home - on Exmoor. At a zoological conference held in Leicester on 30 April 1994, horse geneticists and evolutionists learnt that paleontological comparisons of fossilised horse bones unearthed in Alaska and modern-day bones of the sturdy Exmoor pony had provided persuasive evidence for believing that the latter breed was a direct, scarcely-altered descendant of the ancient race of hill ponies which had entered southern Britain from continental Europe at least 130,000 years ago, long before the English Channel's formation. To quote zoologist Dr Alma Swan, founder of the British Pony Centre at Gaddesby in Leicestershire: "What we have here is effectively a living fossil." Anyone who feels compelled to trek through remote swamps and scale treacherous plateaux in search of prehistoric survivors, please take note!

SELECTED REFERENCES

ANON. (1994). Pony a living fossil – claim. *Express and Star* (Wolverhampton), 30 April.
BILLE, Matthew A. (ed.) (1996). Recent discoveries: unknown horses. *Exotic Zoology*, 3 (January-February): 1-2.
LOWRY, Suzanne (1995). Explorer backs Tibetan dark horse in the history stakes. *Daily Telegraph* (London), 16 November.
PEISSEL, Michel (2002). *Tibet: The Secret Continent*. Macmillan (London).
SANDERSON, Ivan T. (1941). *Living Treasure*. Viking Press (New York).
SIMONS, Marlise (1995). A stone-age horse still roams a Tibetan plateau. *New York Times* (New York), 12 November.
SIMONS, Marlise (1995). A stone age horse in Tibet. *Times Higher Educational Supplement* (London), 16 November.

Exmoor pony at the Valley of the Rocks, Lynton, Devon (Richard Pharo)

2002

LURKING ON LUZON. Five mysterious water monsters, first sighted on 5 November 2001, are causing consternation among families of the Aeta tribe living in the village of Buhawen, on the island of Luzon in the Philippines, whose oral history has no mention of such creatures. The monsters have appeared in the Tikis River, are jet black in colour, and elongate in outline. One of them was estimated by eyewitnesses to measure approximately 7 ft long and 3 ft wide, another was said to be the size of a jeep, but all remain unidentified, as they have never revealed their heads or tails. The villagers are so concerned for their own safety that they no longer catch fish in the river, have forbidden their children from bathing there, and are appealing to scientists to visit the area in the hope of solving the mystery of this bizarre quintet. http://www.inq7.net/nat/2002/jan/14/nat_9-1.htm 14 January 2002 [no longer online].

R.I.P. TASEK BERA. Still with Asian water monsters: in my book *In Search of Prehistoric Survivors* (1995), I recalled the huge 'golden serpents' or ular tedong of Tasek Bera - a remote lake in the Malaysian state of Pahang, which in 1951 was visited by explorer Stewart Wavell in the hope of spying the mystifying 'neodinosaurian' cryptids claimed to inhabit its waters by the local Semelai people. Sadly, he failed to do so, though he did hear an extraordinary trumpeting sound said to be the cry of these creatures. Since then, however, they have been largely forgotten - and now, regrettably, it would seem that any future opportunity to examine the cryptozoological secrets of Tasek Bera has been lost. For in a recent e-mail, correspondent Todd Jurasek informed me that a combination of choking weeds and palm-oil plantations nearby have transformed this once-thriving lake into little more than a cesspool, thus rendering the likelihood of any sizeable cryptids surviving here improbable in the extreme. **Todd Jurasek, pers. comm., 16 January 2002.**

Alien Zoo

Slender-billed curlew (centre)

A VIP VAGRANT IN BRITAIN. Almost as incredible, but at last confirmed as official after three years of debate, was the sighting back in 1998 of a slender-billed curlew *Numenius tenuirostris* at Druridge Bay, in Northumberland. Spied by Tim Cleeves, a Yorkshire-based RSPB conservation officer, who also videoed and photographed it, what makes this report so extraordinary is that the slender-billed curlew is one of the world's rarest species of bird, possibly down to as few as 50 specimens in total globally, and whose nesting site has never been discovered.

Thus Cleeves's video and photos were scrutinised at great length and in meticulous detail by several experts before the British Ornithologists' Union Records Committee finally announced in January 2002 that this critically-endangered species had indeed been accepted onto the official British birds list - making it the rarest bird to be seen in Britain since the great auk *Pinguinus impennis*, which became extinct in 1844. **Wolverhampton Express and Star, 25 January 2002.**

IT'S DINO-BIRD!...OR IS IT? In late January 2002, rumours were circulating on the internet to the effect that the palaeontological community was "abuzz" with news of the discovery of some incredible 'dinosaur-birds' living in seclusion atop a Venezuelan tepui (high isolated plateau) called Aqueputa. According to their alleged discoverer, a Dr José Ramos-Pajaron of Caracas University, who claimed to have observed them with his students, these extraordinary creatures travel in small cooperative family groups, walk upright, stand 2 m tall, are three-toed, and superficially avian, but with a tooth-bearing beak, long stiff bony tail, primitive hair-like feathers, a horny violet crest in the males, long flexible neck, and claw-bearing wings. Yet no such researcher can be traced via online searches; and as noted by veteran fortean researcher Scott Corrales, the name 'Pajaron' actually translates as 'big bird'. Scott feels sure that this whole story is nothing more than an internet-based hoax, as does palaeontologist Darren Naish - who has swiftly pointed out that the palaeontological community is certainly not "abuzz" with news of this case, and that the creatures themselves seem to be based directly upon the latest reconstruction proposals for Cretaceous coelurosaur dinosaurs. **cz@yahoogroups.com, various postings, 31 January 2002.**

A MERE MONSTER? If the swans on the waters of west Lancashire's Martin Mere nature reserve are rather more wary than most at present, they have good reason to be. After all, it takes a pretty sizeable creature to haul underwater a fully-grown swan, which can weigh up to 13 kg. Yet, incredible as it may seem, just such a scene was witnessed by several visitors

Is the monster of Martin Mere a giant catfish?

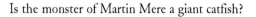

here on 7 February, causing all of the other swans to flee the 20-acre, 4-m-deep lake on which they habitually gather in winter. They had also mysteriously vanished, all 1500-plus of them, on 17 January from this same lake. The cause of their panic (not to mention the abduction of their brethren) remained unseen on both occasions. However, the reserve's manager, Pat Wisniewski, clearly recalls seeing four years ago a mysterious 'something' that appeared to be size of a small car, circling just below the waterline of this same lake. Two favoured contenders are a giant northern pike *Esox lucius* or an oversized wels catfish *Silurus glanis*, both of which could survive in this murky, oxygen-starved water for many years, growing steadily bigger all the time. *Liverpool Echo*, **14 February 2002.**

LION AROUND FLORIDA'S COAST. In January 2002, American newspapers carried reports of an exotic but unwelcome newcomer to the waters off North Carolina. A popular exhibit in tropical marine aquaria due to its vividly-striped, dramatic appearance, the lionfish *Pterois volitans* is (in)famous for the array of highly-venomous, needle-like spines borne upon its long spectacular dorsal, anal, and pelvic fins. Although widely-distributed in the Indo-Pacific, it had not previously been reported from the South Atlantic waters around the southeastern USA. However, it was formally identified after one of several specimens observed near two shipwrecks here by some local scuba divers was collected and sent by the Department of Commerce's National Oceanic and Atmospheric Administration (NOAA) to a lionfish expert. A month later, scientists at the Florida Marine Research Institute of St Petersburg verified this species' existence further south, off the eastern coast of Florida, where a number of specimens had been recorded. Research is now planned by the NOAA to determine the lionfish's precise numbers and range in southeastern American waters, and whether it is reproducing. Warnings to divers and bathers about its spines' dangerous nature are being publicised too. One theory regarding the lionfish's unexpected origin here is that captive specimens may have been purposefully released by aquarists. *Constituent* **(North Carolina), 10 January 2002; http://story.news.yahoo.com/news?tmpl=story&u=/nm/20020216/sc_nm/health_fish_dc_1 16 February 2002 [no longer online].**

Engraving of lionfish

THE (UN)USUAL SUSPECTS. Two decidedly suspect claims have come to my notice recently. One is the briefest of mentions: a colleague in passing stated that during the 1980s a small pterodactyl was reputedly killed and displayed in a store front in Queensland, Australia, until the decomposing carcase was discarded. No further details are available, but if such a specimen did exist, it may well have been a

Alien Zoo

Fake stuffed skunk ape head on eBay, February 2002

variation upon the modified 'Jenny Haniver' or composite 'Feejee mermaid' theme of fake fauna. The same surely applies to the supposed stuffed 'skunk ape' head with fearsome tusks that was for sale on eBay during mid-February 2002. Commenting upon this specimen, American cryptozoologist Chad Arment noted that he had previously seen pictures of a similar taxiderm creation labelled as an 'Ozark mountain monkey', with the same kind of red and white hair colouration but without tusks. **cz@yahoogroups.com, various postings, 17 February 2002.**

SMALL BUT SENSATIONAL. A notable episode in the modern-day history of the giant squid *Architeuthis* took place a little earlier this year, when in March 2002 an international team of scientists led by marine biologist Dr Steve O'Shea announced that they had become the first researchers ever to capture living specimens of juvenile giant squid. Measuring a hardly giant-sized 9-13 mm, the seven 'squidlings' in question were captured off the coast of New Zealand and were briefly filmed by The Discovery Channel before (despite the scientists' best efforts at maintaining them alive in captivity) the newsworthy septet died. **http://www.ananova.com/mews/story/sm_534339.html 2 March 2002 [no longer online].**

BIGFOOT GOES EAST. A lesser-known hairy man-beast that hit the crypto-news headlines lately is a huge, bipedal, shaggy entity of northeastern India's West Garo Hills, known locally as the mande burung. In February 2002, Nebilson Sangma claimed that he and his brother had encountered one of these creatures while hunting in the jungles here, and had even spied its nest-like home - from where, while they watched discreetly at a safe distance, it emerged on three consecutive days to feast upon bananas growing in a nearby grove. However, the Sangma brothers were not widely believed - until the mande burung's nesting place was videoed just a few days later by Dipu Marak. Moreover, huge footprints, measuring some 20 in long, were also discovered here, by a team of wildlife researchers. Suggestions by forest officials that it may have been a bear or even an elephant were swiftly dismissed by its two observers. ***Strait Times*, 17 March 2002.**

Alien Zoo

WHATEVER HAPPENED TO THE TYGOMELIA? Every so often, a mystery beast comes along that, even by cryptozoology's standards, is quite exceptional.

One such creature is the tygomelia, brought to my attention lately by Edmonton correspondent Kevin Stewart via a 19th-Century Canadian newspaper article, which reads as follows:

> Sir John E. Packenham, an officer in the English army, who has been spending the last year in her Majesty's northern provinces, arrived at Fort Buford with an animal of rare beauty, and never before caught on this continent, nor has it been known till late years that the species existed in this country. It is of the same family as the giraffe, or camelopard, of Africa, and is known to naturalists as the tygomelia. They are known to inhibit the high table lands of Cashmere and Hindoo Kush, but are more frequently seen on the high peaks of the Himalaya Mountains. The animal was taken when quite young, and is thoroughly domesticated, and follows its keeper like a dog. It is only four months old, and ordinarily stands about five feet high, but is capable of raising its head two feet, which makes the animal seven feet when standing erect. It is of a dark brown mouse color, large projecting eyes, with slight indications of horns growing out. The wonderful animal was caught north of Lake Athabasca, on the water of the McKenzie's River. It has a craw similar to the pelican, by which means it can carry subsistence for several days. It was very fleet, being able to outfoot the fastest horse in the country. The black dapper spots on the rich brown color make it one of the most beautiful animals in existence, more beautiful than the leopard of the Chinese jungle. Sir John did not consider it safe to transport this pet by water down the Mississippi River, fearing the uncertain navigation and the great change of climate from the Manitoba to the sunny south. He has, therefore, wisely concluded to go by way of St. Paul, Minnesota. The commander of Fort Buford furnishes him with an escort for the trip. He will then proceed through Canada to Montreal, where he will ship his cargo to England.

As is so often the case, however, nothing further seems to have been reported about this singular animal, whose description, if accurate, does not recall any known species. Moreover, there is certainly no living giraffid native either to North America or to anywhere in Asia. So could it simply be a journalistic whimsy, so common at that time, or should we be leafing back through the New World newspaper archives for the late 1800s in search of extra details? Kevin Stewart speculated that perhaps this animal was a tick-infected moose *Alces alces*, but if so then surely its state would have been swiftly noted by its owner and anyone else who observed it closely.

If there is anyone out there who knows more about the tantalising tygomelia, please let me know. (More recently, Kevin brought to my attention a remarkable freak piebald moose in Kenai, Alaska, whose striking mottled pelage is indeed superficially giraffine in appearance, so could the tygomelia have been another, younger example of this unusual but highly distinctive moose aberration?) *Ottawa Times*, **22 November 1870; Kevin Stewart, pers. comms, 21 March 2002, 29 July 2010; http://moosewatch.com/?p=310 accessed 29 July 2010.**

Wels catfish, a favoured identity for some Scandinavian lake monsters, but seemingly not for Norway's foal-headed Mjoesorm (Wikipedia/public domain)

Alien Zoo

ON THE BEACH. Sometimes, aquatic mystery beasts are not content to lurk in lakes, but actually come out onto land - much to the amazement of anyone observing them. Such was the hitherto-undocumented but recently-unearthed case lately revealed by Norwegian cryptozoologist Erik Knatterud, which took place at noon one hot sunny midsummer day at Norway's Lake Mjoesa. This was when a farmer and his wife who maintain a small farm near the lake's shore spied a bizarre-looking creature emerging from the lake onto a gravel beach. When the farmer bravely walked down to the shore for a better view, he saw the long slender front half of a strange beast resting out of the water, stretched out like a pole. He estimated that this portion alone was about 10 m long and 40-50 cm wide, blackish-brown in colour. Its head was foal-like but had no visible ears and was broader than the neck. When the creature saw the man, it raised its head, coiled backwards, and rushed into the water, swimming at the surface briefly before diving. Full details are held by Knatterud, who plans to release them in a future book documenting the Lake Mjoesa monster, or Mjoesorm. **Erik Knatterud, cz@yahoogroups.com 25 March 2002.**

OGGY-OGLING IN OGSTON. Ornithologists visiting Ogston Reservoir in Derbyshire may conceivably spot something far more exotic than anything of the feathered variety - if an anonymous posting in the forum section of the Ogston Bird Club's website were to be believed. On 8 April 2002, the unnamed person in question claimed that just before 5.00 am that morning, while visiting the reservoir, he/she heard a commotion at the south end, and saw a flock of gulls flying just above the water surface but diving at it, seemingly agitated by something that was creating a large bow wave moving from right to left at the surface. As the observer watched, two dark humps broke the surface, each about 6 ft long and rising about 18 inches high, in line with each other and emerging twice before the bow wave subsided. The observer claimed that these humps looked almost slimy "like the skin of an eel, similar in colour too - but whatever it was was far too large to be an eel". Many of the responses subsequently posted adopted a jocular stance, although a few, including longstanding Nessie-seeker Steve Feltham (responsible for dubbing this enigmatic entity Oggy), treated the matter soberly at first, but it was subsequently outed as a hoax. **Forum, Ogston Bird Club, http://www.ogstonbirdclub.co.uk/ postings from 8 April 2002 onwards.**

A SHINGLE-SCUTTLING SURPRISE. Of particular interest to English entomologists is a new species of small black scuttle fly, previously unknown to science, that has been unexpectedly discovered living up to 3 ft deep in shingle in East Sussex. Christened *Megaselia yatesi*, it probably avoided prior detection because it not only breeds but also feeds out of sight within the shingle, and only made its belated scientific debut when specimens were captured in a special trap invented and set by Dr Barry Yates (after whom this species has been named) at Rye Harbour Nature Reserve. *Wolverhampton Express and Star*, 13 April 2002.

DEADLY NEW JELLYFISH? Robert King may lay claim to the dubious honour of having been killed by a species of jellyfish hitherto unknown to science. After being stung in March 2002 while swimming off Port Douglas in Queensland, Australia, the 44-year-old American tourist died a month later in Townsville Hospital. Skin scrapings were taken for examination, which revealed the presence of nematocysts (stinging cells) from a cubomedusan or box jellyfish, yet differing from those of any known species. This find prompted the investigator in question, jellyfish expert Dr Jamie Seymour, to announce that in his view the culprit may be a species new to science. **http://abc.net.au/ra/newstories/ RANewsStories_532107.htm 17 April 2002 [no longer online].**

JUST WHEN YOU THOUGHT IT WAS SAFE TO GO BACK IN THE WATER... Zoologist Dr Alan Rabinowitz's recent book, *Beyond the Last Village*, featuring his explorations of Myanmar (formerly Burma), contains some tantalising details of a hitherto-unpublicised mystery beast: "In Putao

Alien Zoo

[a northern village here], there were stories about a giant water snake, the bu-rin, 40 to 50 feet long, that attacked swimmers or even small boats. Sounding somewhat like a larger, aquatic version of the Myanmar [Burmese] python, this snake was considered incredibly hostile and dangerous. No one had first-hand knowledge of the creature, yet because of it, children were often discouraged from spending long periods of time in the water". Just an extra-large, or exaggerated, python - or something quite different? **Matthew Bille, pers. comm., 19 April 2002.**

OBSERVING OGOPOGO. Fact is often stranger than fiction, and a film company staging a recreation of a Lake Okanagan monster (Ogopogo) sighting 24 years ago would certainly vouch for that! On 18 April, while a crew from Tripod Film and Video Productions was recording a scene reconstructing a 1978 sighting by Bill Steciuk of an alleged Ogopogo at the west of Okanagan Lake Bridge, with Bill's son Rob playing the role of Bill, three humps suddenly appeared about 200 yards away, above an otherwise totally calm expanse of water. They were undulating in and out of the water, at one point clearly revealing a space between one hump and the water, and were witnessed by a group of 14 observers, including Bill Steciuk himself, who was acting as an advisor for the film. Fortunately, the humps were filmed, yielding a clip lasting about 90 seconds and of excellent quality. One eyewitness, Renee Boucher, described what she saw as "something moving in a very slow and calm manner. It was moving up and down so gracefully...It was huge, black and shiny, very slick looking. It was very, very long". **http://www.ok.bc.ca/archive/2002/04/22/stories/2241_full.html 22 April 2002 [no longer online].**

Ogopogo (Richard Svensson)

Alien Zoo

BINTURONG, BINTURONG, WHERE HAVE YOU BEEN? The binturong *Arctictis binturong* is a black, shaggy-haired, long-bodied, short-legged, nocturnal, arboreal civet, generally up to 2.5 ft long plus a slightly shorter semi-prehensile tail. Normally, it is native exclusively to the tropical jungles of southeastern Asia, which is why Gerry and Joan States of Economy, Pennsylvania, were alarmed to say the least when they found a mature female specimen, subsequently identified conclusively as a binturong - and a none-too-happy one at that - curled up under a bench on their front doorstep on the morning of 21 April 2002. After being captured, it spent the following evening in a cage within a cage for security purposes at the McKees Rocks kennel of Triangle Pet Control Service. Where it had originally come from, conversely, is another matter entirely. After all, I would not normally have imagined it to be among the more likely animals to be kept as a pet. Having said that, however, news articles covering this incident did mention that one of the experts who identified it as a binturong was Wendy Looker, described as a binturong breeder. So if there are people in the States actually breeding these exotic creatures, perhaps there is a call for them as pets (or alternatively, possibly, for their fur?) after all. Anyone with knowledge of why binturongs are being bred in the States is definitely encouraged to share their news with *FT*. http://www.post-gazette.com/neigh_west/20020422critter0422p2.asp 22 April 2002.

Binturong (Jastrow/Wikipedia)

HAIR TODAY...? Hair clippings obtained in Kentucky may be from a bona fide bigfoot, according to Ohio-based geneticist John Lewis who recently analysed them. Moreover, not only does he assert that the samples are from a primate, he also claims that they are closer to the bonobo or pygmy chimpanzee *Pan paniscus* than they are to humans, and suggests that they may therefore constitute evidence for the existence of a still-undiscovered species of American primate. Nor is the bigfoot the only cryptid of interest to Lewis, as he is presently planning an expedition to Manaus, Brazil, in search of the equally elusive mapinguary - variously claimed by cryptozoologists to be a giant ape-like primate or a surviving species of ground sloth. **http://downtoearth.ncbuy.com/newscenter/weirdnews.html 20 May 2002.**

Engraving of a wolverine

AN UNEXPECTED TAIL? I am greatly indebted to correspondent Larry Tribula for sharing the following intriguing discovery with me. While recently dining at a rustic restaurant called 'Antlers' in the Upper Peninsula of Michigan, which was filled with taxiderm exhibits, Larry noticed one very unusual specimen. He describes it as being a bear on all fours, displayed near to some black bears. However, this particular individual, roughly the size of a juvenile black bear, was golden brown in colour and, departing significantly from the typical ursine form, sported a long bushy tail, resembling that of anteaters but not quite as large. Larry felt that this odd creature looked like a giant-sized wolverine *Gulo gulo*, noting that a true wolverine was mounted close to it but was much smaller, only a third or quarter the size of this strange 'tailed bear'. Larry did not consider it to be a hoax, and hopes to take a picture of it, should he ever return there. **Larry Tribula, pers. comm., 21 June 2002.**

PYGMIES ABOUNDING. During a recent re-examination of preserved 'baby' octopuses at California's Santa Barbara Museum of Natural History, each weighing only tenths of one gram, Dr Eric Hochberg was stunned to discover that they were not babies at all. Instead, they were all fully mature specimens - and represented dozens of new, valid, yet hitherto-undescribed species of pygmy octopus. Moreover, they had not been collected from just a single location, but over the years had been acquired from a wide range of tropical provenances - as far apart as the Philippines and Mexico. Hochberg and colleagues now hope to collect enough material from these minuscule creatures for DNA analysis to be feasible, which would greatly assist their precise classification. **UPI Science News, 4 July 2002.**

IN A FLAP OVER A MUSCULAR MYSTERY. During the third weekend in July 2002, a very distinctive (albeit dead) specimen of giant squid was discovered washed ashore on a beach in Hobart, Tasmania. Weighing 250 kg and lacking the two extra-long tentacles that distinguish squids from octopuses in terms of limb count, what made this particular giant squid so noteworthy were the long thin flaps or keels of muscular tissue attached to each of its eight arms. According to a statement attributed in newspaper accounts to Dr David Pemberton, senior zoology curator at the Tasmanian Museum and

Art Gallery, which received the squid's carcase: "What we've seen on this animal we haven't seen on other squid, and it's a significant feature. It's basically like having a pile of muscles on your own body that nobody else has". As a result, speculation arose that the flap-armed squid may represent a new species of giant squid. On 23 July, however, further news reports announced that another expert was claiming this creature to be nothing more dramatic than a damaged specimen of the known species of giant squid, not a taxonomic novelty after all. http://news.bbc.co.uk/hi/english/world/asia-pacific/newsid_214400/2144379.stm 22 July 2002; ITV Teletext news service, p. 318, 10.20 pm, 23 July 2002.

WATER ELEPHANTS AHOY? As documented in Dr Bernard Heuvelmans's book *On the Track of Unknown Animals* (1958) and more recently reassessed within my own *In Search of Prehistoric Survivors* (1995), one of the most intriguing of Africa's many mammalian cryptids must surely be that amphibious anomaly generally dubbed the water elephant. Possessing a noticeably long, ovoid skull and a relatively short trunk according to eyewitness accounts, it seems to bear more than a passing resemblance to some of the earliest, long-extinct members of the proboscidean (elephant) lineage, but has not been seriously investigated in the field for several decades. However, in mid-2003, Bill Gibbons, a veteran seeker of cryptozoological curiosities, plans to visit the Democratic Republic of Congo (formerly Zaire) with a Belgian helicopter company operating there, in order to pursue claims by the company's president and CEO that a military helicopter flying over Lake Tumba spied a herd of very strange-looking elephants that the helicopter's pilots thought may be the legendary water elephants. According to Bill, the producer of a French TV documentary company is keen to film the expedition, so we wish everyone associated with this project the best of luck, and await further developments with interest. **Bill Gibbons, pers. comm., 25 July 2002.**

Water elephant, based upon eyewitness descriptions (Markus Bühler)

MONITORING COLORADO'S RIVER DINOSAUR. Since 1982, reports have emerged from several locations in Colorado of an elusive, strange-sounding reptile that has been dubbed the river dinosaur. According to a dossier of accounts gathered by investigating cryptozoologist Nick Sucik from Minnesota, it is a graceful swift mover that walks exclusively on its hind legs, stands about 3 ft tall, has a long neck and arm-like appendages rather than forelegs, and is most commonly sighted close to a wet environment. One eyewitness from Mesa Verde added that it had a cone-shaped nose, a 2-ft-long tail, and came from a pond area. Jeff Thulin from the local Reptile Reserve suggested to this eyewitness that the creature may have been a monitor lizard or varanid. Although not native to North America, several species of large water-associated monitor are

Alien Zoo

frequently kept as exotic pets in the States and elsewhere, and reports of escapees/releases are quite commonplace (even in Britain). Moreover, not only are monitors very fast runners, but when moving swiftly they also often rise up onto their hind legs and can run bipedally for long distances, using their tail for balancing purposes, and while doing so bear at least a passing resemblance to small dinosaurs. http://www.cortezjournal.com/asp-bin/article_generation.asp?article_type=news&article_path=/news/news020730_3.htm 30 July 2002 [no longer online].

European bee-eater, painted by John Gould

MAKING A BEE-LINE FOR COUNTY DURHAM. Twitchers near to home have been privileged to witness a remarkable comeback - thanks to the first confirmed nesting and rearing of young in Britain for 47 years by a pair of European bee-eaters *Merops apiaster*. These slim, rainbow-hued cousins of the kingfisher are more typically seen perching on telegraph lines and other lofty but visible vantage-points in Andalusia and similar sunny areas elsewhere in southern Europe. However, in June 2002 a pair was observed performing courtship displays in a disused quarry at Bishop Middleham, County Durham. The Durham Wildlife Trust and the RSPB promptly mounted a 24-hour guard on the birds and the nest burrow that they duly constructed in the quarry's cliff face, and on 20 August at least two chicks were seen and photographed, coming to the mouth of the burrow to be fed by their parents. Ornithologists hope that global warming may result in this exotic species becoming a regular breeder in Britain. **http://www.ananova.com/news/story/sm_608343.html 15 June 2002 [no longer online];** *Daily Mail*, **25 July 2002;** *Wolverhampton Express and Star*, **20 August 2002.**

ITCHING FOR ANSWERS. One of the weirdest mystery beasts reported of late must surely be the muhnochwa or scratch monster of Lakhimour Kheri in India. Claimed to be a luminous aerial entity that severely scratches its human victims, leaving distinctive lacerations in their skin, for quite some time it eluded all attempts at capture. However, a supposed muhnochwa was finally caught in August 2002 and brought to Lucknow University for formal identification. Here, zoologist Prof. K.C. Pandey later announced that it was a specimen of a rare, 3-inch-long species of cricket, known scientifically (and aptly, as it turns out) as *Schizodactylus monstrosus*. Its most noteworthy feature, in view of its alleged synonymity with the scratch monster, is the series of sharp spikes on its hind limbs. Yet although these may conceivably produce a slight mark on the skin if the insect

Engraving from 1914 of *Schizodactylus monstrosus*

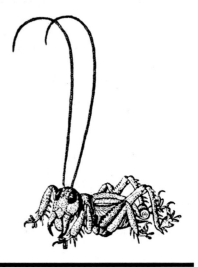

Alien Zoo

is handled, they do not seem powerful enough to inflict the kind of wounds blamed upon the muhnochwa - which leaves the experts scratching their heads, and the monster's victims still scratching their wounds. *The Times of India*, **24 August 2002.**

RIDDLE OF A RAIL. What may well be a still-undescribed bird species was photographed on the island of Malaita in the Solomons in July 2002 but remains unresolved. A rail, it closely resembles the flightless Woodford's rail *Nesoclopeus woodfordi*, also of the Solomons, but can be differentiated by its browner overall colouration as well as by having more pale spots on its wings. A photograph of this mystery Malaita rail was published in *World Birdwatch*, **September 2002.**

A PERFECT PICTURE OF MYSTERY. As I've often noted, mystery animals can turn up in the most unlikely places - and one of these just so happens to be the American version of eBay, the online auction house. From 16 to 26 September 2002, a most extraordinary painting was offered on there, as Item #907237942, by a female seller called Steph, from Old Town, in north central Florida, with the user name ectopistes (interestingly, the generic name of the now-extinct passenger pigeon), and requiring an undisclosed reserve price. According to the seller's description of this painting, it was an original unframed water-colour, measuring 22" by 30", bearing the signature Canzanella, and it depicted a pair

Canzanella's mysterious painting of two equally mysterious mammals (Canzanella)

Alien Zoo

of very odd-looking mammals. The one in the foreground was described by the seller as white with a very slight pinkish hue (as the seller's video-camera photo of the painting had made it seem yellow). The seller had owned the painting for roughly 28 years after paying a thousand dollars for it, but had no information as to what the animals in it were. Emphasising their curious appearance, she had entitled the item 'Strange Cryptozoology Animal Painting'. Certainly, the creatures depicted are decidedly unusual, even sinister, and I cannot readily identify them with any known species. There are certain similarities to badgers (though the portrayed animals' legs seem too long for all but perhaps the Asian hog badger *Arctonyx collaris*), and others to various viverrids. The pink-hued white fur of the foreground specimen may indicate albinism. In any event, the painting is reproduced here, and we'd welcome any opinions or information concerning Canzanella's mystery animals, and also concerning Canzanella him/herself - the equally mysterious artist responsible for depicting them. http://www.ebay.com 16-26 September 2002 [no longer online].

Engraving of sugar glider

LOOK OUT, LOOK OUT, THERE'S A SUGAR GLIDER ABOUT! One of the most extraordinary cases of out-of-place animals en masse occurred on 30 September 2002, when an explosion caused by a suspected gas leak blew apart a sizeable storage container in Auburn, Seattle - releasing into the nearby woods hundreds of Australian sugar gliders *Petaurus breviceps*. These enchanting little marsupials closely resemble flying squirrels - indeed, they are the marsupial world's direct ecological counterpart of the latter gliding rodents - and were being sold as pets by the container's owner, Sharon Massena, who was critically injured by the blast. About 120 sugar gliders died, another 350 or so were rescued, but around 500 escaped. Local conservationists fear that if these escapees survive in the woods, they may compete with native species (sugar gliders have a wide-ranging diet - from flowers, nectar, buds, and sap, to insects and even small birds) and may even establish a colony - which would certainly constitute one of the most exotic and unexpected recent examples of mammalian naturalisation anywhere. *Tacoma News Tribune*, 1 October 2002.

IT'S SUPER SPONGE! Unlike their maritime kin, freshwater sponges are typically small, rarely exceeding an inch or two in length. Until recently, the only major exception has been a 3-ft-high species in Siberia's Lake Baikal, where it grows at a depth of around 120 ft. However, an unexpected rival has now been discovered in the Staffelsee, a freshwater lake in Upper Bavaria. Moreover, not only is this giant specimen 3 ft tall too, but it is far more readily visible than its Siberian counterpart, having wrapped itself round the timber support of a ruined bridge - and it is still growing! Researchers hope to film their super-sized sponge soon.
http://www.iol.co.za/index.php?click_id=143&art_id=qw103347720040B265&set_id=1 1 October 2002.

A STING IN THE TALE? In mid-October 2002, a box jellyfish (cubomedusan) was found at The Strand in Townsville, Queensland, Australia, that has caused even more than the usual stir which occurs whenever one of these highly venomous sea creatures turns up. Beaches are often closed and stinger

Alien Zoo

nets set out if a quantity of box jellies appear, but what makes this particular (dead) specimen especially startling is that it does not seem to match any known cubomedusan species anywhere in the world. According to cubomedusan expert Dr Jamie Seymour at James Cook University's School of Tropical Biology, its nematocysts (stinging cells) are different in shape, its tentacles are shorter, its general shape and size are different too, and it doesn't have as many dots on its body. Clearly, therefore, as noted by Dr Seymour, "this highlights the fact that we know so very little about stingers. We are really at the very, very beginning of a long journey". http://townsvillebulletin.news.com.au/common/story_page/0,7034,5316713%255E14787,00.html 19 October 2002 [no longer online].

SOUTH AUSTRALIA'S ELUSIVE GREY PEOPLE. In late October 2002, announcing her forthcoming Quest Trek 2002 expedition in search of putative surviving thylacoleonids or marsupial lions in the Flinders Range region of South Australia, Aussie cryptozoologist Debbie Hynes also referred to a much less familiar crypto-subject, the Grey People. This is the name given by Westerners here to a mysterious race of very small, furry, black-skinned humanoids, which walk upright but are only 1.0-1.5 m tall, with sloping brows, pronounced eyebrow ridges, and ape-like faces. Long spoken of in the local desert aboriginals' legends, they also have their own aboriginal name, and they appear in ancient rock art, but are still being reported today too. According to Debbie, in 2001 an American back-packer startled one of these beings when he returned to his camp and discovered it stealing his belongings. Debbie's own guide for her expedition was a trapper back in the 1950s and 1960s, who mentioned to her that on one occasion he'd been followed by some Grey People, hoping to steal rabbits from his traps, and during that trip he'd found one sprung trap containing a fresh human-like fingernail but no bigger than the nail of a young child's little finger; he assumes that in springing it, a hand of one such entity had been caught by the trap. Debbie hopes to visit an area in the heart of Grey People territory, so let's keep our fingers crossed that she sees one of these enigmatic beings. **Debbie Hynes, cz@yahoogroups.com 26 & 27 October 2002.**

A MONSTROUS DELAY? Unicorns have yet to be added to the wrong kind of leaves or snow as reasons for causing travel and transport hold-ups over here in Britain. Yet a comparable situation has recently occurred in New Zealand. As revealed in November 2002, work on a major new highway in this southern hemisphere country has been halted, at least temporarily, because a local Maori tribe called the Ngati Ngaho has claimed the path of the new road would lead directly to a guardian spirit entity known as a taniwha, which takes the form of a huge lizard-like river or swamp monster. As taniwhas are sacred to the Maori people, the claim of the Ngati Ngaho is being taken very seriously by the road building agency involved, Transit New Zealand, and meetings between the agency and Maori elders are being planned, to see if any mutually-acceptable compromise can be reached. **McConnell Media Group, 6 November 2002; http://www.xzone-radio.com/strange_news/948474.htm [no longer online].**

Ureia, a guardian taniwha depicted in a carved poupou (house post) from the interior of Hotunui, a carved meeting house (built in 1878) of the Ngati Maru people, Thames, New Zealand (Wikipedia)

Alien Zoo

TIGER TALES. The last recorded sighting of a tiger in South Korea took place in Gyeonggi Province's Daedok Mountains, way back in 1921. However, this has not prevented Lim Sun-nam from devoting every available spare moment of the past seven years to searching for evidence that this dramatic felid still exists in his country. Having seen claw marks in a tree that he is convinced were made by a tiger, as well as 30 or so tiger-like footprints during February 1998 in snow on the remote Hanbuk Mountains, he returns here each week in a truck containing a sound machine that emits the mating call of a male tiger as a lure, and checks a camera set up with a live rabbit as bait. So far, however, no tiger has responded to either the aural or the living lure, but Lim Sun-nam believes that at least ten tigers do still exist in his country. He also claims to have emails from United States soldiers stationed here who have described tiger noises and smells, and that there is even a videotape of a tiger pawing a fence along the 38th Parallel. Meanwhile in Indonesia, Didik Raharyono, a biology graduate from Gadjah Mada University, is equally adamant that tigers still survive in Java, even though the Javan tiger *Panthera tigris sondaica* was formally declared extinct during the 1980s. Yet sightings continue to this day among local villagers and poachers, who have been interviewed at length by Didik. They have even given him tiger teeth, skin, and whiskers, some of which may be from tigers killed only months earlier. And he has collected faeces and spied footprints in caves and other localities that tigers allegedly still frequent. Since beginning his search in 1997, Didik claims to have seen a tiger himself once, albeit briefly, during a 14-day survey in Meru Betiri National Park. Moreover, hair samples have apparently been formally identified as Javan tiger in origin by the Indonesian Institute of Sciences. Notwithstanding such evidence, finding sponsors to support these two tiger researchers' quests in their respective lands, however, is proving to be as great a problem as finding the tigers themselves. http://english.joins.com/Article.asp?aid=20021115001926&sid=600 15 November 2002 [no longer online], http://www.thejakartapost.com/detailfeatures.asp?fileid=20021030.S03&irec=2 30 October 2002 [no longer online].

Two Javan tigers at Berlin Zoo, c.1905 (John Edwards)

Alien Zoo

MISSING MONK SEALS? There are a number of mystifying historical records on file of an alleged but unidentified species of seal formerly inhabiting the tropical waters of the western Indian Ocean and Red Sea. Attempts have been made to classify this inconnu species variously as a monk seal *Monachus* sp., an oop southern elephant seal *Mirounga leonina*, and even a dugong *Dugong dugon*. Now, a new account of this abiding mystery has appeared, authored by François Moutou of the French Mammal Society, who has offered up two additional pieces of hitherto-unconsidered historical data for discussion. One is the discovery of two authentic 3000-4700-year-old sites of dugong butchery in the United Arab Emirate of Umm al Qiwain, thereby adding weight to the dugong identity. Conversely, the other concerns the sighting of 20 or so "sea cows" lying asleep on a beach on Mamelles islet in the Seychelles by a Marion-Dufresne expedition on 4 October 1768. Although the term 'sea cow' normally alludes to dugongs, these and other sirenians do not come ashore to lie on beaches, whereas seals commonly do. Thus, the riddle of the Indian Ocean cryptid remains unsolved. ***Monachus Guardian*, 5 (November 2002); http://www.monachus-guardian.org/mguard10/1032scien.htm**

Classic image of the unicorn

DEM BONES, DEM BONES, DEM UNICORN BONES. Customs officers may claim that they've seen it all as far as the smuggling of exotic items is concerned, but Auckland Harbour officers recently claimed a first when, in November 2002, they confiscated from an Auckland furniture importer an undeclared crate. For the crate, shipped in by him from Indonesia, was found to contain bones that the furniture importer genuinely believed to be from a unicorn! He had bought them, apparently in good faith regarding their identity, as a special Christmas present. When the bones were examined at Auckland Museum, however, they were identified as originating from a cow or water buffalo, and had been coated with a thin layer of cement to make them appear fossilised. ***Dominion Post*, 11 December 2002.**

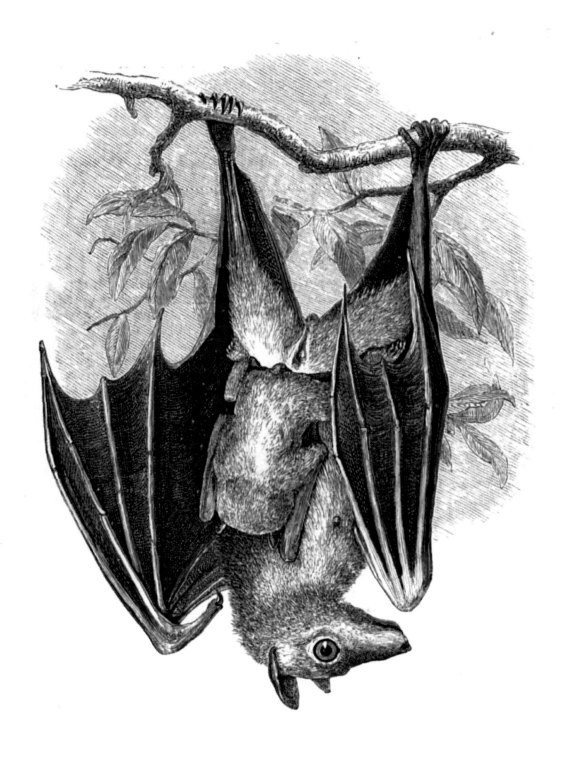

The Lost Ark

OUT-FOXED BY OUR HIGH-FLYING COUSINS?

But when he brushes up against a screen,
We are afraid of what our eyes have seen:

For something is amiss or out of place
When mice with wings can wear a human face.

Theodore Roethke – 'The Bat'

The fortean literature contains reports of some exceedingly bizarre entities, but few are any stranger than the various bat-winged humanoids spasmodically reported from certain corners of the world. These include such aerial anomalies as the Vietnamese 'bat-woman' soberly described by three American Marines in 1969, the letayuschiy chelovek ('flying human') reputedly frequenting the enormous taiga forest within far-eastern Russia's Primorskiy Kray Territory, and the child-abducting orang bati from the Indonesian island of Seram (see separate chapter in this present book).[1]

Zoologists have traditionally averted their eyes from such heretical horrors as these, but in a classic 'fact is stranger than fiction' scenario, a remarkable evolutionary theory has lately re-emerged that unites humans and bats in a wholly unexpected evolutionary manner.

FLYING FOXES AS WINGED PRIMATES?

As far back as 1910, W.K. Gregory proposed that bats were closely related to primates - the order of mammals containing the lemurs, monkeys, apes, and humans.[2] More recently, Dr Alan Walker revealed that dental features of a supposed fossil primate christened *Propotto leakeyi* in 1967 by American zoologist Prof. George Gaylord Simpson indicated that it was not a primate at all, but actually a species of fruit bat.[3]

In 1986, however, Queensland University neurobiologist Dr John D. Pettigrew took this whole issue of apparent bat-primate affinity one very significant step further, by providing thought-provoking evidence for believing that the fruit bats may be more than just relatives of primates - that, in reality, these winged mammals *are* primates!

All species of bat are traditionally grouped together within the taxonomic order of mammals known as Chiroptera. Within that order, however, they are split into two well-defined suborders. The fruit bats or flying foxes belong to the suborder Megachiroptera ('big bats'), and are therefore colloquially termed mega-bats. All of the other bats belong to the second suborder, Microchiroptera ('small bats'), and hence are termed micro-bats.

MACRO-BATS AND MICRO-BATS – NOT SEEING EYE TO EYE?

As a neurobiologist, Dr Pettigrew had been interested in determining the degree of similarity between the nervous systems of mega-bats and micro-bats. In particular, he sought to compare the pattern of connections linking the retina of the eyes with a portion of the mid-brain called the tectum, or superior colliculus. He used specimens of three *Pteropus* species of fruit bat to represent the mega-bats. And to obtain the most effective comparison with these, he chose for his micro-bat representatives some specimens of the Australian ghost bat *Macroderma gigas* - one of the world's largest micro-bats. Ideally suited for this purpose because its visual system is better developed than that of many other micro-bats, it has large eyes like those of fruit bats, and retinas with a similar positional arrangement.

Pettigrew's examination of all of these specimens revealed that the pattern of retinotectal neural connections was very different between mega-bats and micro-bats, but far more important was the precise *manner* in which they differed - providing a radically new insight not merely into bat evolution but also into the family tree of humanity.

Reporting his remarkable findings in 1986[4], Pettigrew announced that the retinotectal pattern of connections in fruit bats was very similar to the highly-advanced version possessed by primates. That fact was made even more astounding by the knowledge that until this discovery, the primate pattern had been unique. In other words, it had unambiguously distinguished primates not only from all other mammals (including the micro-bats) but also from all other vertebrates, i.e. fishes, amphibians, reptiles, and birds - all of which have a quite different, more primitive pattern. Suddenly, the fruit bats were in taxonomic turmoil.

NOT SUCH A FLIGHT OF FANCY?

Until now, the fact that micro-bats and mega-bats all possessed wings and were capable of controlled flight had been considered sufficient proof that they were directly related, because it seemed unlikely that true flight could have evolved in two totally independent groups of mammals. Gliding, via extensible membranes of skin, had evolved several times (e.g. in the scalytail rodents, the 'flying' squirrels, three different groups of 'flying' marsupial phalanger, and the peculiar colugos or 'flying lemurs'), but

Alien Zoo

this did not require such anatomical specialisations as the evolution of bona fide, flapping wings for true flight.

Yet it seemed even less likely that the advanced retinotectal pathway displayed by primates could have evolved wholly independently in fruit bats.

In short, by exhibiting the latter organisation of neural connections, fruit bats now provided persuasive reasons for zoologists to consider seriously the quite extraordinary possibility that these winged mammals were not bats at all, in the sense of being relatives of the micro-bats. Instead, they were nothing less than flying primates!

Moreover, as Pettigrew noted in his paper, even the wings of mega-bats and micro-bats are not as similar as commonly thought. On the contrary, they show certain consistent skeletal differences, which point once again to separate evolutionary lines. And even that is not all - thanks to Dermoptera, that tiny taxonomic order of gliding mammals known somewhat haplessly as the flying lemurs (bearing in mind that they are not lemurs, and do not fly!) or, more suitably, as the colugos.

For by combining previously-disclosed similarities in blood proteins between the primates and the flying lemurs with the structural and neural homology apparent between the flying lemurs' gliding membranes and the wings of the mega-bats, extra evidence is obtained for a direct evolutionary link between fruit bats, primates, *and* the flying lemurs - thus resurrecting another possibility that had been suggested by researchers in the past.

FACING UP TO THE FACTS

One of the most familiar external differences between mega-bats and micro-bats is the basic shape of their face. Whereas the face of most fruit bats is surprisingly vulpine (hence 'flying fox') or even lemurine, in many micro-bats it is flatter in shape - though in some species, evolution has superimposed

LEFT: The uniquely grotesque face of the aptly-named Antillean ghost-faced bat *Mormoops blainvillii*, a species of micro-bat native to the West Indies RIGHT: The remarkably lemur-like face of a *Pteropus* fruit bat

upon this shape all manner of grotesque flaps and projections.

The lemur-like shape exhibited by the face of many fruit bats has traditionally been dismissed as evolutionary convergence, engendered merely by these two mammalian groups' comparable frugivorous tendencies.

Judging from Pettigrew's revelations, however, there may now be good reason to believe that such a similarity is a manifestation of a genuine taxonomic relationship between lemurs and fruit bats. The faces of the flying lemurs are also very lemurine (hence their name), which ties in with the above-noted serological evidence for a direct, flying lemur-primate link.

A colugo or flying lemur, again sporting a very lemurine face

Thought-provoking indeed is the evidence for believing that fruit bats are legitimate, albeit aerially-modified, offshoots from the fundamental family tree of the primates.

As Pettigrew pointed out, it is highly implausible that the reverse theory is true (i.e. that the fruit bats gave rise to the primates), because fruit bats seem to be relatively recent species, first evolving long after the primate link had emerged.

EVIDENCE FOR AND AGAINST FLYING PRIMATES

Inevitably, no theory as radical as one implying primate parentage for the fruit bats will remain unchallenged for very long. In 1992, for instance, molecular biologist Dr Wendy Bailey and two other colleagues from Detroit's Wayne State University School of Medicine announced that DNA analysis of the epsilon(e)-globin gene of both groups of bats, primates, and a selection of other mammals implies that the two bat groups are more closely related to one another than either is to any other mammalian group.[5] This finding would therefore seem to support the traditional bat classification[6], but as noted by proponents of Pettigrew's ideas, it does not explain the extraordinary development by fruit bats of the primates' diagnostic visual pathway. Consequently, this tantalising physiological riddle currently remains unanswered.

Moreover, in a comparative immunological study whose results were published during 1994, Drs Arnd Schreiber, Doris Erker, and Klausdieter Bauer from Heidelberg University showed that proteins in the blood serum of fruit bats and primates share enough features to suggest a close taxonomic relationship between these two mammalian groups after all - thus bringing this continuing controversy full circle.[7]

Alien Zoo

Many primitive tribes believe that fruit bats are the spirits of their long-departed ancestors. In view of the fascinating disclosures reported here, these tribes could be closer to the truth than they realise!

UPDATE

The above article of mine was originally published by *FT* in its April 1997 issue (and is reprinted in unchanged form above). Yet despite the initially encouraging research documented in it, the passage of time following its publication did not prove kind to the flying primates hypothesis. In more recent years, sufficient evidence against its veracity as obtained via comparative DNA analysis with primates, mega-bats, and micro-bats has been proffered for it to be largely (though not entirely) discounted nowadays by mainstream workers. A detailed examination of this evidence is presented online in palaeontologist Dr Darren Naish's Tetrapod Zoology blog.[8] Nevertheless, even though the notion of fruit bats as our winged cousins may have been grounded, zoologically speaking it remains of undeniable historical interest, and was such a charming novelty while it lasted that I couldn't resist recalling it here.

SELECTED REFERENCES

1 SHUKER, Karl P.N. (1996). *The Unexplained: An Illustrated Guide to the World's Natural and Paranormal Mysteries.* Carlton (London).

2 GREGORY, W.K. (1910). The orders of mammals. *Bulletin of the American Museum of Natural History*, 27: 1-524.

3 WALKER, Alan (1969). 'True affinities of *Propotto leakeyi* Simpson 1967. *Nature*, 223 (9 August): 647-8.

4 PETTIGREW, John D. (1986). Flying primates? Megabats have the advanced pathway from eye to midbrain. *Science*, 231 (14 March): 1304-6.

5 BAILEY, Wendy J., SLIGHTOM, Jerry L., & GOODMAN, Morris (1992). Rejection of the "flying primate" hypothesis by phylogenetic evidence from the e-globin gene. *Science*, 256 (3 April): 86-9.

6 GIBBONS, Ann (1992). Is "flying primate" hypothesis headed for a crash landing? *Science*, 256 (13 April): 34.

7 SCHREIBER, Arnd, ERKER, Doris, & BAUER, Klausdieter. (1994). The eutherian phylogeny from a primate perspective. *Biological Journal of the Linnaean Society*, 51: 359-76.

8 NAISH, Darren (2009). We flightless primates. http://scienceblogs.com/tetrapodzoology/2009/07/we_flightless_primates.php (posted 13 July).

The dodu (William Rebsamen)

2003

ALIEN CATTLE ALERT! At the end of 2002, during a six-week foray in the area, a team of scientists from Edinburgh's Royal Botanic Gardens was visiting the hitherto-unexplored Gaoligong Shan mountain range on the border of China and Myanmar (formerly Burma) when its members encountered some highly unusual 'alien' cattle, of a kind previously unreported by scientists. Inhabiting a huge forest and numbering around a dozen, the imposing beasts each weighed nearly a ton and were very muscular, but by far their most eyecatching features were their horns. Almost a foot long and sticking straight out to the sides, the sharp horns emerged from a huge bony dome covering the whole of the cow's head from its brows upwards, making it look, at least in the scientists' opinion, as if it were wearing a Viking's war helmet! According to the expedition's leader, Mark Watson, these odd cattle are a semi-wild, feral variety, unafraid of humans, but there only seems to be a small number in existence, confined entirely to this extremely remote, uncharted wilderness. *Aberdeen Press & Journal*, 3 January 2003; *Scotsman*, 3 January 2003.

CAMEROON'S DEADLY DODU. Explorer Bill Gibbons has brought several hitherto-obscure mystery beasts to widespread cryptozoological attention over the years, but his latest 'find' must surely be one of the most interesting to date. As only recently revealed by him, while exploring southern Cameroon in search of reports of mokele-mbembe-type cryptids during April 2000, Bill was informed by the Baka pygmies and Bantus there of a dangerous primate known to them as the dodu. They claimed that it is dark grey in colour, stands up to 6 ft tall, is mostly bipedal but will sometimes knuckle-walk on all fours, and, of particular note, has only three fingers on each hand and just three clawed toes on each foot. (This provides an unexpected parallel with the puzzling reports of three-toed bigfoot prints sometimes encountered in North America.) Highly aggressive, a dodu will attack gorillas, and has an unusual dietary predilection. After killing an antelope or some other sizeable prey, it does not touch the carcase. Instead, the dodu abandons it for a while, leaving the rotting carcase to fill with maggots, after which

Alien Zoo

the dodu returns, scoops the grubs out, and eats them in quantity. It is also well known for leaving piles of sticks on the forest floor, which, as Bill speculates, may be a form of territorial marking behaviour. Bill has collected several native reports of encounters with dodus, but by far the most remarkable was one that he heard when returning to Cameroon in 2001. While visiting the lower Boumba region, he learnt that a few months earlier, a group of white men, accompanied by pygmy trackers, had allegedly captured a live dodu, which was seen by residents of a town called Moloundou. Bill suspects that its captors were loggers, but where this unique specimen is now (assuming the report's validity) is unknown. **Bill Gibbons, cryptolist@yahoogroups.com 1 & 5 January 2003.**

ALL A-TWITCH OVER THE ALBATWITCHERS. The following highly interesting request for information was publicly aired on the cz@yahoogroups.com list recently by veteran American cryptozoologist Ron Schaffner. He had received it on 9 January 2003, from a person signing himself as Jim J (who can be reached at AlmanacJim@aol.com). Its subject was new to me, and will no doubt be of interest to *FT* readers:

I'm trying to track down a source story about the Albatwitchers - small, hairy, bipedal, tree-living creatures that at one time supposedly lived in Pennsylvania along creeks, ate fruit and insects and were about the size of a ten year old. I encountered a story last year somewhere on the internet while doing a story for an Almanac I publish on the Little Red Men (similar creatures that at one time lived between the Ohio and Arkansas rivers) who had similar food tastes and liked stealing red items from river travelers. The Indians told the earliest settlers about these creatures but claimed they were nearly extinct (this in the 1780s). I was wondering if you know of any site with information about them. **Ron Schaffner, cz@yahoogroups.com 9 January 2003.**

FATHER OF THE JACKALOPE, R.I.P. Today, stuffed jackalopes - rabbits ingeniously but fraudulently gifted with antelope horns - are popular tourist souvenirs and equally rich sources of tourist gullibility throughout the American West. On 6 January 2003, however, the widely-claimed 'father' of the jackalope, Douglas Herrick, died, aged 82, in Casper, Wyoming. It is he who is popularly deemed to have engendered this incongruous beast when, in the 1930s, he and his brother Ralph had the idea of attaching a pair of pronghorn antelope horns to a dead jack rabbit in their shop. And thus a legend was born. *New York Times*, 19 January 2003.

A stuffed jackalope (CFZ)

174

Alien Zoo

FROM CHILE TO CHILLED. One of the most extraordinary cases of an oop (out-of-place) animal to hit the headlines lately featured a Patagonian toothfish *Dissostichus eleginoides*, also known as the Chilean sea bass, discovered alive and well but thousands of miles north of its typical sub-antarctic South American waters. Normally unknown north of Chile in the Pacific and Uruguay in the Atlantic, this particular specimen, measuring 5.9 ft long and weighing roughly 154 lb, was caught in the Davis Strait, off Greenland! Its perplexed discoverer, a commercial halibut fisherman called Olaf Sólsker, froze it and brought it ashore, where it was eventually shown to Copenhagen University zoologist Dr Peter Rask Møller, who positively identified the long-distance wanderer. He suspects that the secret of this intrepid toothfish's marathon inter-polar journey is that it migrated through the tropics by swimming in very deep, cold ocean waters, beneath the tropics' warm water - in which it ordinarily could not survive. *Nature*, 6 February 2003; *National Geographic News*, 5 February 2003.

A Patagonian toothfish

BEWARE THE COCKATRICE, MY SON! It's always nice to hear that the world's favourite legendary animals have not vanished totally without issue. On four consecutive days during early February 2003, a cockerel owned by Dibakara Jena, a farmer from Tikanapur village in Orissa, India, startled its owner and the other villagers alike by laying eggs - as confirmed by Laxman Behera, the chief veterinary officer of Kendrapada district, who visited Jena to see his feathered wonder for himself.

A cockatrice

Though undeniably unusual, egg-laying cockerels are far from unknown, generally resulting from a hormonal imbalance - often they are, in reality, hens that have developed secondary male sexual characteristics, such as a coxcomb and a crowing voice, rather than being genuine cockerels - and gave rise in earlier times to belief in the monstrous but wholly mythological cockatrice. **Ananova, 13 February 2003.**

CHAD'S STRANGE ARK IS NOW OFFICIALLY AFLOAT! Much as I enjoy cryptozoology, I am also, like many other *FT* readers, very interested in a wide range of other animal mysteries and anomalies, but these are often very difficult to track down on the internet because there are no sites de-

Alien Zoo

voted to this loosely-knit field of zoological forteana. Happily, however, that is no longer true, thanks to American cryptozoologist Chad Arment, who has instigated an online newsletter list entitled 'Strange Ark', which members can join and forward news through directly to him for posting. It can be accessed at: http://groups.yahoo.com/group/strangeark [see also http://www.strangeark.com]. **Chad Arment, cz@yahoogroups.com 22 April 2003.**

COLT-KILLER ON THE LOOSE! For at least two years, eyewitnesses in the rainforested area of Arroyo Salado, in northern Argentina's Salta province, have reported seeing a bipedal manbeast-like entity, standing 6.5 ft tall, covered in black fur, sporting very large eyes and ears, armed with sharp powerful claws on its hands, emitting ear-splitting howls, and allegedly responsible for killing and devouring a number of fairly large animals, as well as scaring or even occasionally attacking humans who have encountered it. However, this daunting entity finally became headline news in mid-February 2003, following the discovery of a killed colt's chewed skeleton, surrounded by large bigfoot-type prints. Veterinarian Luis Calderon confirmed that the colt's bones had been gnawed by something with extremely powerful jaws and sharp teeth, and plaster casts were made of the prints by Jose Exequiel Alvarez, head of the Juan Carlos Rivas Archaeological and Paleontological Group. The mystery beast is now the subject of an official enquiry by the Thirty-first Sheriff's Office, at Rosario de la Frontera, led by Rene Humberto Tacacho. *El Tribuno*, **21 February 2003.**

Western tragopan painted by D.J. Elliot in 1872.

MÜSHMURGH IS A MISH-MASH. Palaeontologist Darren Naish recently informed me of a mysterious bird, brought to his attention by A.D.H. Bivar. While serving in the pre-partition Indian army, Bivar visited the guest rooms of the Wali of Swat at Saidu Sharif in the North-West Frontier Province, whose foyer contained two stuffed birds. One was a familiar monal pheasant, but the other was a very unfamiliar specimen. Sporting red foreparts, and grey posterior plumage, it was labelled as a tragopan (squat, short-tailed pheasant), but unlike all known tragopans it exhibited a lyre-shaped tail. Moreover, it was clearly distinct from the western tragopan *Tragopan melanocephalus*, the only species native to this region.

Intriguingly, Bivar then alluded to a now-vanished species of bird from Iran, called the müshmurgh, whose flesh made particularly tasty eating. Could this be one and the same as the stuffed bird that he had spied? Darren wondered whether the latter specimen might conceivably have been a taxiderm composite, noting that the only lyre-tailed galliform bird in this whole area is the Caucasian black grouse *Tetrao mlokosiewiczi*, whose remaining plumage is very different from

Alien Zoo

Bivar's bird, but Bivar dismissed that possibility. So if this curious specimen still exists, it may well be worth a detailed examination by an ornithologist versed in pheasant taxonomy. **Darren Naish, pers. comm., 26 February 2003.**

SURPRISES IN SUMATRA? Cryptozoologist Richard Freeman from the Centre for Fortean Zoology, alongside Chris Clarke and Jon Hare, will be taking part in a three-week expedition to Sumatra in June of this year, seeking two of its most notable cryptids - the reclusive orang pendek, and an equally elusive mystery cat called the cigau. The expedition will scale Mount Kerinci to Lake Kerinci, 750 m above sea level, where it will establish a series of base camps. Although other teams have looked for the orang pendek, this is the first to make a purposeful search for the cigau. Described as having yellow or tan-coloured unpatterned fur, a short tail, and a ruff encircling its neck, it is allegedly slightly smaller than the Sumatran tiger but is extremely aggressive and thus greatly feared by the locals. **Jon Downes/ CFZ, cz@yahoogroups.com 16 April 2003.**

SCIMITAR CAT SURVIVAL. The extraordinary scimitar cat *Homotherium latidens* is traditionally deemed to have become extinct in Europe around 300,000 years BP. However, as I noted in my book *Mystery Cats of the World* (1989), there is some enigmatic iconographical evidence to suggest that it still survived as late, geologically speaking, as 30,000 years ago. Now, moreover, there is palaeontological back-up for this prospect. As revealed in a newly-published paper by Jelle Reumer and colleagues, in March 2000 a partial lower jaw of *H. latidens* was trawled from an area southeast of the Brown Bank in the North Sea by the fishing vessel UK33. When tested by radiocarbon analysis, it was found to date to 28,000 BP, i.e. roughly the same age as the iconographical items, which most notably include a small statuette found in 1896 inside Isturitz cave, southwestern France. ***Journal of Paleontology*, 23 (no. 1); Darren Naish, cz@yahoogroups.com 1 May 2003.**

Scimitar cat skull at the Paleozoological Museum of China (Captmondo/Wikipedia)

A MADAGASCAN MYSTERY. Although Madagascar is home to several endemic mammalian carnivores, including the very cat-like fossa *Cryptoprocta ferox*, it has no native species of true cat - or does it? As spotted by Darren Naish while watching a National Geographic documentary in their 'Out There' series one weekend in May 2003, during some studies in northwestern Madagascar's Ankarafantsika National Park Tennessee University fossa researcher Luke Dollar trapped what looked like a wildcat, the second such creature that he had caught there. Moreover, instead of resembling a feral domestic cat (the simplest identity for this ostensibly oop felid), it seemed exactly like the African wildcat *Felis lybica*. In the programme, Dollar hinted that it may be either a valid new record for Madagascar, or a bona fide new species. A blood sample from this intriguing specimen, a pregnant juvenile, was taken for examination. However, as I noted in a cz@yahoogroups.com posting (of 19 May 2003), this incident is not without precedent. As far back as 1967, in his book *The Life, History,*

Alien Zoo

and Magic of the Cat, Fernand Mery included the following pertinent snippet:

> The Malagasy Academy possesses a specimen of a magnificent tabby cat, larger than a domestic cat. Details of its capture on Madagascar are uncertain, but of interest is that in the local Malagasy language, *pisu* f domestic cat, with *kary* used to denote 'wild cats', even though wildcats do not officially exist on the island.

Mery felt that this lent support for the probable existence of wildcats on Madagascar. Who knows - after more than 35 years, the blood sample obtained by Dollar may belatedly vindicate him. **Darren Naish, cz@yahoogroups.com 19 May 2003.**

CRYPTIC IN CAMEROON. Ebo Forest is a remote area within Bakossiland, a mountainous region of Cameroon, where, traditionally, no gorillas were supposed to exist. However, during a recent expedition to Ebo Forest in search of one of the world's largest monkeys, the drill *Mandrillus leucophaeus* (related to the mandrill *M. sphinx* - famous for the male's blue face...and other body regions), researchers Bethan J. Morgan and Chris Wild from the Zoological Society of San Diego were very surprised to espy a troop of seven gorillas. Plans have now been made to test genetically some hairs obtained from nests made by these gorillas, to determine whether they belong to a previously-known subspecies of gorilla or whether they represent an unknown version. **http://www.heraldtribune.com/apps/pbcs.dll/article?AID=/20030521/APN/305210917&ca 21 May 2003 [no longer online].**

HE SAW A SEA SERPENT(?). One of the most controversial photos of an alleged sea monster was snapped during the 1960s off Hook Island, in Queensland's Great Barrier Reef, by Robert le Serrec, depicting what looks somewhat like a colossal tadpole, but which certain investigators have dismissed as a hoax. Since the 1960s, the whereabouts of le Serrec have remained as mysterious as the identity of the object in his photo, but recently GUST (Global Underwater Search Team), the organisation headed by cryptozoologist Jan-Ove Sundberg, located him. Aged 75, le Serrec now lives in Asia, and GUST is hoping to interview him in the near future. **Terry W. Colvin, forteana@yahoogroups.com 1 June 2003.**

SKUNK APE CELEBRATIONS. The first Everglades Skunk Ape Festival, held on 14 June 2003 at Trail Lakes Campground in Ochopee, Florida, was a great success, and was held in honour of Bill Mitchell. He in turn had launched the first-ever skunk ape festival 25 years previously, commemorating a tourist's earlier encounter with Florida's most reclusive (and odiferous) inhabitant. *Naples* **(Florida)** *Daily News,* **15 June 2003.**

A MIAOWING MYSTERY ON MAUI? Although ABC (alien big cat) reports have regularly emerged for decades from most states of the USA, such activity in Hawaii has tended to be conspicuous by its absence - but no longer. Since December 2002, several sightings of a dark-furred, large-headed, long-tailed feline mystery beast said to be at least as large as a labrador dog have been reported in the lower Olinda area of Maui. Accordingly, being well aware of the danger that this creature may pose to the island's endemic fauna (not to mention humans and livestock?), attempts to capture it using two large box traps are now underway, involving the state's Department of Land & Natural Resources.

Interestingly, this is not the first such ABC flap in the Hawaiian archipelago. A mystery cat was sighted leaping from a ditch onto a road on Kaui in May 1984 but was never captured; a suspected puma was sought but again eluded capture in the hills above Aina Haina in February 1988 - where, four months

Alien Zoo

later, a large white felid was also reported; and in August 1993 a big tawny wildcat was spied by locals in Palolo Valley. *Honolulu Star-Bulletin*, 14 & 17 June 2003.

ARMOURED SHRIMPS AND PRICKLY SHARKS. Add to those two afore-mentioned marine novelties some gelatinous sea cucumbers, giant sea spiders the size of dinner plates, and several new species of blue ray, and you have just a very select sample of the 1800+ species of fishes and invertebrates recorded and photographed during a month-long marine expedition in May-June aboard *Tangaroa* - the research vessel of New Zealand's National Institute of Water and Atmospheric Research (NIWA). Surveying the deep waters around Lord Howe Island and Norfolk Island down to 2 km, the team found that out of their vast haul of species, more than 100 of the fishes were new to science, and up to 300 of the invertebrates were too. **http://www.stuff.co.nz/stuff/0,2106,2534982a11,00.html 10 June 2003 [no longer online]; http://www.abc.net.au/news/newsitems/s884828.htm 20 June 2003 [no longer online].**

An 1890s photo of the Florida globster

A NEW HOME FOR THE FLORIDA GLOBSTER - or what's left of it. Since it was washed ashore in Florida during 1896, tissue samples taken from this anomalous object, allegedly the remains of a colossal octopus, have been subjected to many examinations in an attempt to determine its zoological identity, with conflicting results. Moreover, most of the samples have ultimately been lost or discarded - an all-too-common (and frustrating) occurrence long associated with cryptozoological specimens. Hence it is important to keep up to date our records of the whereabouts of any such specimens that do still survive - and these include some samples of the Florida globster analysed by veteran cryptozoologist Prof. Roy P. Mackal. Thus, as now revealed by Bill Gibbons, Prof. Mackal has recently donated these precious remains to the Institute of Creation Research in El Cajon, California, where they have been received by the Institute's biology professor, Dr Kenneth Cummings. **Bill Gibbons, cryptolist@yahoogroups.com 29 June 2003.**

AN EEL OF A DISCOVERY. New species, and even new genera, of fishes are still described quite frequently by zoologists, but the creation of an entirely new taxonomic family to house a new fish is a far, far rarer occurrence. Yet it may be happening soon, thanks to a peculiar eel-like fish discovered in 1999 by researcher Ilse Walker. Working with Brazil's National Amazon Research Institute (INPA) in flood plains near Manaus, she came upon an unfamiliar long-bodied 15-cm fish that was later found to combine characteristics from a range of quite separate fish groups. Among its most notable features are its full set of fins (unlike true eels), a tail resembling that of the huge freshwater arapaima (pirarucu) *Arapaima gigas*, and up to 10 air chambers (instead of the usual 2-3 in other fishes, used for maintaining their position underwater). This latter feature has incited researchers to speculate that their still-

Alien Zoo

undescribed 'mystery fish' (as they call it) may breathe at the water surface. They plan to describe and officially name this dramatically different species by the end of 2003. **http://www.sun-sentinel.com/news/custom/fringe/sfl-73mysteryfish,0,1152499.story?coll=sfla-news-fringe/ 3 July 2003.**

CHAMP SPEAKS! Over at Lake Champlain, controversy concerning the existence of Nessie's New World rival, Champ, has been particularly audible lately - and literally! In June, while conducting sonar research in the lake's Button Bay area for a television show on Champ being produced by the Discovery Channel, a team from Fauna Communications Research led by Elizabeth von Muggenthaler detected a series of biosonar readings similar to those emitted by a beluga whale or dolphin when seeking food underwater. Ten times louder than any known fish species in the lake, too irregular to be produced by a mechanical device or fish finder, and recorded on multiple instruments on three separate days (3, 4, and 10 June), some readings lasted up to 10 minutes, and von Muggenthaler believes that on one occasion the unidentified creature approached to within 30 ft of their boat. Renowned for her bioacoustic work on giraffe communication, von Muggenthaler has remained non-committal as to the likely identity of the sounds' originator, but once her readings are fully analysed (by National Instruments Inc. of Austin,

Has Champ been sounding out in search of prey beneath Lake Champlain's water surface? (William Rebsamen)

Alien Zoo

Texas) she hopes to be able to reveal how fast the creature was moving. Not surprisingly, cryptozoologists hail her findings as a major development in the ongoing debate as to whether or not Champ exists, and we all await the results with keen interest. ***Burlington* (Vermont) *Free Press*, 20 July 2003.**

NESSIE'S BACK...OR JUST A BIT OF IT? It was the find that cryptozoologists had been praying for - a piece of plesiosaur backbone found in Loch Ness! Inevitably, however, things were not quite that simple. Yes, the alga-covered 12-in-long object discovered in shallow water at the loch by pensioner Gerald McSorley in July 2003 was indeed identified conclusively by palaeontologist Dr Lyall Anderson at Edinburgh's National Museum of Scotland as a series of four vertebrae from a plesiosaur - but they were fossil vertebrae, approximately 150 million years old. Could they therefore be from Nessie's far-distant loch-dwelling ancestor? Sadly, no - if only because Loch Ness only came into existence a mere 12,000 years ago, at the end of the last Ice Age. Moreover, the fossil had been intensely drilled by marine sponges, whereas the loch is freshwater; and it is embedded in a grey Jurassic-aged limestone, yet rocks in the Loch Ness area are much older and are all crystalline, igneous, and metamorphic. According to Anderson, the nearest match for the fossil's limestone is some 30 miles away, at Eathie on the Black Isle. Accordingly, it has been suggested that the fossil may have been deliberately planted at Loch Ness as a hoax, the latest in a long line associated with Nessie over the years, in the hope that someone would eventually come along and find it, as has now happened. Conversely, plesiosaur expert Dr Richard Forrest from Leicester's New Walk

A plesiosaur vertebra from my zoological collection
(Dr Karl Shuker)

Museum has been quoted as suggesting that the fossil may have reached Loch Ness by more natural means - transported by glaciers. Also, there is a telling argument against the hoax theory that I have not seen mentioned in print before: namely, plesiosaur fossils of this type are neither cheap nor commonplace. As someone who regularly peruses mineral and fossil fairs, I have seen single plesiosaur vertebrae on sale for around £40 (the sum that I myself paid for one such specimen, pictured below). Hence I would expect a four-vertebra example (should such a specimen actually be available to buy) to cost rather more than £100 - which seems an expensive item to discard purposefully in Loch Ness just on the off-chance that it would one day be found. ***Inverness Courier*, 18 July 2003; Darren Naish, cz@yahoogroups.com 22 July 2003; *National Geographic News*, 29 July 2003.**

HAIR TODAY, KNOWN TOMORROW? Another piece of cryptozoology-allied evidence in the headlines recently was the hair sample allegedly from an orang pendek that was obtained in Sumatra by a British expedition featuring Adam Davies, Keith Townley, and Andrew Sanderson back in September 2001. Since then, the sample's DNA has been rigorously analysed by internationally-acclaimed hair expert Dr Hans Brunner from Melbourne's Deakin University in conjunction with Cambridge Univer-

Alien Zoo

sity primatologist Dr David Chivers, and in a soon-to-be-published scientific paper they confirm that they have been unable to match it with the DNA of any known animal species. A footprint also said to be from an orang pendek and obtained during the same expedition cannot be identified with any known species either. At last, there may well be hard evidence for the existence of the orang pendek or 'short man' as a species undescribed by science - a man-beast on the brink of zoological acceptance. **Manchester News, 19 July 2003; Evening Chronicle, 9 August 2003.**

GLOBSTER GENETICS. And speaking of DNA: two independent DNA analyses conducted on tissue samples taken from the globster washed ashore in Chile during June 2003 have confirmed that its gelatinous remains are indeed from a sperm whale *Physeter macrocephalus* - the identity promoted in scientific circles following the discovery of spermaceti organ tissue amid this decomposing mass by Chilean zoologists in July. The DNA analyses were performed by Auckland University-based Chilean scientist Carlos Olavarria and by marine biologist Dr Sidney Pierce at the University of South Florida. *National Geographic News*, **25 August 2003.**

LANDLOCKED IN GASPARILLA. During late August 2003, news emerged of a 'monster' inhabiting Florida's Gasparilla Lake. And not just any old monster either - for according to local eyewitnesses Tom Farrish and Kristine Barr who have seen it several times during August, chasing fishes to the shore, the creature in question is none other than a dolphin. The only problem with this identity, however, is how it comes to be here - bearing in mind that Gasparilla Lake is totally landlocked, with no access to the Gulf of Mexico. Not surprisingly, therefore, experts have discounted a dolphin identity in favour of either a large tarpon or an otter. However, Barr, who lives on the lake and spends plenty of time on her porch, remains adamant that what she has seen virtually every day for a fortnight is a bona fide dolphin, and is now armed with a camera to snap the photos that she hopes will confirm her claim. *Boca Beacon*, **29 August 2003.**

GREEN GROW THE LIZARDS-O! Generally up to a foot long, the western green lizard *Lacerta bilineata* is much larger than any native species of British lizard - always, assuming, of course, that it is not itself of UK nationality. Although common on the European continent and also the Channel Islands, attempts in the past to establish colonies on mainland Britain have failed, presumably because our cooler climate is not to this species' liking. Now, however, a discovery has been made that could shatter both of these fondly-held preconceptions. At the end of August 2003, reptile expert Dr Chris Gleed-Owen revealed that he had found what appears to be a thriving, breeding population of green lizards that he estimates may

Western green lizards

number in the hundreds, living in a 200-square-yard area of cliff in the Bournemouth region of Dorset. Ironically, he found them only 100 yards or so from the Herpetological Conservation Trust's office, yet they had not been reported there before. Their origin, however, remains controversial. It may be that they are the descendants of a few pet specimens abandoned or deliberately released by persons unknown some time ago. More excitingly, conversely, is the possibility that they represent a hitherto-overlooked native population, dating back 10,000 years. ***Daily Mail*, 29 August 2003.**

STRICTLY FOR THE GOONS - THE GUNNI. Australia is famous for marsupial mammals that have evolved by convergence to resemble ecological counterparts elsewhere in the world, e.g. marsupial mole, marsupial wolf or thylacine, marsupial mice (albeit more like shrews), flying phalangers (closely paralleling flying squirrels). Now, however, it seems that Down Under can even boast a marsupial counterpart to that most infamous of fraudulent fauna, the jackalope. Known as the gunni (and pronounced 'goon-eye'), this horned wombat-like beast is proudly represented by an ingenious taxiderm specimen on display in the visitors' information centre at the tourist town of Marysville, Victoria. Its wombat body is additionally adorned with stripes on its back and hindquarters, plus a tail, and it bears deer antlers on its head. It was recently presented to the centre by local ranger Miles Stewart-Howie as a private project, along with a detailed account of the gunni's fictitious history, which is now also displayed by the centre alongside their newest and certainly most entertaining wildlife exhibit. **http://www.theage.com.au/articles/2003/10/04/1064988455505.html 4 October 2003.**

A RELIC FROM EDEN'S SERPENT? One of the most enigmatic yet hitherto-obscure zoological relics held in any scientific establishment must surely be the 8-inch by 4-inch piece of scaly rusty-red leathery skin contained inside Archive Box #1920.1714 within the very sizeable collection of the Chicago Historical Society. For according to its yellowing French label, this is supposedly a genuine piece of skin from the very serpent that tempted Eve in the Garden of Eden! Indeed, the label goes on to say that the serpent was killed by Adam on the day after its treachery to Eve, using a stake whose traces can be seen on this skin sample, which was preserved by his family in Asia. Affixed to the skin is a document written on velum or similar hide in an Asian script. The society purchased this mystifying exhibit, along with many other items, in 1920 from the eclectic collection of Chicago confectioner Charles F. Gunther - a grand collector of curiosities. Although the society's chief curator, Olivia Mahoney, has no doubt that it is a fraud (as opposed to a bona fide piece of snakeskin dating back to the dawn of time), no research has ever been conducted on it to ascertain what it really is. Moreover, Mahoney is very reluctant to permit any, in case the skin is damaged, and also because in her view it is so evidently a fake. That may well be, but it still doesn't answer what - if not a sample of skin from the Eden serpent - this anomalous object is. As noted by the *Chicago Sun-Times*'s religion writer, Cathleen Falsani - who viewed and wrote about this biblical(?) relic in October 2003 - after watching it being carried back in its box to the society's archives: "I couldn't help thinking about that scene from 'Raiders of the Lost Ark', where the Ark of the Covenant, and all of its

'The Fall of Adam and Eve' by Dutch painter Hugo van der Goes, c.1470, depicting the pre-cursed serpent of Eden as a bipedal humanoid reptile.

power, is crated up and wheeled into a military warehouse among thousands of other generic crates. I wonder what else might be hiding anonymously in a quiet corner of a museum archive somewhere else, waiting to shock us with its mystery". What else indeed? *Chicago Sun-Times*, **10 October 2003**.

The great bustard

BRING BACK THE BUSTARD! A very startling addition to the British fauna may well be one of the world's heaviest flying birds! Once a common species over much of Britain, the crane-related but much heftier, burlier great bustard *Otis tarda* was a popular target for hunting parties, especially as its roasted flesh was tasty and plentiful enough in an adult male (weighing up to 2.5 stones) to feed up to 16 diners. Coupling this with habitat destruction due to farming, its days were numbered, and by 1832 the great bustard had been wiped out in Britain, though it still survives in mainland Europe.

Starting in 1970, a plan was launched to re-establish it, via a semi-captive population maintained on Wiltshire's Salisbury Plain, but this ultimately failed. Now, however, a second attempt is to be made, again on Salisbury Plain, after the British government granted a licence for a trial reintroduction.

Next May, 40 chicks from Russia will be shipped to the UK, where they will be quarantined and then reared in protective pens on the plain, before being released in autumn once they can fly. Forty more could be introduced in the same way each year afterwards for the next decade. *Daily Mail, Wolverhampton Express & Star*, **4 November 2003**.

KELLAS CATS GO WEST? The Kellas cat of northern Scotland, first publicly reported during the early 1980s, proved to be an introgressive hybrid (i.e. resulting from several generations of crossbreeding) of domestic cat *Felis catus* and Scottish wildcat *F. silvestris*. Intriguingly, however, some odd feral domestics have been reported from North Carolina, USA, that sound very reminiscent of the Caledonian Kellas, even though there are obviously no Scottish wildcats living in the wild over there. In November 2003, I learnt from Ben Willis via the MysteryCats@yahoogroups.com online discussion group that from the early 1990s onwards, he has encountered a number of black Kellas-like felids around coastal N.C. He estimated one such specimen to be twice the size of an ordinary domestic cat, and he was even able to capture and rear a second one from kittenhood, which he described as being "considerably larger than a domestic, with white guard-hairs, a kinked tail, and extraordinarily large canine teeth". The white guard hairs and unusually large canine teeth are familiar Kellas cat features, and the kinked tail suggests that this particular specimen may have had Siamese cat ancestry - which has also been mooted in the past for the Kellas cat. Ben states that there is another of these odd cats presently roaming the woods near his home, which he describes as having "the same coat as the others, and has a small white star on his chest" - as do the Kellas cats. The existence of such creatures in an area bereft of Scottish wildcats indicates that the Kellas cat's distinctive features may owe more to its domestic (as opposed to its Scottish wildcat) ancestors than previously supposed. **Ben Willis, Mystery-Cats@yahoogroups.com 12 November 2003.**

PILTDOWN ON PARADE. One of the most infamous scientific frauds of the 20th Century must surely be the Piltdown man, whose remains (which included the jawbone of an orang utan) were exposed as a hoax in 1953, though the perpetrator has never been conclusively identified. On 21 November 2003, marking the exact 50th anniversary of their denouement, these fake fossils were placed on public display at the Natural History Museum in London, where they will remain on show until 20 February 2004. So if you want to see some notorious bones of contention that quite literally made an ape out of the scientific community for a fair number of years after their initial discovery, this is your chance! *Wolverhampton Express & Star*, **20 November 2003.**

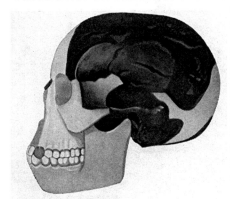

Piltdown man - skull reconstruction from 1922

HAIR OF THE CAT. During summer 2003, CFZ cryptozoologist Richard Freeman, together with Dr Chris Clarke and Jon Hare, spent three weeks in Sumatra searching for the elusive orang pendek or 'short man'. Although they obtained photos of possible orang pendek footprints, the hair samples that they brought back turned out to be feline in origin. However, these are still very interesting, because when later checked by Danish zoologist Lars Thomas against all but one of the known cat species from the region in which the team had been exploring, no positive match could made. The only known felid now left to check with these intriguing hairs is the Asian golden cat *Catopuma* [=*Felis*] *temminckii*, and if there is still no correspondence the team is hoping that they may have obtained the first physical evidence for the reality of Sumatra's leonine mystery cat, the cigau - a fierce maned felid reported by natives from the Mount Kerinci area of Sumatra. *CFZ Newsletter*, **December 2003.**

Statues in Kings Park, Perth, Western Australia, of *Bullockornis,* a Miocene relative of *Genyornis* (Seanmack/Wikipedia)

The Lost Ark

CZ IN OZ
- AN ARK OF NEW MYSTERY BEASTS DOWN UNDER.

At any rate, we hope that the evidence presented in this book is strong enough to convince you, the reader, that the existence of some of these creatures is at least a possibility. We suspect, however, that no matter how good the footprint casts or testimonial evidence may be, you, like ourselves, will never fully believe in any of these creatures until you see one yourself.

*But who knows? Perhaps next time you're driving down a dark and lonely country road you'll turn a corner and there, right in front of you, something – some looming, hairy nightmare – will step
...out of the shadows.*

Tony Healy and Paul Cropper – *Out of the Shadows: Mystery Animals of Australia*

Australia is famed as the continent with the most unusual and unexpected known fauna, but it must also be in strong contention for the most bizarre array of unknown fauna - everything from elusive panthers, giant kangaroos, and marsupial tigers to bunyips, yowies, and even the odd (sometimes very odd) neo-dinosaur or two. Now, as if these were not sufficient to occupy Aussie cryptozoologists for the foreseeable future, I have pleasure in introducing a selection of additional anomalies, whose details have not previously been published. I am extremely grateful to Australian herpetologist Richard Wells for sharing his data with me and for very kindly permitting me to include it in this chapter.

Alien Zoo

A MEGA-PYTHON IN ARNHEM LAND?

We begin with a python of truly Pythonic proportions. Back in the 1970s, while Richard was working for the Northern Territory Museum in Darwin, an amateur naturalist brought in an extremely sizeable python, roughly 4 m long, of a previously unknown species, which was later formally described and named *Liasis* [now *Morelia*] *oenpelliensis*, the Oenpelli python. Subsequent specimens of this new snake have been recorded that vie in size with the mighty amethystine python *L. amethystinus* - Australia's largest species of snake, and the fourth largest in the world, which can attain a total length of up to 7 m.

It seems remarkable that a snake as sizeable as *M. oenpelliensis* could remain undiscovered by science until as recently as the 1970s, and even more so if local aboriginal claims are correct - namely, that specimens greatly in excess of 4 m exist. Nor is the evidence for this dramatic prospect wholly anecdotal. According to Richard: "...the obvious tracks of a huge python have been found within vast underground caverns under Arnhem Land, and these were so large as to indicate that it may attain around 10 metres in length".

Carpet (diamond) python

Moreover, Richard has received consistent reports from local tribes of a massive aquatic python-like snake, whose description leads him to consider that a bona fide prehistoric survivor may be the answer - *Wonambi naracoortensis*. This remarkable species was a giant 5-m-long constricting species of snake, but belonging to an ancient, primitive group known as the madtsoiids, and deemed to have been at least partially aquatic. At one time, madtsoiids existed on all continents formerly part of the southern supercontinent Gondwanaland, but they all died out around 55 million years ago - except in Australia, where they continued to diversify, culminating in *Wonambi*, which became extinct only within the last 50,000 years, inciting speculation that it may have been contemporary with the first humans to reach Australia, and might even have inspired Dream-Time myths of the great Rainbow Serpent.

But could it still exist today? The concept of surviving madtsoiids is not restricted to Australian cryptids. As noted in my book *The Beasts That Hide From Man* (2003), English cryptozoologist Richard Freeman has recently offered this identity as an explanation for reports of gigantic water snakes inhabiting the Mekong River comprising Thailand's northeastern border with Laos, and known as nagas (not to be confused with the ancient snake deities of the same name), but until now there had been no reports of comparable aquatic mystery snakes in Australia.

Conversely, when I communicated with palaeontologist Dr Ralph Molnar, formerly Curator of the Queensland Museum and an expert in Australian fossil reptiles, concerning these cryptids, he favoured the modern-day carpet (diamond) python *Morelia spilota* as a more plausible candidate taxonomically, though he is not surprised to learn that reports of extra-large constrictors have emerged here. Personally, I look towards the amethystine python, a northern species already known to attain immense lengths, which is also a good swimmer and usually occurs near water. The prospect of exceptionally-large aquatic specimens, their huge size buoyed by their liquid medium, is not untenable.

MARY RIVER'S MYSTERY CROCODILE

Does a gharial-like mystery beast inhabit Australia's Mary River system? (Dr Karl Shuker)

Alien Zoo

Today, only two species of crocodile are known from Australia - the monstrous estuarine or saltwater crocodile *Crocodylus porosus*, and the smaller freshwater or Johnston's crocodile *C. johnstoni*. However, when Richard Wells lived in the Northern Territory, he received several consistent reports from a variety of informants, including fish poachers, old ex-crocodile hunters, and aboriginals, indicating that a third, very large but very different and seemingly scientifically-unknown crocodilian existed Down Under, inhabiting at least the tidal parts of the Northern Territory's Mary River system. According to Richard's data:

> [it] would appear to be totally aquatic with an elongated jaw with numerous exposed teeth more in keeping with that of some kind of gharial (*Gavialis*) - but it appears to reach a larger size, has paddle-like limbs, and is of nocturnal behaviour. Most reports of the creature have been dismissed as representing sawfish or crocodiles, but all the people who reported it were very familiar with sawfish and crocs, but were adamant that it is some sort of crocodile-like reptile and it scared the hell [out] of them.

Gavialis gangeticus, the only known modern-day species of *Gavialis* gharial, is not native to modern-day Australia. However, crocodiles feeding on a low-protein diet (as sometimes in zoos) are prone to developing exposed teeth that point laterally in the jaws. Could this explain the dental peculiarity of the Mary River mystery croc - is it based merely upon sightings of some abnormal specimens of a known species? (Johnston's crocodile does possess a noticeably long, thin, superficially gharial-like snout.)

Possibly - were it not for the claim that its limbs are paddle-like. As Ralph stated to me, this feature immediately calls to mind those highly-specialised, wholly aquatic prehistoric crocodilians known as thalattosuchians. However, they were a relatively short-lived group, and became extinct around 110 million years ago. In contrast, although its limbs were not paddled (which may be just an exaggeration on the part of the eyewitnesses - perhaps the Mary River beast's limbs are simply more rounded than those of Australia's two known crocodiles, creating the illusion of being paddle-shaped?), less than two million years ago the Solomon Sea just above southeastern New Guinea was hunted in by a very gharial-like species, known as the Murua crocodile and dubbed *Gavialis papuensis*.

Some authorities have challenged this creature's gharial classification, favouring a thoracosaur affinity, but Ralph, although initially sceptical, now informs me that it may indeed have been a gharial. Its elongated snout is certainly reminiscent of one, and it is believed to have been a fish-eater, again like modern-day gharials. A round-limbed present-day representative of this species would correspond very closely to the Mary River cryptid.

AN ENIGMATIC BIRD FROM EL SHARANA

Richard's final unknown creature from the Northern Territory: "...represents a large mainly nocturnal or crepuscular flightless bird at the southern end of Arnhem Land. I presumed that it may be [a] sort of tropical emu, or a hitherto unknown population of cassowary, but whatever it is, it certainly does exist, because I found its tracks near El Sharana (an abandoned uranium mine) in 1977".

The most romantic identity for this creature would be a living mihirung *Genyornis newtoni* - a superficially ostrich-like Australian bird (but actually more closely related to waterfowl) that officially died out around 15,000 years ago. As I documented in *The Beasts That Hide From Man*, there is at least one

Alien Zoo

Newton's mihirung *Genyornis newtoni* (Tim Morris)

tantalising modern-day sighting of a giant mystery bird here that recalls a mihirung, but Richard's account does not contain enough detail to attempt any identification of the El Sharana bird.

All that we can say for this, and for the other cryptids reported here, is that it would surely bear investigating by anyone fancying themselves as a cryptozoological wizard in Oz (or even a wizard of Aus)!

Is the El Sharana mystery bird a species of cassowary?

2004

SEEKING THE DEATH WORM. I have recently learnt from Czech explorer Ivan Mackerle that he is planning to return this summer to the Gobi Desert to search once again for its most elusive, and deadly, cryptozoological enigma - the electrifying Mongolian death worm or allghoi khorkhoi. This time, to facilitate covering a greater search area within the Gobi's vast terrain, Ivan is planning to pilot an ultralight aircraft, from which he hopes to spot one of these vermiform mystery beasts basking on the surface of the sands, which it is said to do during the months of June and July. **Ivan Mackerle, pers. comm., 16 January 2004.**

NOT SO EXTINCT? In 2000, writing off the world's rarest monkey as extinct may prove to have been a load of colobus - Miss Waldron's red colobus *Piliocolobus badius waldronae*, to be precise. A small black-furred monkey with reddish brow and thighs, this enigmatic form was once abundant in the canopy rainforests of eastern Ivory Coast and western Ghana. But plummeting numbers caused by hunting and especially by forest destruction meant that by 2000, it seemed to have vanished; several years of searching in vain for it by Ohio State University anthropologist Prof. Scott McGraw and colleagues finally resulted in their publishing a paper formally consigning Miss Waldron's colobus to zoological oblivion - the first monkey to have become extinct in over 200 years.

Two years ago, however, what seemed to be the skin of a recently-killed specimen was handed to McGraw by an Ivory Coast hunter, and in a separate incident during that same year another Ivory Coast hunter gave McGraw a still-fresh tail whose DNA was later confirmed to be from a red colobus.

FACING PAGE: In the lair of the Mongolian death worm (Andy Paciorek)

Alien Zoo

Miss Waldron's red colobus

McGraw now hopes to return to Ivory Coast, notwithstanding this country's recent civil war and associated political turmoil, to seek out conclusive evidence that Miss Waldron's simian namesake has indeed survived. **http://www.innovations-report.com/ html/reports/environment_sciences/ report-25365.html 3 February 2004.**

AN ANGOLAN ANOMALY. There have been brief media reports lately alluding to a mysterious 'giant' black panther whose traces have apparently been found in Kangandala National Park, in Angola's northern Malanje province. Indeed, the province's governor, Cristovao da Cunha, has personally dispatched a four-man expedition to investigate these claims. What is particularly noteworthy about this matter is that whereas black panthers are very common in Asia, they are decidedly rare in Africa. Moreover, whereas the Asian black panthers are the familiar melanistic leopard variety (whose spots are hidden by the animals' abnormally dark background fur colouration), African black panthers tend to be the much more unusual pseudomelanistic variant - in which the background fur colour is the normal golden shade but the spots are abnormal, having coalesced into one another so that the golden background colour is almost obscured, yielding predominantly black-furred cats with just an occasional streak of gold, except for their normal, pale-coloured underparts. If any of these cats are alive and well in Angola, it would be extremely interesting to monitor them in the wild state, as virtually all previously known pseudomelanistic specimens are preserved trophy pelts or a very occasional captive specimen. **Angola Press Agency, 11 April 2004.**

A PICTURE OF PARADISE? Painter Errol Fuller, in his recent, lavishly-illustrated book *The Lost Birds of Paradise*, put forward a bold but compelling argument promoting the possibility that certain controversial museum specimens of bird of paradise dismissed since the 1930s as hybrids actually represented valid (but presumably now-extinct) species in their own right. Now he has uncovered further evidence to support his thesis, in the form of certain 18th-Century bird paintings by French artist Jacques Barraband. Known for the accuracy of his avian illustrations, most of Barraband's subjects (which include several birds of paradise) can be readily identified with known species - but there are a few enigmatic, mystery birds of paradise painted by him that do not resemble any recorded species. One of these paintings, published by *New Scientist*, shows a

Barraband's mystery black-breasted 12-wired bird of paradise (Errol Fuller)

black-breasted 12-wired bird of paradise, whereas the typical version has a yellow breast. Could Barraband's paintings thus include depictions of unknown species, or merely a selection of curious crossbreeds? *New Scientist*, **15 May 2004.**

CURELOMS AND CUMOMS. Cryptozoologically, the Holy Bible is famous for mentioning two very mystifying creatures - the leviathan and the behemoth. Less well-known, however, is that another sacred work also contains a pair of highly mysterious animals. The following quote is from Ether 9:19 in the Book of Mormon: "And they also had horses, and asses, and there were elephants and curelom and cumoms; all of which were useful unto man, and more especially the elephants and curelom and cumoms". But what exactly are, or were, these latter two beasts? Surfing the internet in search of answers, the best that I could find was the curelom entry in Wikipedia, which notes that Mormons have suggested various possibilities. These include: late-surviving mastodons or mammoths; a yet-undiscovered but probably now-extinct species; or some Central American species with which Joseph Smith was unfamiliar, such as the tapir or jaguar. **http://en.wikipedia.org/wiki/Curelom accessed 21 May 2004.**

Was the curelom a late-surviving mastodon?

THE HORNS OF A DILEMMA? A veteran cryptozoologist with some intriguing news is Bill Gibbons. During his most recent Cameroon expedition (of November 2003) in search of its equivalent to the Congolese mokele-mbembe, he learnt of a previously-unreported and highly perplexing mystery beast known to the Baka people as the ngoubou. Described by them as being the size of an elephant, it has a substantial neck frill (smaller in females than in males), a long thin tail, and either three horns (female) or four horns (male) - with the female in particular seeming to bear an incredible resemblance

Alien Zoo

to the famous prehistoric ceratopsian dinosaur *Triceratops*, and the male likened to the multi-horned ceratopsian *Styracosaurus*. Moreover, there appear to be two distinct types - one is semi-aquatic, the other a savannah-dwelling herd animal (which is shot for food by the Bantu). According to Timbo, the Baka chief, a female of the semi-aquatic variety was snared and killed by the tribe in an elephant trap set near the Boumba River in 1996.

And in September 2000, Bill's field director, Pierre Sima, visited a village whose hunters had just shot a male specimen of the savannah version, whose four-horned carcase he saw (and dined upon - it tasted like pork). Bill hopes to revisit Cameroon in January 2005, and will be publishing a full report of his Cameroon discoveries in due course - which also include accounts of a truly gargantuan type of ground-dwelling spider called the j'ba fofi ('great spider') that reputedly boasts a 4-5-ft leg span! **Bill Gibbons, cryptolist@yahoogroups.com 24 May 2004.**

Styracosaurus, reputedly resembling the male ngoubou (Dr Karl Shuker/reproduction courtesy of Drayton Manor Park and Zoo)

SNAKES ON THE WING. For over a year, American cryptozoologist Nick Sucik has been conducting research into a hitherto-obscure but fascinating cryptid - the Navajo flying snake. This extraordi-

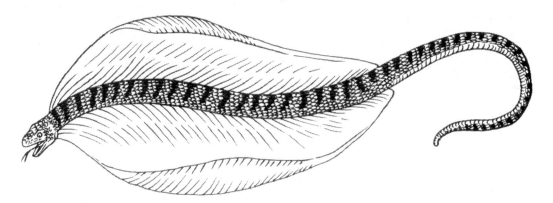

Reconstruction of Navajo flying snake (Tim Morris)

Alien Zoo

nary reptile is apparently well-known to the native Americans of the region, who describe it as being serpentine but with a pair of lateral wing-like membranes running down the anterior portion of its body that it uses for flight purposes. I was privileged to receive from Nick a copy of his extensive research report a while ago, and encouraged him to make it publicly available. This he has now done, with the assistance of fellow cryptozoologist Chad Arment, so I am pleased to announce that it can be accessed at: http://www.voicesofthepeople.org/DragonsOfTheDine.pdf [now at http://www.azcentral.com/12news/pics/dragonsofthedine.pdf]. **Nick Sucik, pers. comm., 26 May 2004.**

GOING FOR GOLD. A few months ago, I reported that an expedition to Sumatra by British cryptozoologist Richard Freeman and colleagues had succeeded in bringing back some fur samples that may be from Sumatra's leonine mystery cat, the cigau. Comparison of the fur with samples from all but one of the known felid species from this area of Asia had failed to find a match; all that remained was to compare the fur with that of the final felid on the list - the Asian golden cat *Catopuma temminckii*. News has now emerged that Copenhagen University zoologist Lars Thomas has conducted this last, all-important comparison - and the result?

Sadly for cryptozoology, the golden cat samples and those of the alleged cigau corresponded. This means either that the mystery fur was not from a cigau but merely from a golden cat, or that the cigau IS a golden cat. However, if local reports are correct, the cigau is far bigger than the latter felid. This recalls an equivalent African scenario, whereby Tanzania's own mystery cat, the huge and ferocious brindle-furred mngwa or nunda, was suggested by Dr Bernard Heuvelmans to be a gigantic variety of the African golden cat *Profelis aurata*. Who knows, perhaps the cigau may prove to be an extra-large version of the Asiatic golden cat - or something even more exciting? Based upon interviews with local eyewitnesses, Richard is considering the dramatic possibility that the cigau is a surviving species of scimitar cat *Homotherium*, allied to the famous but officially-demised sabre-tooths. *CFZ Newsletter*, **30 May 2004.**

CLOSE ENCOUNTERS OF THE LEAFY KIND. One of my many correspondents is Juan Cabana, who skilfully constructs and sells as curiosities amazing stuffed/preserved fake monsters, especially mermaids - some of which can be seen on his website [now at http://www.thefeejeemermaid.com]. During one e-mail conversation, however, he informed me of a truly extraordinary kind of entity that I had never previously heard of. Juan mentioned that he had once seen a couple of examples of so-called plant people - small mummified humanoids, each about 12 inches tall and originating from southeast Asia, that were made from plant tissue and, according to local tradition, grew upon a plant, just like a fruit! Juan examined them very closely and was astonished at the degree of detail - such that they did not appear to have been hand-crafted.

He now believes that they had been made by growing some type of fruit or vegetable inside a human-shaped glass vessel until the plant tissue had totally filled and thus acquired the container's shape, whereupon the vessel was broken, and the resulting humanoid plant tissue dried out to resemble a leafy corpse. **Juan Cabana, pers. comm., 31 May 2004.**

CHILLING OUT WITH LIVING TRILOBITES? Superficially resembling aquatic woodlice (though sometimes far bigger in size) but constituting a totally separate arthropod lineage, trilobites officially became extinct during the mass extinction of the Permian Period, roughly 250 million years ago. However, in a recent media report, no less eminent a mainstream figure than Ron O'Dor, chief scientist with the $1-billion, 53-nation, 10-year global Census of Marine Life (CoML), was quoted as saying that the chilling, barely-unexplored waters of the Arctic Ocean may conceal living trilobites. He

Alien Zoo

Have the trilobites truly vanished?

was speaking at the time about a new survey of life in the depths of this ocean, and what may well still await discovery there, including new species of jellyfishes, giant squid, and other exciting marine fauna. **Yahoo! News, 24 June 2004.**

AN IMR INTANGIBLE. Very mysterious is the 1-ft-long worm-like creature with vivid red anterior and long, curved, sand-buried posterior photographed while crawling around the sea bottom at a depth of 6500 ft, during the first comprehensive deepsea probes of the Mid-Atlantic Ridge, conducted recently by a 2-month-long Norwegian-led international expedition. The Mid-Atlantic Ridge is a chain of undersea mountains running between Iceland and the Azores, and over 80,000 specimens were collected, but sadly the mystery worm was not one of them, as it defied attempts to bring it to the surface. Scientists from Bergen's Institute of Marine Research (IMR) have likened the creature to the acorn worms or enteropneusts, but concede that it may even represent an entirely new phylum of animals. Looking at its photos, it recalls those equally enigmatic vermiform beasts the pogonophorans or beardworms, which include the giant tube-dwelling, red-tentacled hydrothermal vent worms that were themselves made known to science as recently as the mid-1970s. Another mystery unveiled by this expedition that may or may not be related to this strange worm was the presence of never-before-seen, perfectly straight, evenly-spaced lines of 2-inch-diameter holes that look as if they have been stitched into the seabed at a depth of 6000 ft. Presumably the burrows of some marine animal, what makes them so perplexing is how any creature could create such straight lines. **http://www.wired.com/news/ technology/0,1282,64483,00.html?tw=wn_tophead_3 6 August 2004.**

CRYPTOZOOLOGY COMES TO KAWASAKI. Dispelling visions of motorbike-riding bigfoots, I hasten to explain that the Kawasaki in question here is the Japanese city of Kawasaki, in Kanagawa Prefecture, whose City Museum is currently hosting a fascinating exhibition entitled 'Japan's Mythical

Alien Zoo

Creatures: Accounts of Unidentified Living Organisms'. It comprehensively surveys Japan's diverse mythological and cryptozoological fauna, including the water-dwelling tortoise-shelled kappa, assorted ogres, merfolk, dragons, thunder-precipitated beasts known as raiju, and the serpentine tzuchinoko, and boasts some extraordinary exhibits. One of the most eye-catching of these, on loan from a temple in Osaka (to whom it was supposedly donated during the late 17th Century), is claimed to be a mummified kappa. Its shrivelled body measures 70 cm long, sports four spindly limbs with disproportionately lengthy fingers and toes, and it has a fish-like face. No doubt its true origin is akin to that of the infamous Feejee mermaids, manufactured in bygone times by skilful vendors to sell to unwary or gullible sightseers. A 50-cm merman with a wizened face is also on

A kappa, from a Japanese bestiary illustrated by Katsushika Hokusai (1760-1849) (Wikipedia)

display, as is a mummified ogre whose sitting height is an imposing 1.2 m. Another intriguing exhibit is an elongate fingernail said to be from a tengu – a legendary half-bird, half-humanoid entity of capricious, mercurial temperament; that particular relic has been highly revered by a temple in Saitama Prefecture. This unique exhibition continues until 5 September 2004. **http://www.yomiuri.co.jp/newse/20040826woaa.htm 26 August 2004.**

IT'S THE DINOSAUR KANGAROO OF ARICA! It's always good to learn of a new cryptid, and Chile's Arica Beast is new in every sense! Several different motorists driving along the main road linking Iquique and Arica, through the Atacama Desert, have lately reported witnessing an extraordinary bipedal creature over 2 m tall, with sharp teeth and three-toed footprints, which has been variously likened to a velociraptoresque dinosaur or even a

Is an anachronistic *Velociraptor* chilling out in Chile? (Dr Karl Shuker)

Alien Zoo

'dinosaur kangaroo'! In the words of one eyewitness, Hernan Cuevas: "A weird animal looking like a dinosaur with two legs and huge thighs crossed the road in front of my car". Not surprisingly, the local authorities are very puzzled. **http://www.ananova.com/news/story/sm_1080697.html 27 August 2004 [no longer online].**

A HORSE OF A VERY DIFFERENT COLOUR. In my book *From Flying Toads To Snakes With Wings* (1997), I documented the remarkable history of a completely hairless, blue-skinned horse that, after being discovered in the company of a herd of quagga *Equus quagga* (=*burchelli*) *quagga* in South Africa during 1860, was captured, brought to England, and successfully exhibited in February 1868 at London's famous Crystal Palace. Until very recently, I had never encountered reports of anything even remotely similar to this equine enigma. However, after reading my book a short while ago, Anna Severson from the USA emailed me to inform me that a living horse of near-identical appearance is depicted and documented in the October 1999 issue of *Western Horseman*. Moreover, Anna subscribes to this magazine, but in a spooky twist, when she went to the shelf where she keeps its back issues to check this particular article, she discovered that the October 1999 issue was missing - and despite searching elsewhere for it, so far she has not been able to track it down! **Anna Severson, pers. comm., 6 October 2004.**

A REAL RARA AVIS! It's not every day that one of the world's rarest birds visits Britain, but this is apparently what happened this autumn. To the amazement, but delight, of birdwatchers at Minsmere in Suffolk, one of the UK's most famous RSPB reserves, what may well have been a slender-billed curlew *Numenius tenuirostris* was spied there on 28 September 2004. Only around 270 specimens of this ultra-rare wading bird are thought to be alive worldwide, its breeding grounds have never been discovered, and only 174 sightings around the world have been recorded since 1900. A century ago, it was much more common, and was thought to nest somewhere in Siberia, then migrate to North Africa. But it was decimated by over-hunting, and more than once this enigmatic species has been written off as extinct. It may take months before experts finally decide whether the Minsmere mystery bird really is a slender-billed curlew, using eyewitness descriptions and photos, and possibly also DNA samples from any droppings or feathers left behind by it. But if it is confirmed as such, anyone who saw it can class themselves as one of the most privileged UK 'twitchers' alive today. *Evening Star* **(Suffolk), 30 September 2004;** *Daily Mail***, 6 October 2004.**

HORSE-EELS AHOY! A Japanese five-man television crew is arriving in Ireland during October 2004 to seek out a highly elusive horse-eel...or two. They plan to visit various lakes around Clifden in County Galway, as well as similar sites in Cork and Kerry. Lough Auna and Lough Shanakeever to the north of Clifden are famous for sightings of a mysterious humped water beast, said to have a long flexible eel-like body but a horse-like head and neck. So too is Lough Mask, just north of Oughterard. Critics claim that such lakes are not large or deep enough to sustain such creatures. However, believers suggest that the horse-eels do not live permanently in any one lake, but migrate overland like true eels from one body of water to another. *Galway Advertiser***, 7 October 2004.**

DIPPY OVER *DIPLOCAULUS*. In September 2004, *FT* forwarded to me a short note from reader Stuart Pike enquiring about an attached photo of a bizarre-looking mystery beast, labelled as a hammerhead lizard.

Not long afterwards, Maltese journalist Tonio Galea independently contacted me, requesting details about this same photo (of unknown origin), whose creature, according to local Maltese rumour, had lately been discovered alive on a rocky beach at Il-Maghluq, Marsascala, in the south of the island. Since then, I have received several more enquiries from other correspondents, and so too, it transpires,

Alien Zoo

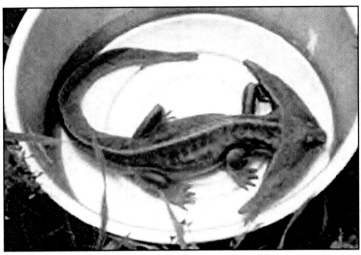

The so-called hammerhead lizard of Malta

have various other scientists, including Malta University biologist Prof. Patrick J. Schembri. In reality, what this intriguing photo depicts is an apparently plastic or resin model of a ancient prehistoric American amphibian called *Diplocaulus*, whose fossils date back to the lower Permian Period (roughly 270 million years ago).

Belonging to an extinct group of salamander-like amphibians termed nectridians, *Diplocaulus* was characterised by its head's two enormous lateral horns, out of all proportion to its body.

In short, if, as seems to be so, this photo is being put forward as a picture of a real, living beast, then it is a hoax, but *Diplocaulus* itself is a genuine albeit long-extinct animal. **Tonio Galea, various pers. comms;** *Times* **(Malta), 21 November 2004.**

STRANGE SONG OF THE DEEP. For the past 12 years, a highly mysterious baleen whale has been cruising the Northern Pacific Ocean every autumn and winter, undergoing a migration pattern unlike that of any known species of whale. Even more remarkable, however, is that its calls - or song - do not match those of any known species either. As confirmed by the US navy, which has been monitoring its song and those of other whales, utilising equipment normally used for tracking submarines, it sings at a frequency of around 52 hertz, whereas other baleens usually call at much lower frequencies, around 15-20 hertz. Since first noted in 1992, its call has deepened slightly with age, but is still very different from those of typical baleens, making its identity a subject for speculation - a previously unknown species, or merely a vocal freak with wanderlust? **Reuters, 8 December 2004.**

THE RIDDLE OF THE SANDS CONTINUES. In 2004, Czech explorer Ivan Mackerle launched his latest expedition in search of Mongolia's most extraordinary mystery beast - the death worm of the Gobi desert, claimed by locals to possess the power of electrocution, or something very like it. This time, Mackerle attempted a very different method of seeking this vast sandy expanse's most elusive inhabitant - employing a pilot in an ultra-light plane who spent many hours soaring above the desert terrain, thereby covering a much greater area than would be possible on foot, in the hope of spying a death worm resting on the surface of a dune. Sadly, however, none was spotted, but bearing in mind that the Gobi covers an area of roughly 500,000 square miles, words such as 'needle' and 'haystack' naturally come to mind, and Ivan should be congratulated for succeeding in monitoring as much of the terrain as he did. **Ivan Mackerle, pers. comm., 8 December 2004.**

KEEPING TRACK OF THE DUENDE. Reported from the Central American country of Belize, the duende or dwendi is said to be a small bigfoot-like bipedal man-beast, inhabiting Belize's dense forests. In December 2004, field investigator Ken Gerhard for the American Primate Conservation Alliance (APCA) returned from a search for this crypto-primate, and revealed that he had obtained four well-formed tracks that fit the duende's profile. APCA founder Chester Moore was sufficiently impressed by

Alien Zoo

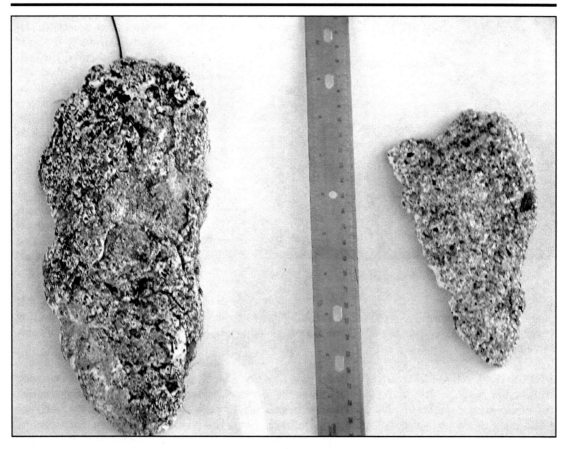

Possible duende tracks obtained by Ken Gerhard in Belize (Ken Gerhard)

the tracks to announce that he will be following up Gerhard's expedition with one of his own, now that there is evidence as to where such entities are living. **http://science-and-research.press-world.com/v/65359/belize-bigfoot-expedition-a-success.html 16 December 2004 [no longer online].**

TOMORROW THE PINK-HEAD? During a survey of wetlands in northern Myanmar between 29 November and 16 December 2004, members of the survey team - drawn from BirdLife International and a local non-government organisation called the Biodiversity and Nature Conservation Association (BANCA) - made a possible sighting of a pink-headed duck *Rhodonessa caryophyllacea*, a highly distinctive species deemed extinct since the early 1940s. Although the sighting lasted for no more than 3 minutes, and was in bright light, with the bird in flight, its observers discerned a bright pink head and neck. One of the eyewitnesses, waterfowl veteran Tim Appleton, is 99 per cent convinced that it was indeed a pink-head, and the others too were sufficiently impressed to decide to renew their search here in November 2005. **BirdLife International press release 20 January 2005, http://www.rutlandtimes.co.uk/detail.asp?aid=723 23 December 2004 [no longer online].**

The Lost Ark

THE ART OF CRYPTO-TWITCHING.

In Hiva-Oa, one of the islands of the Marquesas archipelago, in the middle of the Pacific Ocean, French painter Paul Gauguin and Belgish [sic] singer Jacques Brel spent the last years of their lives, and their graves can be found on that island. But cryptozoologists should also know it, as well as people fond of art, for in it dwells a bird unknown to science.

Michel Raynal – 'Paul Gauguin's Mystery Bird',
in *Cryptozoology and the
Investigation of Lesser-
Known Mystery Animals*

Twitchers seek to increase their life-lists of known bird species that they have seen by scouring the great outdoors with their binoculars. Crypto-twitchers, conversely, increase their lists of unknown bird species by much more eclectic means – such as perusing obscure periodicals and long-forgotten explorers' journals in libraries, questioning native tribes, and even - as demonstrated here - scrutinising the artworks of famous painters.

AUDUBON'S MISSING WARBLERS AND A LONG-LOST EAGLE

As for bird painters, they don't come much more famous than John James Audubon (1785-1851). His paintings are among the most spectacular, most documented, and most valuable bird illustrations of all time.

So it may be surprising to learn that his work includes some decidedly mysterious forms whose status

Alien Zoo

Audubon's painting of the small-headed flycatcher

remains contentious to this day. Three of these are the Blue Mountain warbler *Sylvicola montana* (an olive and yellow species of parulid or New World wood warbler), the carbonated swamp-warbler *Helinaia carbonata* (another parulid), and the small-headed flycatcher *Sylvania microcephala* (now also deemed to have been a parulid).

Bearing in mind that Audubon normally painted from preserved specimens, wired wherever possible in life-like poses, there must have been at least a single specimen of each of these enigmatic birds in existence at one time, but where these are now is unknown. The same, sadly, is also true of whatever species, subspecies, uncommon hybrid, or colour phase that these three birds respectively represented, as none has been spied since Audubon's paintings were created.

A fourth Audubon mystery bird only just escapes this ignominy. The reality of Townsend's bunting *Spiza townsendi* is based solely upon a single individual collected in New Garden, Chester County, Pennsylvania, in the 1830s, described by Audubon, and now held at the Smithsonian.

Audubon's painting of Townsend's bunting

No other specimen of this white-throated grey bird is known, and some ornithologists believe it to be either a hybrid of the blue grosbeak *Passerina caerulea* and the dickcissel *Spiza americana* or a colour mutant of the dickcissel.

Even more mystifying is Audubon's long-lost eagle, the magnificent Washington's eagle *Haliaeetus washingtonii*. Uniformly brown in colour and sporting a huge 10-ft wingspan, it readily surpassed in size both of North America's known eagle species – the golden eagle *Aquila chrysaetos* and the bald eagle *H. leucocephalus* – and the adult male individual painted by Audubon was one that he had shot after spying it scavenging at a pig slaughter near the village of Henderson, Kentucky, in or around the 1820s. The reality of Washington's eagle was initially accepted by naturalists, but by the time of Audubon's death in 1851 the first doubts had already begun to be voiced. Eventually, it was dismissed as a non-existent form based upon misidentified juvenile bald eagles (which lack the familiar white head of adults) – even though Washington's eagle was far bigger than any juvenile bald eagle. It also differed with regard to certain morphological features, such as the appearance of its cere (a fleshy swelling at the base of its

Audubon's painting of the Blue Mountain warbler

Audubon's painting of the carbonated swamp-warbler

Alien Zoo

Audubon's painting of his Washington's eagle

beak's upper mandible, containing the nares or nostrils) and the uniform scutellation (scaling) on its tarsi (lower legs). Even its flight, feeding behaviour, and nesting preferences were all markedly different. Nevertheless, today this spectacular bird is largely forgotten, but is this justified, and, whatever its taxonomic status, might it still exist in the more remote, little-visited backwaters of North America?

As recently as November 2009, I received a fascinating email concerning a modern-day sighting in a harbour around the Aleutian Islands off western Alaska of what the eyewitness (one of three) referred to as a gigantic bald eagle, far bigger than the other, normal-sized specimens also present there.

He didn't say whether it had a white head, like adult bald eagles, or was all-brown, like juveniles – but if the latter, this enigmatic bird would have borne more than a passing resemblance to Washington's eagle (see my Washington's eagle article in *FT262*, May 2010, for other recent sightings and full details of this impressive bird's chequered history).

GAUGUIN'S MAGICAL MYSTERY BIRD

No less celebrated an artist than Audubon is Paul Gauguin (1848-1903), though his fame lies far more with paintings of dusky South Sea Island maidens than with ornithological subjects. Having said that, however, it may well be that one of his paintings has considerable crypto-twitching significance.

One of his last works was painted in 1902 while on the small Pacific island of Hiva Oa in French Polynesia's Marquesas group, and is now at the Museum of Modern and Contemporary Art in Liege, Belgium. Entitled 'The Sorcerer of Hiva Oa', it depicts a tall man in a striking red cape standing near a forest - but what is most intriguing from a cryptozoological viewpoint is the brightly-plumaged bird portrayed in the painting's bottom-right corner, and seemingly held in place by a dog, which is seizing it with its jaws.

Remarkably, this bird looks very like the famous New Zealand takahe *Porphyrio mantelli*, the large

Gauguin's painting 'The Sorcerer of Hiva Oa'

Alien Zoo

The New Zealand takahe (Dr Karl Shuker)

flightless gallinule thought to be extinct until rediscovered on South Island in 1948. In recent years, specimens of this greatly-endangered species have been transferred to, and have successfully bred on, the small island bird sanctuary of Tiritiri Matangi, which I visited in November 2006 and where I was greatly privileged to see wild takahes at close range. Hence I can confirm that Gauguin's bird does indeed look very like - though not identical to - a takahe; the main difference is that the mystery bird's head is green, whereas the takahe's is dark blue.

But what could any such bird be doing far from New Zealand, on the tiny South Pacific island of Hiva Oa? No such species is known to exist here.

Nevertheless, two aspects of the painting clearly indicate that the bird was indeed native to this island. Firstly, the sorcerer depicted by it in 1902 was a famous Hiva Oa local of that time called Haapuani. Secondly, there is a distinct suggestion that the bird had been newly-captured on the island during a hunt, because Gauguin depicted it gripped by the jaws of a hunting-type dog.

What makes this painting so important cryptozoologically, as brought to attention by French researcher Michel Raynal in a number of his writings, is that Gauguin's rara avis compares very closely with descriptions of a still-undescribed, uncaptured species of bird reported on several occasions from Hiva Oa (it was even briefly spied there by the famous Norwegian voyager Thor Heyerdahl in 1937) and known locally here as the koao. Moreover, sub-fossil remains of an officially-extinct gallinule, *Porphyrio paepae*, have been uncovered on Hiva Oa, leading to the exciting possibility that this species and the elusive koao are one and the same. And perhaps, unknowingly, Gauguin has left us a unique portrait of this bird.

Close-up of the mystery bird in Gauguin's painting

Alien Zoo

SEEING RED ABOUT A PERPLEXING PARROT

George Edwards's painting of a Jamaican mystery parrot

George Edwards (1694-1773) was a very talented English bird painter and author, but his claim to cryptozoological fame lies with his painting of a most unusual parrot. Even the painting itself was unknown until fellow bird artist-author Errol Fuller spied it on the stall of an art dealer standing at the annual Olympia Antiques Fair in London during June 1996. Although very familiar with Edwards's work, Fuller did not recognise this painting, and after purchasing it he was able to confirm that it was indeed hitherto undocumented (in 1999, my book *Mysteries of Planet Earth* became the first publication in the world to reproduce it).

It portrayed a very mysterious 9-inch parrot, vaguely resembling in basic form one of the Amazon parrots, but primarily red (instead of green) in colour, and unlike any species known to science. At the bottom of the painting was an inscription:

> A very uncommon parrot from Jamaica.
> Drawn from Nature of the size of life by
> G. Edwards. July 1764.

A more detailed note penned by Edwards at that same time was present on the painting's reverse, revealing that the bird had been shot in Jamaica, and brought to England in dried form, after which it was loaned to Edwards by Dr Alexander Russell who now owned it. Moreover, even the native Jamaicans had never seen a parrot like this one before.

Tragically, the current fate of Russell's collection of zoological specimens, including his seemingly unique parrot, is unknown - though as taxiderm techniques were generally very unsatisfactory in those far-off days, his mystery parrot may well have decomposed long ago. But what was it? Could it have been a wholly novel, unknown species – or alternatively, in view of its apparent singularity, was it a freak red (erythristic) mutant of one of the familiar green species of Amazon parrot? Moreover, and regardless of its taxonomy, it may not even have been native to Jamaica; perhaps it was a pet bird brought by someone to Jamaica from elsewhere - hence the Jamaican people's unfamiliarity with it. Regrettably, we'll probably never know.

A COUPLE OF MYSTIFYING MACAWS

Staying with George Edwards and mystery parrots: I have long been intrigued by another painting associated with (though not prepared by) George Edwards - one that portrays a bird far removed from any parrot, yet which also has previously-unrecognised significance of a crypto-psittacine nature. One of the most famous depictions of the dodo *Raphus cucullatus* is the beautiful oil painting by Flemish artist Roelandt Savery (1576-1639), painted in Holland in 1626, which was once owned by Edwards, and is now at London's Natural History Museum. This great work has become virtually the 'standard' notion of

Alien Zoo

what the dodo looked like in life (though in more recent times, the veracity of its chubby form has been questioned), and has thus attracted great attention. In stark contrast, almost entirely ignored from an ornithological standpoint are the two very striking, colourful parrots that Savery painted to the dodo's immediate left and top-right in this same painting. Judging from their size and form, they are evidently macaws, but they do not resemble any known species.

Apart from its green facial skin, the left-hand macaw is almost totally red; and the top-right macaw is entirely green and yellow except for a black half-collar and white facial skin. Interestingly, Japan's dodo expert Masauji Hachisuka owned a later copy of Savery's painting that I have seen and which is a near-identical mirror image of it, except that the colours of the macaws are more distinct (and the dodo itself is more brown than grey), emphasising the red plumage of the one bird and (especially) the green and yellow of the other.

If these macaws were meant to be real, it suggests that they may be undescribed, lost species, or un-

Engraving based upon Savery's dodo painting and also featuring its two mystery macaws

Alien Zoo

Close-up of the mysterious red macaw in Savery's dodo painting

usual freak/hybrid individuals (but what happened to the specimens that they were based upon?). Or could they have been 'invented' by Savery purely as colourful support for the more prosaic plumage of the dodo? Or perhaps they did exist but Savery's paintings of them were based not upon physical specimens but instead upon inaccurate verbal descriptions of known species? Oddly, the top-right macaw looks very like a blue-and-yellow macaw *Ara ararauna* would look if its blue hues were completely replaced by green, so might it have been a rare colour mutant or even a preserved, sun-faded skin?

Although there is no species of macaw with almost exclusively green and yellow plumage alive today, two so-called green-and-yellow macaws, both now extinct, have been described and named from the West Indies. Having said that, one of these, the Jamaican green-and-yellow macaw *Ara erythrocephala*, which became extinct around 1842, also had a red head, blue wings, and a red-and-blue tail! But the other one, the Dominican green-and-yellow macaw *Ara atwoodi*, believed to have died out around 1800, was indeed predominantly green and yellow. So could it be that at least one such bird was brought back to Holland, perhaps as a pet, over two centuries earlier? And what of Savery's other mystery macaw? Various scientifically-unnamed red macaws seemingly existed several centuries ago on Guadeloupe and Hispaniola. Could a specimen of one of these brought back to Holland have been the model for his red macaw?

Several West Indian species of macaw became extinct during the late 1700s and 1800s. There may have been others (certainly one, the mysterious macaw of St Croix, known only from a single leg bone) that died out before their physical appearance had even been documented - except, perhaps, for a couple of perplexing portraits alongside, ironically, the most famous demised bird of all, the dodo?

SELECTED REFERENCES

DAY, David (1981). *The Doomsday Book of Animals.* Ebury (London).
FULLER, Errol (1999). Pers. comm., 22 February.
PARKES, Kenneth C. (1985). Audubon's mystery birds. *Natural History*, 94 (April): 88-93.
RAYNAL, Michel (1995). The mystery bird from Hiva-Oa. *In:* DOWNES, Jonathan (ed.) (1995). *CFZ Yearbook 1996.* CFZ (Exeter), pp. 63-74.
RAYNAL, Michel (2006). Paul Gauguin's mystery bird. *In:* ARMENT, Chad (ed.) (2006). *Cryptozoology and the Investigation of Lesser-Known Mystery Animals.* Coachwhip Publications (Landisville – Pennsylvania), pp. 115-36.
SHUKER, Karl P.N. (1996). Mystery bird of the Marquesas. *Wild About Animals*, 8 (September): 11.
SHUKER, Karl P.N. (1999). *Mysteries of Planet Earth: An Encyclopedia of the Inexplicable.* Carlton (London).
SHUKER, Karl P.N. (2010). Whatever happened to Washington's eagle? *Fortean Times*, no. 262 (May): 44-6.
WEIDENSAUL, Scott (1994). *Mountains of the Heart: A Natural History of the Appalachians.* Fulcrum Publishing (Golden – Colorado).

2005

Close-up of European eagle owl (Dr Karl Shuker)

EAGLE OWLS THRIVING IN ENGLAND. Despite a regular stream of reports over the years of European eagle owls *Bubo bubo* escaping from captivity in the UK yet rarely being recaptured, the possibility of these non-native giant owls - sporting a wingspan of nearly 6 ft and standing 2.5 ft high - breeding in the wild here has always been flatly denied by ornithologists, until now.

In January 2005, photographer Rachel Smith confirmed that a family of this species had been discovered living in secluded

Alien Zoo

woodland near Alnwick, Northumberland.

She had first spotted them in May 2004, and has been monitoring them ever since. The owl family comprised a female and two babies, living in a tree hollow. By July, the youngsters were starting to fly, and by now they will be nearing adulthood. ***Daily Mail*, 5 January 2005.**

IT'S ALL IN THE MOVES. Bearing in mind that mainstream science is all too often wary of showing even tacit interest in cryptozoological cases, it is a delight when such a case is formally examined by a mainstream authority, and especially so when reportable results emerge. An excellent recent example features Jürgen Konczak, associate professor in the School of Kinesiology (the mechanics of body movement) and director of the Human Sensorimotor Control laboratories at Minnesota University. Last winter, Prof. Konczak agreed to examine the famous footage shot by Roger Patterson of an alleged bigfoot striding across Bluff Creek, California. Although not confirming that the entity was an unknown species, he commented in the university's magazine, *Link*, that its walk was not that of a typical ape, as it exhibited some human qualities, yet was not a normal human walk either, concluding: "If it was a guy in an ape suit, he certainly did a good job trying to be peculiar".

Such a statement from someone so eminently qualified and experienced in kinesiology as Prof. Konczak is of particular note, not merely because it independently corroborates similar statements made in the past by Russian cryptozoologists and by the late Prof. Grover Krantz, a veteran bigfoot investigator, but also because, unlike these latter figures, Konczak has no professional interest in the subject of bigfoot. Thus his comments stem from a wholly unbiased, objective standpoint, based solely upon the evidence examined by him and thereby uninfluenced by anything else on record appertaining to bigfoot. Konczak's views were also included in a bigfoot documentary produced by local film-maker Doug Hajicek and aired last autumn on the Outdoor Life Network; it may be aired again this coming spring. ***Pioneer Press* (St Paul, Minnesota), 7 February 2005.**

THEY DUNG HIM WRONG. Sadly, however, other examples of scientific involvement in cryptozoological cases sometimes lead to a decidedly negative, disquieting outcome - as experienced lately by cryptozoologist Mike Williams, who for some time has been investigating reports of ABCs (alien big cats) across the Hawkesbury Range, in New South Wales, Australia. On several occasions, scat samples reputed to originate from these mystery cats have been obtained and sent for analysis to various scientific laboratories. Every time, however, they have been resolutely identified as canine, not feline.

Accordingly, Mike decided to put the accuracy (or otherwise) of such testing to a real test. He obtained (with permission and assistance) some scats produced by a black panther (melanistic leopard) *Panthera pardus* held in captivity at Bullens' Animal World in Warragamba, then sent them off as usual for identification (having deliberately refrained from stating what animal they were from), to two different laboratories - one of which, moreover, is a centre preferred by the State Government in its own investigations of such material. And when the results came back - yes, you've guessed it, both laboratories had identified the scats as canine.

In addition, fur samples obtained from the black panther, which Mike sent off to a third laboratory, were identified there as domestic cat. Not surprisingly, faced with such an extraordinary outcome, Mike now believes that the only hope of obtaining undeniable evidence for the existence of big cats in Australia is DNA analysis, concluding: "I can understand the fur result. Fur can be subjective...But with the scats, they had a stench of cat urine and gigantic fur balls with bones. For whatever reason, the experts used by the State Government are incapable of seeing the difference between felid and canid". ***Hawkesbury Gazette*, 17 February 2005.**

A bigfoot (Richard Svensson)

Alien Zoo

MERMAN OF THE MOONLIT SKIN. According to Iranian media sources, for the past two years a merman has been spied by Iranian sailors and residents of coastal regions around the southern and southwestern Caspian Sea. Referred to by sailors as the runan-shah ('master of the sea and rivers'), it was said to be roughly 5 ft 5 in tall and of strong build, with a protruding stomach, relatively squat legs and arms, flipper-like feet, four webbed fingers on each hand, black and green hair on its head, a sharply-pointed nose, large orbicular eyes, a fairly large mouth with a projecting upper jaw but no chin, and skin the colour of moonlight(!). It has also been supposedly sighted by Azerbaijani fishermen in waters linking the cities of Astara and Lenkoran during May 2004. ***Pravda*, 25 March 2005.**

SEA SERPENTS RISING TO THE OCCASION? One of Dr Bernard Heuvelmans's many postulated categories of sea serpent was the super-otter (which he considered to be a highly primitive, limbed whale), and one of the most important documents cited by him in support of its reality was an encounter in 1734 off the coast of Greenland by missionary's son Poul Egede with an erect cylindrical sea serpent, rising up out of the water. In a recently-published paper reappraising this classic sighting, however, Drs Charles Paxton and Sharon Hedley from St Andrews University's Centre for Research into Ecological and Environmental Modelling, in conjunction with Norwegian cryptozoological researcher Erik Knatterud, have proposed a very different, and memorable, alternative explanation. Namely, that what Egede actually spied was a known species of baleen whale but exhibiting a state of sexual arousal, with its prominent genitalia misidentified by Egede as the flukeless tail of a sea monster! Not a super-otter after all, therefore, but merely another case of mistaken identity - a veritable cryptozoological cock-up, in fact. ***Archives of Natural History*, 32: 1-9 (2005); *Courier* (Fife), 5 April 2005.**

Egede's super-otter

GIANT EELS AND DEADLY WORMS. English cryptozoologist Richard Freeman of the CFZ has been busy of late. In April, Richard, Lisa Dowley, and David Curtis spent some days at Loch Morar in

---

FACING PAGE 'Swamp Dweller' – Jade Bengco's spectacular merman painting (Jade Bengco)

Alien Zoo

A moray eel impersonating Nessie underwater! (Dr Karl Shuker)

the hope of enticing its cryptic monster Morag into view. Their bait was a mixture of mussels, fish guts, herring, cow liver, and a fish-attracting chemical called Van Den Eynd Predator Plus, placed inside permeable Hessian sacks, floating 20 ft or so beneath the water surface, attached to flotation devices. Richard believes that reports of monsters in various Scottish lochs, including Morar and Ness, may be due to the existence of extra-large, abnormal eels nicknamed eunuch eels because they are sterile and, instead of migrating to the sea to breed, remain permanently in freshwater where they continue to grow, attaining great sizes. However, although the flotation devices were successful, the bait was not touched. Undaunted, Richard is now pursuing a very different kind of vermiform cryptid, as the leader of 'Operation Deathworm' - a four-man month-long expedition in May to Mongolia's southern Gobi Desert to seek out, and perhaps even capture, some specimens of the legendary and reputedly lethal allghoi khorkhoi or Mongolian death worm.

As I discussed in my extensive *Fortean Studies* death worm article (Vol 4) back in 1998, one likely candidate for this mystery beast is a giant species of amphisbaenian or worm-lizard. These reptiles are actually harmless, but many other innocuous serpentiform creatures have been imbued with all sorts of fearsome if totally fictitious capabilities by superstitious locals, so it is possible that the death worm's formidable powers of venom-squirting and electrocution are merely apocryphal. Let's hope so - Richard and co plan to create localised floods by damming small streams in order to flush the worms to the surface of the sands. *CFZ Newsletter*, **24 April 2005, http://www.deathworm.co.uk [now http://cryptoworld.co.uk/projects/operation-deathworm-1/] 28 April 2005.**

NEW FIND FOXES SCIENTISTS. The discovery of a hitherto-unknown species of large mammalian carnivore is a rare find indeed, which is why scientists are so intrigued by claims that a new species of fox may have been uncovered in Kalimantan (Indonesian Borneo).

Snapped twice in 2003 by an automatic infra-red camera placed within the forests of the Kayam

Alien Zoo

Menterong National Park by WWF scientists studying zoological diversity here (but with the photos only made public now), the animal has a bushy tail but is foxy-red all over, with no typically vulpine white markings at all, and its back legs are slightly extended - leading to speculation that it may be at least partly arboreal. Scientists who have seen the photos, including staff at the Jakarta Natural History Museum, believe that this Bornean mystery mammal does constitute a new species, but more evidence is needed for confirmation. *The Independent*, **8 May 2005.**

A SECOND TYPE OF DEATH WORM? A most interesting snippet of information from its current Mongolian expedition seeking the Gobi's deadly death worm has reached the CFZ back home in Devon. Reporting from Dalanzadgad, team member Dave Churchill notes:

> We have also been told that there are possibly two different types of Death-worm, the second is smaller and grey in colour and is referred to by locals as Temeenii suul, literally translated as Camel's Tail. This could be a second species. But more likely the young or sexual dimorphism of the Deathworm.

Despite several expeditions to the Gobi in pursuit of the death worm, Czech explorer Ivan Mackerle has never reported the temeenii suul, so this is a very intriguing report, well worth further investigation by the team if the opportunity arises. *CFZ Newsletter*, **8 May 2005.**

KAYADI, NEW GUINEA'S FORGOTTEN BIGFOOT. Reports of large, hairy man-beasts have emerged from many regions of the globe, but not from the otherwise cryptozoologically-fecund island of New Guinea – until now, that is. In autumn 2002, during a visit to Papua New Guinea, American cryptozoologist Todd Jurasek stayed at the village of Siawi, roughly 18 miles from the border with Irian Jaya (Indonesian New Guinea), and whilst there he questioned the villagers concerning a number of cryptids. I am greatly indebted to Todd for kindly sharing with me recently the following information that he gleaned there concerning a hitherto-undocumented Bigfoot-like entity:

> This bigfoot-type animal was described by individuals of both the Siawi and Amto tribes. What's even more remarkable, is the fact that there are no known primates in either Irian Jaya or Papua New Guinea. The creature is basically described as a man size or larger (most natives are under 5'5"), bi-pedal, hairy primate which can climb trees very fast and can "physically throw people". Members of the Amto tribe called the beast, "Kayadi". An Amto man told me his uncle had startled a "wild hairy tall man" or Kayadi once while digging for pigeon eggs in cave near his village in 1981. A similar beast apparently kidnapped a local girl at some point. Needless to say, this was exciting news for someone interested in cryptozoology. Natives also told me of giant monitor lizards and snakes, however I was unable to isolate any specific useable material. The only other mysterious creatures I was able to solidly identify were not reptiles but mammals. These included a lion sized feline (there are no known cats in PNG) and a possible large canine species.

I have previously heard of reports from New Guinea of a feline mystery beast resembling the yarri or Queensland tiger, and the suspiciously thylacine-like dobsegna of Irian Jaya has been documented by me in several previous publications, but the only mystery primate reports from New Guinea that I had previously encountered were of small gnome-like entities, akin to the Little People of western tradition.

Alien Zoo

Wooden Papuan figurine of an orang utan-like entity at the Museum für Völkerkunde, Kulturen der Südsee, in Kiel, Germany (Markus Bühler)

Having said that, however, I do have on file a photograph snapped and sent to me by German cryptozoological enthusiast Markus Bühler of a mystifying wooden figurine from Papua New Guinea that depicts an orang utan-like entity (reproduced above). Could this be a representation of the kayadi? **Todd Jurasek, pers. comm., 24 June 2005.**

MONSTROUS FROM THE DEEP. An extraordinary 'monster' carcase was discovered on 20 July 2005 washed ashore near Yangshashan of Chunxiao Town in Beilun District, within eastern China's Zhejiang Province. Roughly 36 ft long and weighing approximately 2 tons, it was already decomposed, with its spine exposed, and characterised by a long thin head and a snout about 3 ft long, said to be crocodilian in appearance, with hair on its body, and orange stripes across its 10-13-ft-wide belly, but lacking a tail or lower limbs. The heavy skull and what has been referred to as the spine's coccyx had already separated from the rest of the body, and a large sample of tissue from the corpse was taken away by some young people. Hairy carcases are generally rotting sharks, with the 'hair' being nothing more than strands of exposed muscle fibres and connective tissue, but its orange-striped belly and crocodilian head do not lend obvious support to this identity. Instead, rather more exotic suggestions, such as an extra-large estuarine crocodile, or even a vagrant elephant seal, have been put forward. Having said that, an unnamed authority from Ningbo University's sea creature research centre has since discounted both of these - but without actually examining or even seeing the carcase himself. Surely,

Alien Zoo

this odd specimen would make a prime candidate for DNA studies, to end the controversy regarding its status. *Today Morning Express* **(China), 29 July 2005.**

FISHING FOR THE TRUTH? Staying in China, in mid-August an expedition was launched by Xinjiang Ecological Association to explore the long-running mystery of the monsters reputedly inhabiting the deep alpine waters of Lake Kanasi (aka Hanas) in China's northwestern Xinjiang Uygur Autonomous Region. Said to devour horses, cattle, and sheep that stray too close to the lake's shores, these monsters were first investigated back in the mid-1980s, by a team from the Xinjiang University Department of Biology. According to this team, they were found to be an enormous version of the taimen *Hucho taimen* (a species of salmon), growing to immense, previously-unsuspected lengths - sometimes exceeding 40 ft, and weighing over 4 tons. If confirmed, this would make them the world's largest freshwater fish, but so far the scientific world, at least beyond China, has not recognised their claim. Perhaps this latest expedition will bring back conclusive evidence - ideally a specimen - that will finally convince everyone. *China View*, **29 July 2005.**

A MOOSE ON THE LOOSE...IN NZ! Scotland has its Nessie, the USA has its bigfoot, and New Zealand has its moose - the only difference being that New Zealand's mystery beast is a mystery no longer. Back in 1910, ten specimens of moose *Alces alces* shipped here from Canada were released in what is today the remote Fjordland National Park of South Island. The hope was that they would multiply and spread like the imported European red deer *Cervus elaphus* had done, to provide good hunting for the country's sportsmen. However, these giant deer did not fare so well, for although they did initially breed, the last confirmed sighting of a NZ moose was in 1952, and even that specimen was shot dead. Nevertheless, there have been rumours of continued moose existence here ever since, paralleling the thylacine *Thylacinus cynocephalus* survival scenario across the waters in mainland Australia and Tasmania, but no conclusive evidence has ever been obtained - until now. Longstanding moose investigator Ken Tustin, who has been pursuing claims of this species' survival in Fjordland since the early 1970s when he found a pair of moose antlers here, recently discovered some moose-like hair snagged at waist height on the bark of a tree in the deepest, most inaccessible reaches of this region. Ruling out the highly unlikely possibility that someone had taken the considerable trouble to place it there deliberately, as a hoax, Tustin submitted it to Trent University's Wildlife Forensic DNA Laboratory for analysis, where it has now been announced that the sample has

Could there be a moose on the loose in Fjordland?

Alien Zoo

indeed tested positive for the presence of moose genes. Consequently, there may be a small but viable population of moose lingering on in this most remote, impenetrable locality after all, as it seems implausible that just a single moose is responsible for every report since 1952, over 50 years ago. ***Edmonton Sun*, 1 November 2005.**

A FATAL FLIGHT OF FANCY? North of St Petersburg, Russia, in the Vepskaya Heights is a small but much-dreaded, dense area of forested marshland, infamous for the discovery here of 16 naked human corpses since 1993, whose cause of death remains unknown. They show no signs of violence, but they appear to go mad before dying, as they strip all their clothes off, eat dirt, throw away edible foodstuffs that they have been gathering in the forests, and replace that food in their baskets with their own clothes. Several possible explanations have been aired by baffled investigators, including a serial killer, some strange form of fever, or the effects of ingesting toxic mushrooms or some other comparably deadly items while collecting their berries, fungi, etc. Perhaps the most intriguing suggestion of all, however, is that the dead people are the victims of a controversial, still-unidentified species of snake said to inhabit the bogs in this area.

According to local testimony collected since the 1980s, such snakes are readily distinguished by a fleshy growth on their head in male specimens that resembles a rooster's cockscomb (curiously, the equally mysterious crowing crested cobra of tropical Africa is also stated to possess this crest). In addition, they have a lethally venomous bite, are often found in the trees here, and can leap great distances from tree to tree, so that they are nicknamed the flying monsters. Only one known species of snake can glide in this manner - the so-called flying snake *Chrysopelea ornata* of southeast Asia. Suddenly, one unsolved mystery - the explanation for the dead bodies - has become two, with a creature of cryptozoology having been cited as that selfsame explanation. **http://english.pravda.ru/science/19/94/378/16500_snake.html 21 November 2005.**

A STING IN THE TRAIL. Eco-tourists visiting Australia often declare an interest in going on the trail of such dangerous native fauna as crocodiles and sharks. Now, at least according to Andy Dunstan, manager and research co-ordinator of Undersea Explorer, a Port Douglas-based tourism company that combines diving with scientific research on the Great Barrier Reef, they may well be enticed here in the future by a creature even deadlier and totally new - in every sense. For in the outer regions of the Reef, a hitherto unknown species of box jellyfish has been discovered, whose closest relatives include the deadliest multicellular animal species known to mankind. At least one person is already believed to have been a victim of this recently-exposed creature - a 27-year-old snorkeller who was stung on the face by a box jelly in late November 2005 at Ribbon Reefs in the far north of Queensland and soon afterwards suffered a combination of symptoms characteristic of the effects of a box jelly's sting. Normally, box jellies are only a problem near coastal regions, but with the discovery of this latest species, the open reef has now become a potential danger zone too for unwary divers. Up to 80 cm across, the new box jelly has tentacles that can attain a length of up to 1.5 m, and it is native not only to Australia but also to Hawaii. ***Advertiser* (Adelaide) 30 November 2005.**

A TANZANIAN TIGER? British scientist Tim Davenport made zoological headlines earlier in 2005 with his co-discovery of a hitherto unknown species of mangabey monkey in Tanzania - a creature well known, incidentally, to the local people who refer to it as the kipunji, but which was previously dismissed by scientists as a wholly mythical spirit beast. Moreover, he could well repeat his success this year, if he succeeds in uncovering the identity of another alleged spirit beast here, dubbed the Rongwe tiger. According to the locals, it is a large striped animal, a description not matching that of any known species from this area of Tanzania. Davenport concedes that it could be a striped hyaena *Hyaena hyaena* or aardwolf *Proteles cristatus* - albeit way out of its known range - but does not discount that it

Alien Zoo

might be a still-unknown species. After all, he has only to look at the kipunji to know that such a prospect is far from unrealistic in this remote African locality. ***The Guardian*, 7 December 2005.**

IT'S ALL IN THE NAME. As readers of my books *The Lost Ark: New and Rediscovered Animals of the 20th Century* (1993) and its successor *The New Zoo* (2002) may recall, loriciferan discoverer Prof. Reinhardt Møbjerg Kristensen (the loriciferans were the first of two entirely new phyla of microscopic multicellular animals revealed by him!) very kindly promised to name a new species of loriciferan after me - there were several new species that were still awaiting formal names and descriptions. I was recently asked by a correspondent whether 'my' loriciferan had ever been named and described, and I am happy to confirm that it has. It is *Pliciloricus shukeri*, and in case anyone wishes to read about Shuker's loriciferan, the bibliographical reference to the paper in which its description appears is as follows: **Iben Heiner & Reinhardt Møbjerg Kristensen (2005). Two new species of the genus *Pliciloricus* (Loricifera, Pliciloricidae) from the Faroe Bank, North Atlantic. *Zoologischer Anzeiger*, 243 (no. 3): 121-38.**

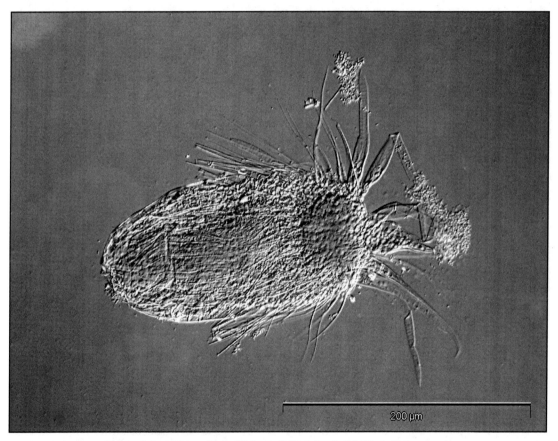

Shuker's loriciferan (Iben Heimer & Reinhardt Møbjerg Kristensen)

Not all deadly spiders are of the giant sci-fi monster variety seen here (Dr Karl Shuker)

The Lost Ark

ALONG CAME A SPIDER
- BEWARE THE PHANTOM FLESH-ROTTER!

Ah, venom mouth and shaggy thighs
And paunch grown sleek with sacrifice,
Thy dolphin back and shoulders round
Coarse-hairy, as some goblin hound
Whom a hag rides to sabbath on,
While shuddering stars in fear grown wan.

Lord de Tabley – 'The Study of a Spider'

As every ophiophobe will readily affirm, Australia is the only continent containing more species of venomous than non-venomous snake. Other equally dubious zoological delights facing the average Antipodean include the world's most venomous fish (stonefish), most venomous octopus (blue-ringed), and most venomous jellyfish (Flecker's sea wasp), plus the deadly funnel-web spider. More recently, yet another example of what happens when Mother Nature has a bad hair day Down Under has been attracting attention - an invisible horror dubbed the phantom flesh-rotter.

NECROTISING ARACHNIDISM – SEEKING THE SOURCE

Australia is home to many species of venomous spider. For some years, however, there have been widespread reports of a mysterious spider whose bite is even more devastating - producing ulcerating, cyanotic (discoloured bluish-grey) lesions that progress in some cases to such profound necrosis (rotting of skin and flesh) that even the underlying bone is exposed. Indeed, the spider's victim may

Alien Zoo

need to undergo plastic surgery in order for the extensive, disfiguring tissue damage stemming from the initial bite wound to be repaired.

This horrific condition is known medically as necrotising arachnidism, but recent research suggests that the Aussie spider clan may well have been taking the blame for the misdeeds of a much more insidious miscreant - i.e. the 'real' phantom flesh-rotter.

RED HERRINGS WITH EIGHT LEGS

Until as recently as the early 1990s, Australian zoologists placed the white-tailed spider *Lampona cylindrata* at the head of their list of suspects responsible for causing necrotising arachnidism. This was because the condition is widespread and so too is this species - a very common spider with a grey or black hairy body and white tip, which occurs in most parts of Australia, and is often found running around inside houses. In other words, a species likely to come into frequent contact with humans, and one which is also well known for biting.

White-tailed spider *Lampona cylindrata* (Chylld/Wikipedia)

Unfortunately for this convenient but unconfirmed identification, however, when victims arrived at hospitals or called into museums and described the spider that had inflicted their bites, their accounts did not recall this species' behaviour. Equally, with instances of spider bites where the white-tailed spi-

Alien Zoo

der was *known* to have been the species responsible (as in cases when the victim had shown the presence of mind to capture the offending arachnid and bring it along for identification), any effects resulting from such bites invariably proved to be short-lived and hardly visible. It thus became clear from these findings (and was duly confirmed via biochemical analyses, plus a study of 130 bites caused by this species published in 2003) that the white-tailed spider's venom was very weak, and could not induce the devastating tissue damage that epitomises necrotising arachnidism.

So what was the identity of the phantom flesh-rotter - could it be a species of spider still undiscovered by science? One of the world's leading experts on spiders and spider venom is Dr Struan Sutherland, who notes in his book *Venomous Creatures of Australia* (revised 1994) that new species of Australian wolf spider are regularly being discovered. In addition, the venomous fiddleback spider *Loxosceles rufescens* remained undescribed by science until as recently as 1974...when it was discovered in a western suburb of Adelaide, South Australia's capital city! Consequently, such a possibility would certainly not be out of the question.

In view of his professional interest in spiders, Dr Sutherland was well aware of the mystery surrounding the species inducing necrotising arachnidism, and he had referred to this matter at a congress of the International Toxicology Society held in Brisbane on 14 July 1982. Eight years later, moreover, he made an unexpected discovery that offered a highly significant yet hitherto-unsuspected insight into the condition's (very) secret agent.

THE MYSTERY SPIDER THAT NEVER WAS!

In 1990, Sutherland published a report revealing that after examining a necrotising wound allegedly inflicted by a white-tailed spider, he had successfully isolated from that wound a species of bacterium called *Mycobacterium ulcerans*, often found in soil. What makes this achievement so important in relation to the long-running saga of the phantom flesh-rotter is that if *M. ulcerans* infects a wound, the outcome is a large, expanding area of necrosis - whose appearance matches the outcome of necrotising arachnidism.

It was as if an opaque veil had suddenly been lifted. There was no unknown species of deadly spider awaiting exposure, because spider venom was not responsible for necrotising arachnidism after all. Instead, spiders were simply the vectors, the eight-limbed modes of transportation, for this dreadful condition - the true culprit was *M. ulcerans*.

Any species of spider dwelling in soily terrain is likely to harbour *M. ulcerans* on its large chelicerae (biting jawparts). Consequently, if a soil-inhabiting spider bites someone, this virulent bacterium will be transferred directly from the spider's chelicerae into the victim's bloodstream and tissues, duly initiating its all-too-familiar pattern of necrosis.

Wolf spiders (*Lycosa* spp.) are soil dwellers, and it is known that they have indeed been responsible for bites that subsequently developed necrotising arachnidism. Conversely, as the white-tailed spider is most commonly encountered by humans inside their homes, that may well explain why bites from this species rarely (if ever?) become gangrenous. After years of undeserved notoriety and suspicion, *Lampona cylindrata* finally seems set to be officially exonerated of the heinous crimes committed by *M. ulcerans* - the phantom flesh-rotter unmasked at last!

I wish to offer my especial thanks to arachnologists Dr Paul Hillyard from the Natural History Museum in London and Dr Robert J. Raven from the Queensland Museum for kindly supplying me with much-appreciated data and opinions relating to this subject.

SELECTED REFERENCES

ANON. (1989). Spiders inside houses. Queensland Museum information sheet, no. 39.

HILLYARD, Paul (1991). Pers. comms, 18 & 20 March.

ISBISTER, G.K. & GRAY, M.R. (2003). White-tail spider bite: a prospective study of 130 definite bites by *Lampona* species. *Medical Journal of Australia*, 179 (no. 4; 18 August): 199–202.

NICHOL, John (1989). *Bites and Stings: The World of Venomous Animals*. David & Charles (Newton Abbot).

RAVEN, Robert J. (undated, <1991). White-tailed spiders and the mystery necrotic lesions. Information sheet.

RAVEN, Robert J. (1991). Pers. comm., 4 July.

SHUKER, Karl P.N. (2001). *The Hidden Powers of Animals*. Marshall Editions (London).

SUTHERLAND, Struan K. (1994). *Venomous Creatures of Australia* (3rd rev. edit.). Oxford University Press (Melbourne).

2006

WORMING ITS WAY BACK. Not everyone may be overcome with joy to learn that a giant white earthworm measuring over 1 m long, able to dig 5-m-deep burrows and also spit (though not at the same time!), and smelling strangely like a lily is not extinct after all.

For earthworm fans and all conservationists, however, such news is extremely welcome - yes, the giant Palouse worm *Driloleirus americanus*, native to earth tracts on the Washington-Idaho border, has recently been confirmed to exist, after previously being written off as long-demised. In May 2005, Idaho University graduate student Yaniria Sanchez-de Leon dug a specimen up (inadvertently but tragically cutting it in two when doing so) while studying earthworms in the Smoot Hill Ecological Preserve near Palouse, Washington State. This is only the fourth specimen of the giant Palouse worm recorded during the past 30 years. *Spokesman-Review*, 1 February 2006.

LOOK OUT FOR THE LAVELLAN. For all its perceived faults and critics, *Wikipedia*, the free online encyclopedia, is a veritable treasure trove of information, including some fascinating crypto-snippets.

One such example that I chanced upon only recently is its short but tantalisingly-intriguing entry on a truly obscure cryptid known as the lavellan. Hailing from northern Scotland, and generally considered to be either a form of rodent or some kind of giant water shrew, it was said to be larger than a rat, lived in rivers and deep pools, and was very noxious - able to cause injury to cattle more than 100 ft away, albeit in an unspecified manner. Paradoxically, however, if a lavellan was killed or found dead, its skin

Alien Zoo

Was the lavellan some kind of giant water shrew?

would be preserved and used as a cure for sick beasts, by giving them the water into which the lavellan's skin had been dipped. This mystifying beast was referred to by some notable writers, including Scottish Gaelic poet Rob Donn and eminent naturalist Thomas Pennant. Whether it was real or merely mythical, however, is unknown; certainly, no such animal is apparently reported nowadays. So what - if anything - was the lavellan? *Wikipedia*, **accessed 23 April 2006.**

A STEGOSAUR IN CAMBODIA? Cryptozoological riddles can turn up in the most unlikely places, but few can be as unexpected as Cambodia's perplexing dinosaur carving. One of this country's most beautiful monuments is the jungle temple of Ta Prohm, part of the awe-inspiring temple complex at Angkor Wat and created around 800 years ago. Like others here, it is intricately adorned with images from Buddhist and Hindu mythology, but it also has one truly exceptional glyph unique to itself. Near to the temple's exit is a circular glyph containing the carving of a burly, small-headed, quad-

Cambodia's stegosaur petroglyph (John ┐ Lesley Burke)

rupedal beast bearing a row of diamond-shaped plates along its back - an image irresistibly reminiscent of a stegosaurian dinosaur! This anachronistic animal carving is reputedly popular with local guides, who delight in baffling western tourists by asking them if they believe dinosaurs still existed 800 years ago and then showing this glyph to them. Could it therefore be a modern fake, skilfully carved amid the genuine glyphs by a trickster hoping to fool unsuspecting tourists? Or is it a bona fide 800-year-old artifact just like the others? If so, perhaps it was inspired by the temple's architects having seen some fossilised dinosaur remains? After all, it surely couldn't have been based upon a sighting of a real-life stegosaur...could it? **http://www.unexplainedearth.com/angkor.php accessed 23 April 2006. [More details and photos re this mystifying petroglyph can be found in my book *Dr Shuker's Casebook* (CFZ Press: Bideford, 2008).]**

THE MAINE MINI-MAN. This curious report of a diminutive cryptic humanoid was sent to me by correspondent Ann Harper, from Abbot, in Maine, USA. Visiting Maine's Moxie Mountain during the berry season a few years ago, Ann and her husband Bob spotted what they thought at first was a baby moose or young bear in some bushes on the side of the road close by, eating berries. Suddenly, however, the creature caught sight of them, and abruptly rose up onto its hind legs and ran swiftly for cover into the forest. They only saw it briefly, but estimated that it stood no more than 3-4 ft tall when on its hind legs, and affirmed that it was definitely not a moose, goat, or bear. They have since travelled to Moxie a number of times in the hope of sighting this mysterious entity again, though so far they have not met with any success, but plan to travel there again during the berry season this year. **Ann Harper, pers. comm., 29 April 2005, 19 May 2006.**

BILI APE UNMASKED. After longstanding controversy, the mystery of the Bili ape's zoological identity has finally been resolved. Confined to the remote northern reaches of the Democratic Republic of the Congo (formerly Zaire), these huge, reputedly savage apes mystifyingly combine morphological and behavioural features normally characteristic of the gorilla with others more typical of the chimpanzee. Until now, this had led to speculation that they may represent a hitherto-unrecorded example of hybridisation between these two species, or even a previously undiscovered, totally separate species of great ape in its own right. However, following analysis of mitochondrial DNA taken from faecal samples, Dr Cleve Hicks and other Amsterdam University colleagues, who have also examined these primates' behaviour in the field, have announced that their findings all confirm that the Bili ape belongs to a known subspecies of chimpanzee - *Pan troglodytes schweinfurthii*. Presumably, the Bili ape's distinctive morphological features have evolved through its population's isolation from others of this subspecies but involve relatively little change at the genetic level. *New Scientist*, **30 June 2006.**

Luckily, not even the Bili ape attains these dimensions! (Dr Karl Shuker)

Alien Zoo

A MIXED-UP MYSTERY BEAST. What do you call an elusive creature that resembles a small deer but has a kangaroo-like head, big upright ears, and a long thin rat-like tail? This mixed-up mystery beast has been lurking in the wooded area between Shem Creek and Chuck Dawley Boulevard, South Carolina, for several months, and has even been photographed, but remains unidentified. One intriguing suggestion that has been offered is that it is a Sampson fox – a very rare, freak red fox *Vulpes vulpes* that lacks the normal thick outer layer of guard hair. Whatever it is, however, the local cats keep well away! *Post & Courier* **(Charleston, SC), 7 August 2006**

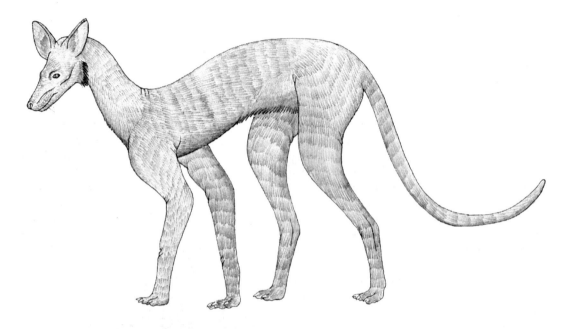

South Carolina's mysterious 'deer-fox', based upon eyewitness descriptions (Tim Morris)

A NETFUL OF SEA SERPENT. The physical capture of cryptids is a relatively rare event, which is why I am indebted to Michael Newton for bringing to my attention the following bygone Canadian newspaper report, published in the *Winnipeg Morning Free Press* on 17 September 1902, but seemingly hitherto-undocumented in the cryptozoological literature. In September 1902, a 48-ft-long sea serpent with 2.5-ft-long horns and 2-ft-long ears allegedly entangled itself in the nets of some fishermen off the coast of Japan. The creature, referred to in the report as a male, bellowed as they attempted to haul it ashore, but when they were unable to do so, they took guns from their schooners and shot it. As it thrashed madly in its death throes, what was claimed to be a female of this same undetermined species (and measuring 39 ft long) rose to the surface and lashed about furiously, until she too had to be killed. Their remains were then taken ashore and exhibited at Osaka. Scientists claimed that these Japanese sea serpents had probably been disturbed by seismic disruptions, causing them to abandon the deep sea for the shallower waters where they were encountered and killed. It would be very interesting to know if any records of these animals exist at Osaka. So if any Japanese *FT* readers are reading this right now... **Michael Newton, pers. comm., 13 August 2006.**

Alien Zoo

AHOY THERE, WINDY! What may well be Britain's newest water monster made its startling debut in August 2006. University lecturer Steve Burnip and his wife Eileen were holidaying in the Lake District, and were looking out from Watbarrow point across Lake Windermere, England's largest lake, when what he later described as a 15-20-ft-long serpentine, eel-like creature with a little head and two small humps following in its wake appeared at the surface. He likened it to a giant eel, and was very shocked to see it, especially as, unlike many Scottish lochs and even a few Welsh lakes, those of England are not generally known for sightings of humped elongate monsters.

Lancaster University fish ecologist Ian Winfield later opined that the creature may have been a wels catfish *Silurus glanis*, which is certainly an extremely large species, but one that in no way resembles an eel. For now, therefore, the mystery of the Lake Windermere monster, or Windy, as I propose that it should henceforth be called, remains unresolved. ***Westmorland Gazette*, 18 August 2006.**

Eyewitness Steve Burnip and a sketch map he drew for CFZ investigators (CFZ)

Alien Zoo

MYSTERY OF MAINE'S 'MUTANT' SOLVED. For many years, locals living in and around Turner in Maine, USA, have claimed that its woodlands are home to a savage mystery beast blamed for the savage killings of pets and other livestock. On 12 August 2006, the carcase of a large grey creature was discovered here, killed by a car while chasing a cat. Claims regarding its identity ranged widely - and wildly - including a wolverine, hyaena, supernatural wendigo beast, Tasmanian devil, mutant monster, and even the legendary chupacabra. When DNA samples taken from its decomposed remains were examined at a Toronto lab, however, it was found to be nothing more remarkable than a big domestic dog. *Lewiston* **(ME)** *Sun Journal*, **26 August 2006.**

The Maine 'Mutant' (Richard Svensson)

A MONSTROUS FILM. In late August 2006, a Japanese team of journalists arrived at Turkey's Lake Van to obtain eyewitness testimony regarding the monster that reputedly inhabits its waters. Just before their fortnight of research was over, however, the team was able to record its members' own testimonies - thanks to an unexpected appearance by the monster! They saw and filmed a large, seemingly animate object swimming 300 m or so offshore. Now they plan to return in 2007 to continue their research. **Anadolu News Agency, 8 September 2006.**

Alien Zoo

PHOTOGRAPHING BIGFOOT. A 'new' alleged photograph of a skunk ape, Florida's version of the legendary bigfoot or sasquatch, was part of an exhibition held in October 2006 at Florida's Museum of the Everglades. Although fuzzy, it appears to show a tall gorilla-like creature walking through Big Cypress National Preserve. The photo was snapped on a hot July afternoon five years ago by Judy Caseley, but she has only now publicly released it, as she was afraid at the time of attracting ridicule and disbelief. **WFTV, Channel 9 (Orlando), 19 October 2006.**

Bigfoot (Richard Svensson)

MARK K. BAYLESS DIES. Expert herpetoculturist and longstanding cryptozoological researcher Mark K. Bayless, of Berkeley, California, died on 1 November 2006 from diabetic complications. He was 46. In addition to his exhaustive knowledge and vast archive re varanid lizards, Mark had moderated an online discussion group re bigfoot, and had also amassed a very sizeable collection of cryptozoological material, especially relating to reptilian mystery beasts. Over the years, he had published a number of articles on such creatures, and he had recently begun work on what was planned to be a major book on this subject. Always more than happy to share his material and assist others in their researches, Mark will be greatly missed by his countless friends worldwide (including me) and by his numerous colleagues and fellow workers. **http://www.cryptomundo.com/cryptozoo-news/bayless-obit/ 3 November 2006.**

Alien Zoo

A MOA NO MORE. It's always good to be able to bring a seemingly-forgotten crypto-case to a close, even if the closing is rather less dramatic than the opening. On 19 May 1992, a supposed pair of German tourists, Franz Christianssen and Hulga Umbreit, visiting the Craigieburn area of New Zealand's Canterbury Range in South Island, wrote entries in its Bealey Hut accommodation's visitors' book claiming that they had been "very surprised to see two moas" in the Cass-Lagoon Saddle Area of Harper Valley. As might be expected, this startling claim soon attracted considerable international media attention, and later featured in many cryptozoological publications. What remained much less well-known until now, conversely, is that when moa researcher Bruce Spittle subsequently investigated these entries, he made a surprising discovery. As he recently revealed to me, two other entries in the Bealey Hut visitors' book, penned by two Canterbury University students, Andrew Reynolds and Jeremy Martin, on the same day as those of the Germans, exhibited very similar handwriting characteristics to the German entries. Consequently, Spittle pursued this curious affair by phoning Jeremy Martin, who admitted that Reynolds had written the German entries as a joke. This exposé was duly published in New Zealand's *Christchurch Press* newspaper on 19 August 1995, but until now had attracted very little overseas cryptozoological notice. **Bruce Spittle, pers. comm., 10 November 2006.**

A pair of moas, by Heinrich Harder

NESSIE IS NUMBER ONE SCOT. In a recent British survey, over 2000 adults across the UK were asked to give their opinion as to Scotland's most famous figure, present or past. Obviously, the surveyors expected someone like Sean Connery, Robert Burns, Ewan McGregor, William Wallace, or Robert the Bruce to top the poll. But no – although all of those names did indeed appear in the list, they were all trounced by a most unexpected name – Nessie, the Loch Ness monster! One can only assume that a fair few of those questioned had cryptozoological leanings! *The Scotsman*, **29 November 2006.**

Alien Zoo

VISITING NEW GUINEA'S JURASSIC PARK. A bold television crew from America's 'Destination Truth' television series headed by longstanding TV producer Neil Mandt is currently (November 2006) exploring the rainforests of New Guinea looking for an assemblage of mystery beasts that may seem more at home in Jurassic Park. These include monstrous reptiles said to resemble the huge dinosaur *Iguanodon*, and a terrifying pterodactyl-like winged beast called the ropen. We wish them well! **Bill Gibbons, pers. comm., December 2006.**

The famous 19th-Century *Iguanodon* statues at Crystal Palace Park in London – zoologically dated, but aesthetically pleasing (Dr Karl Shuker)

SECRET CIVET, OR SURPRISING SQUIRREL? Ever since the WWF released two colour photos in 2005 of a bushy-tailed, foxy-furred mystery mammal originally snapped in 2003 by an automatic infra-red camera within the forests of the Kayam Menterong National Park in Kalimantan (Indonesian Borneo), the question of the creature's likely identity has been hotly debated. At first, it was thought to be some form of carnivoran - either a completely new species of fox not previously recorded by science, or an obscure species of viverrid known as Hose's palm civet *Diplogale hosei*. Although this civet is indeed native to Borneo, its coat colouration is normally darker, however, than that of the mystery beast (unless the latter's red hue was due to the camera's flash?), and its ears are much smaller But now a radically different identity has been proffered. In a *Mammal Review* paper published online this month, Kalimantan-based zoologist Dr Erik Meijaard and two European co-workers have proposed that the Bornean mystery mammal was not a carnivoran of any type. Instead, they suggest, it was more likely to have been a large gliding rodent, most probably Thomas's flying squirrel *Aeromys thomasi* - a rarely-sighted species but whose morphology very closely matches that of the mystery beast. ***Mammal Review*, 36 (no. 4; 12 December 2006): 318-24.**

WHEN TWO HEADS WEREN'T BETTER THAN ONE. Two-headed terrapins, snakes, lambs, and other dicephalous mutants are commonly reported in the fortean literature, but a quite extraordinary example has come light that was around (albeit briefly) long, long before forteana, Charles Fort, or, indeed, *Homo sapiens* itself, had ever come into being. The creature in question is a specimen of an archaic fossil reptile called *Sinohydrosaurus lingyuanensis*, which was found, exquisitely preserved, in the Yixian Formation of northeastern China by a team of scientific researchers, and is approximately 145 million years old – thereby making it the oldest two-headed freak vertebrate currently known to science. Sporting a pair of perfectly-formed necks and heads (a teratological phenomenon know as axial bifurcation), sadly this unique specimen clearly did not benefit from its dual complement, because it appears to have been either a foetus or an individual that died at (or soon after) birth, as it measured a mere 7 cm long; normal adult specimens measured 1 m long. ***Daily Mail*, 20 December 2006.**

Alien Zoo

BIG CATS, BIG CATS, EVERYWHERE! One of the favourite claims made by cryptozoological sceptics against the possibility that ABC (alien big cat) sightings in Britain are due to big cats escaping or being deliberately released from captivity is that few such beasts are kept privately. Now, however, that claim has been totally shot down, thanks to figures newly-obtained under the Freedom of Information Act by the group Big Cats in Britain (BCIB). They show that 12 African lions *Panthera leo*, 14 tigers *Panthera tigris*, and no less than 50 leopards *Panthera pardus* (whose melanistic form is the familiar black panther) are being kept here as pets, plus dozens of specimens of smaller wild species such as servals *Leptailurus serval*, leopard cats *Prionailurus bengalensis*, etc. And these are only the ones being kept legally – obviously, illegally-kept wild pets have not been registered. So clearly there are more than enough captive specimens around for escapes/releases to occur from time to time. **Big Cats in Britain press release, 26 December 2006.**

A leopard cat (Dr Karl Shuker)

A serval, just one of several species of wild cat being kept as pets in Britain (Dr Karl Shuker)

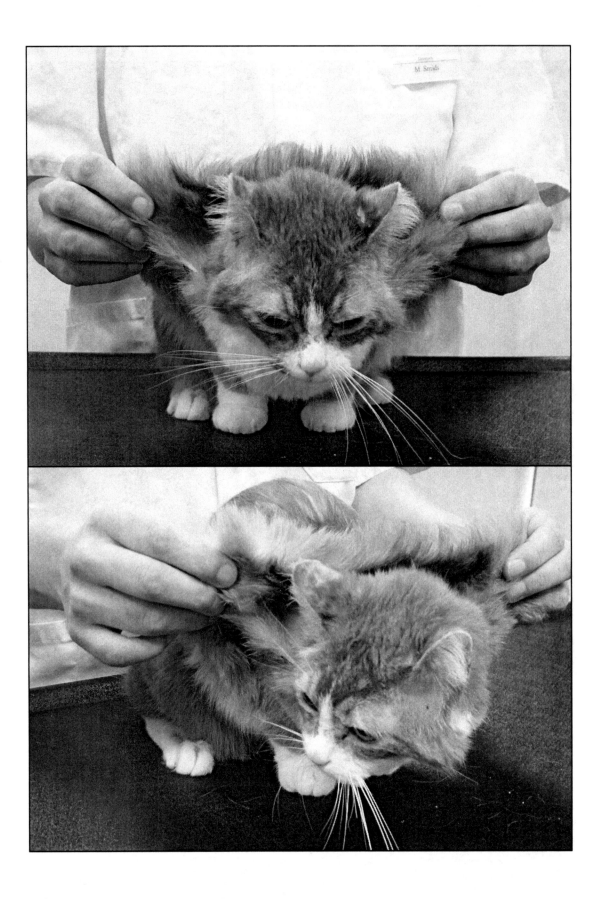

The Lost Ark

BRINGING THE WINGED CATS DOWN TO EARTH.

So Dame Nature does her best to fulfil every cat's ambition to get up amongst the birds and to confirm Darwin's theory about the survival of the fittest...

Sid Birchby – 'Winged Cats', *Fortean Times*, Summer 1981.

The subject of winged cats has fascinated me for many years, and I have researched it very extensively. Indeed, my book *Dr Shuker's Casebook: In Pursuit of Marvels and Mysteries* (2008) contains the most comprehensive survey of these anomalous felids ever compiled and published (a survey that will be updated and expanded even further within one of my forthcoming books, *The Anomalarium of Doctor Shuker*, scheduled for publication within the next 12-18 months). During that time, I have published a number of articles and news reports in *FT* concerning such animals.

Two of these articles, however, are of particular importance – because it was in one of them (published in the December 1994/January 1995 issue) that I exclusively revealed to the world the long-awaited explanation for the development of these cats' wings; and it was in the other one (published in the January 2009 issue) that I presented the hitherto-undocumented but truly extraordinary history of a set of famous and not-so-famous photographs depicting one of the best-known winged cats on file. Conse-

FACING PAGE: Martine Smids's winged cat, Prul, exhibiting FCA (Martine Smids)

quently, I am reproducing both of these articles here, in their original, unedited form (with the addition of a significant later case in the first one), as a unique double-bill of delights for fellow winged cat aficionados everywhere!

CAT FLAPS.

The phenomenon of winged cats - feline wonders with extraordinary wing-like extensions of skin and fur sprouting incongruously from their shoulders, haunches, or back - has been documented in various publications over the years[1], providing the case-histories of several authentic examples, but the explanation for these anomalous animals has never been forthcoming. During my own investigations into the mystery of the winged cats, however, I discovered that the answer has actually been available for many

A typical winged cat

Alien Zoo

years, concealed within the veterinary literature, but not recognised for what it was - until now.

The medical condition responsible for cats with wings is in fact an obscure, inherited skin disorder known as feline cutaneous asthenia (FCA) - literally translated as 'weak skin in cats'. A comparable condition has also been recorded from dogs and mink, as well as from cattle and sheep - in which it is referred to as dermatosparaxis ('torn skin'). There is even a human equivalent - Ehlers-Danlos Syndrome (EDS) - occurring in at least seven different forms.[2]

In cats exhibiting FCA, the skin is hyperextensible. That is to say, it is so elastic that it can readily stretch to yield furry wing-like extensions - thereby creating 'winged cats'. Not only is it hyperextensible, it is also abnormally fragile (friable) - tearing at the slightest contact with anything sharp, such as the cat's own claws when grooming, or any rough surfaces against which the cat might choose to rub itself. Curiously, however, such lacerations do not result in bleeding - instead, the skin simply peels away and drops off.[2] Needless to say, this fully explains those occasions on record when a winged cat has suddenly moulted its 'wings' - in reality, these pieces of stretched skin had simply sloughed off.

In mammals, the skin comprises two principal layers. The surface layer is the epidermis, and is relatively thin. Below this, however, is the dermis, which is much thicker, and contains the creature's connective tissue - providing support and packing, as well as containing the nerves and blood vessels. It consists largely of fibres, and most are composed of a protein known as collagen. This binds the dermis's cells together, but cats exhibiting FCA possess defective collagen in certain areas of their body's skin, which is unable to carry out effectively the connective tissue's packing function.[2]

As a result, the skin of such cats is exceptionally flexible and fragile in these areas - usually the shoulders, back, and haunches, the sites where 'wings' have arisen in the various winged cats on record.[2] Furthermore, if they arise in regions containing sufficient musculature, they can even be moved slightly, thus explaining reports describing cats that could lift their wings. (Conversely, the supposed ability of certain winged cats to become airborne owes more to foaflore than flight mechanics!)

A still-obscure, little-studied condition, only a few FCA cases are currently recorded in veterinary publications. In 1970, Kent vet Peter Fitchie received a female tabby cat, approximately 5 months old, for spaying - but when he attempted to anaesthetise it via intravenous injection, to his horror the skin immediately split, and the same thing happened in its flank when he shaved it. More splits occurred when he tried to sew up the first two, but he eventually succeeded using a round-bodied needle. Despite their dramatic formation, however, these slits all healed in a straightforward manner.[3]

In 1974, the case of a four-year-old tom cat said by its owner to have always had 'fragile skin', which was brought in for treatment to the Small Animal Clinic of Cornell University's New York State Veterinary College, was documented by Dr D.V. Scott. He noted that its skin was exceptionally thin and velvety in texture (typical of FCA specimens), extremely hyperextensible, and marked with a criss-cross network of fine white tear scars upon its body. When fur was clipped from one of its forelegs to take a blood sample for analysis, the skin immediately peeled away, and the same occurred whenever the slightest pressure was applied elsewhere to the cat's skin. Skin studies revealed that its dermis's collagen fibres were fragmented, irregular, and disoriented, with hardly any normal fibres present.[4]

An adult female cat noted in 1975 by W.F. Butler from Bristol University's Anatomy Department also had very fragile body skin (although its head and limbs' skin was much less so), with particularly low collagen levels in the dermis of its back's lumbar region.[5]

Alien Zoo

So far, the genetic nature of skin hyperextensibility and fragility in the cat had not been discovered, but in 1977 Drs Donald F. Patterson and Ronald R. Minor - vets from the University of Pennsylvania's School of Veterinary Medicine - were able to uncover the answer. They had received for treatment a 6-month-old male cat, grey in colour with short hair, that had severely lacerated its skin simply by scratching itself with its paws. When examined, its skin was found to be extraordinarily delicate and elastic, and when the fur upon the dorsal lumbar portion of its back was gently lifted during examination, it extended to a distance above the spine that was equal to 22 per cent of its body's entire length![6]

Their paper describing this remarkable cat contains a photograph depicting that action. The result is a classic winged cat, identical to those documented in fortean reports.

Due to its skin's fragile nature, the cat's owner decided to donate it to the veterinary school, where it afterwards lived and bred with four 1-year-old female cats (long-haired). Studies of the resulting kittens revealed that this skin condition was inherited as a non-sex-linked dominant trait, and all of the affected kittens exhibited packing defects in the collagen of their skin's dermis.[6]

None of the winged cats recorded in fortean publications has been documented in the veterinary literature. Equally, none of the FCA cases in veterinary accounts had previously featured in fortean reports. Happily, my interest in both of these disciplines enabled me to make the crucial connection between them and thereby uncover the winged cats' secret.

In early 2008, I was contacted by Dutch vet Martine Smids, who actually owns a male winged cat with FCA.[7] Below is my original documentation of this highly significant case as it appeared within my earlier book, *Dr Shuker's Casebook* (2008):[8]

In an email to me of 24 January 2008 (the first of several that we subsequently exchanged, discussing her cat, Prul), Martine provided the following information:

> It happens to be, that I own a so called 'winged cat'. It was brought to our practice in November 2005 at the age of 6 months. He was brought in because he had lost the complete skin of his tail, in a fight with a dog. The tail needed to be amputated. The owner didn't want to make a lot of costs and decided to leave the cat with us (this instead of euthanasing the cat immediately). The cat didn't seem to be sick or unhappy, so I decided to keep the cat, although I didn't know what was wrong with him at that point. After weeks of diagnostic investigations and talking to veterinarian dermatologists, the diagnosis of Cutis [sic] Asthenia was made.
>
> Now, 2 years later, the cat is doing fine. Of course he is an indoor cat and he wears most of the time a little baby-sweater to protect his skin. I have 3 more cats and a dog, this doesn't seem to be a problem. But this is probably due to the good character of the cat. Whenever he has skin lesions, I treat them very easy with agraffes (staples), he doesn't need sedation for this and even the largest wounds heal within a week. His wounds don't bleed and don't seem to hurt really. I've stopped giving him antibiotics for every wound that he has, because it didn't seem to make a lot of a difference. In the last 2 years, he only had one infected wound, that formed a small ab-

scess. He also has hip dysplasia. The first months that I had him, he seemed to have problems with this. His hips subluxated spontaneously when he walked. This causes him to walk lame, non responsive to NSAIDs. After a few months, he started to walk normally again and he doesn't really have problems with this ever since. The last year he even seems to be getting stronger, he can jump up much better, although not as good as a normal cat.

Now I am writing a case-report about him for the Dutch Veterinarian Magazine (*Tijdschrift voor Diergeneeskunde*).

In her next email to me, of 6 February, Martine included some additional details:

Prul hasn't always 'wings', he only really has them, when he has been licking on a certain spot for a long time, his skin stretches then into folds,

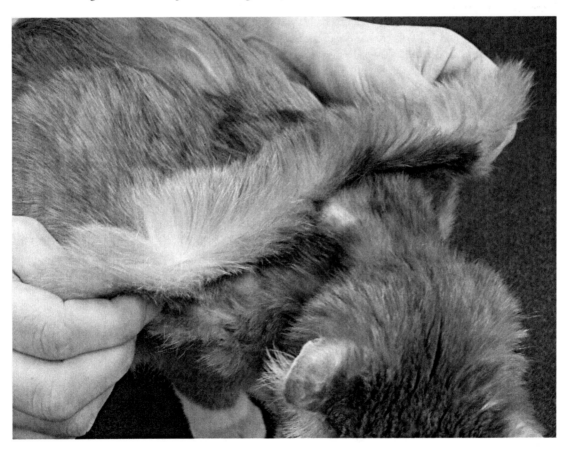

Close-up of Purl's 'wings' (Martine Smids)

Alien Zoo

sometimes his skin tears. These folds usually disappear after a while, when he stops licking. I've made some pictures with his skin stretched out.

In later emails, Martine did indeed enclose a number of photos, including the ones reproduced here with Martine's kind permission.

Also of interest is an old newspaper report of a winged rabbit! Dating from 1911 (and containing a photo by J.E. Stutter of Colchester), it reads:[9]

Mr. William Walton, of Morant Road, Colchester, is the owner of an extraordinary freak of nature - a winged rabbit. This animal, which is about six months old, is a great pet in the home. Its wings, which seem to take no regular form [a clear indication of cutaneous asthenia], proceed from near the two fore feet, and are about seven inches in length and in width about two inches at their widest part. They are fed with blood and resemble pieces of flesh. These wings are covered with white fur similar to the fur of the animal. At times this wonderful rabbit will sit in front of the fire and flap its wings, and it has been known to ascend on to a chair [by becoming hare-borne, no doubt!]. When the fur which covers the head is ruffled, it somewhat resembles an owl.

Finally: the earliest report of a winged cat currently contained within the fortean literature [this is no longer the case – see *Dr Shuker's Casebook* for several earlier reports[8]] appears to be the kitten from Wiveliscombe, Somerset, whose now-familiar photograph was published by the *Strand Magazine* in 1899.[10] During my own researches, however, I uncovered an odd report that pre-dates this by 31 years - assuming, of course, that the strange creature featured in it really is a winged cat. The report is as follows:[11]

The following is extracted from a late paper:-

"A nondescript animal, said to be a flying cat, and called by the Bhells *pauca billee*, has just been shot by Mr. Alexander Gibson, in the Punch Mehali [India]. The dried skin was exhibited at the last meeting of the Bombay Asiatic Society. It measured 18 inches in length, and was quite as broad when extended in the air. Mr. Gibson, who is well known as a member of the Asiatic Society and a contributor to its journal, believes the animal to be really a cat, and not a bat or a flying-fox, as some contend."

Cat or not, it certainly sounds batty to me!

I wish to thank felid geneticist Roy Robinson most sincerely for pointing me in the right direction during my quest for a solution to the winged cat mystery.

REFERENCES

1. MICHELL, John & RICKARD, Robert J.M. (1982). *Living Wonders*. Thames & Hudson (London); WOOD, Gerald L. (1984). *Guinness Book of Pet Records*. Guinness Superlatives (Enville); BORD, Janet & BORD, Colin (1987). *Modern Mysteries of Britain*. Grafton (London).
2. ROBINSON, Roy & PEDERSEN, N.C. (1991). Chapter 2. Normal Genetics, Genetic Disorders, Developmental Anomalies and Breeding Programs. *In:* Neils C. Pedersen (ed.) (1991). *Feline Husbandry: Diseases and Management in the Multiple-Cat Environment*. American Veterinary Publications (Goleta - California).
3. FITCHIE, Peter (1972). The Ehlers-Danlos Syndrome. *The Veterinary Record*, 90 (5 February): 165.
4. SCOTT, D.V. (1974). Cutaneous asthenia in a cat, resembling Ehlers-Danlos syndrome in man. *Veterinary Medicine/Small Animal Clinic*, 69 (October): 1256-8.
5. BUTLER, W.F. (1975). Fragility in the skin of a cat. *Research in Veterinary Science*, 19: 213-16.
6. PATTERSON, Donald F. & MINOR, Ronald R. (1977). Hereditary fragility and hyperextensibility of the skin of cats. A defect in collagen fibrillogenesis. *Laboratory Investigation*, 37: 170-9.
7. SMIDS, Martine (2008). Pers. comms, 24 January, and 6, 7 (several), 13, & 18 February.
8. SHUKER, Karl P.N. (2008). *Dr Shuker's Casebook: In Pursuit of Marvels and Miracles*. CFZ Press (Bideford).
9. ANON. (1911). A rabbit with wings. *Essex County Standard, West Suffolk Gazette & East Counties Advertiser*, 16 December, p. 1.
10. ANON. (1899). Can a cat fly? *Strand Magazine*, 18 (November): 599.
11. R.B.W. (1868). Flying cat. *The Naturalist's Note Book*, p. 318.

WINGED CATS AND FORTEAN PHOTOS.

As noted earlier, one of my most recent books, *Dr Shuker's Casebook* (2008), presently contains the most comprehensive survey of winged cats ever published, and includes the case histories of all of the significant, classic examples. Of these, the most famous one must surely be the Manchester winged cat, whose photographs have appeared in a vast range of publications. However, the extraordinary, decidedly fortean history of those photographs has never been fully revealed, nor has the entire set of photographs themselves ever been published - until now.

I first saw photographs of winged cats back in the 1980s, when I read the concise account of them that featured in *Living Wonders* (1982) by John Michell and *FT*'s very own founder, Bob Rickard. One of the cats documented was the Manchester specimen, accompanied by a photograph (hereafter Photo #1) depicting it sitting between two kneeling men, each holding up one of the cat's 'wings', and with the cat's head pointing to the right. Its history was recalled in the photo's caption:

> A reader's letter to the *Manchester Evening News*, 23 December 1975, requested the editor to print again a photograph which had been shown in the paper some years earlier of a 'winged cat which appeared in the yard of Banister Walton ┐ Co., Trafford Park, Manchester, and made its home

Alien Zoo

there for some years.' The writer of the letter, Mr H. Bootles, said that the people now working with the firm were sceptical about winged cats. The editor put an end to their doubts with this photograph.

A few years later saw the publication of Janet and Colin Bord's book *Modern Mysteries of Britain* (1987), in which they also documented this specimen, and commented that in addition to its wings it was also unusual on account of its noticeably broad, flat tail, which could clearly be seen in the accompanying photograph (hereafter Photo #2). Interestingly, although this photo was very similar to the one (Photo #1) that had appeared in *Living Wonders*, and showed it sitting between two men holding up its wings, it was not identical, because this time the cat's head faced slightly to the left. Moreover, it was this photo (Photo #2) that subsequently appeared in many other publications. In contrast, the only other occurrence of the *Living Wonders* photo (Photo #1) that I am aware of was in Gerald L. Wood's *Guinness Book of Pet Records* (1984).

As *FT* readers will know, by now I had become very interested in the winged cat phenomenon, which at that time was unexplained. Over the years, I went on to publish several updates in *FT* (articles in *FT*78 and *FT*168, plus various Alien Zoo items) and elsewhere as I uncovered more and more cases – including my eventual discovery of the solution to the phenomenon itself (revealed in my *FT*78 article [and reprinted above in this present book]). Namely, true winged cats (as opposed to some that merely have clumps of matted fur) suffer from a rare genetic skin condition called feline cutaneous asthenia (FCA), in which the skin is hyperextensible – i.e. readily stretching out into long wing-like furry extensions on the back, shoulders, or haunches of these cats, and which, on account of containing connective and neuromuscular tissue, can even be actively raised and lowered. Moreover, if (as often happens) these delicate 'wings' eventually break off, there is no bleeding or wounding; instead, it simply looks as if the cat has shed them.

During my ongoing winged cat researches, I sought to obtain as much published evidence of their existence as possible, and so one of the items that I was particularly keen to trace was the original *Manchester Evening News* (hereafter *MEN*) article mentioned by Mr H. Bootles in his letter of 23 December 1975 to that newspaper. Writing to its editor in 1998, I explained my interest and investigations, I provided what information I had concerning that *MEN* article, and then waited to see if this would be enough for it to be uncovered. Happily, I had indeed sent sufficient details, because not long afterwards I received not only a photocopy of the long-awaited article but also some large prints of the two photographs that had been reproduced in it – only to discover that the history of the Manchester winged cat was infinitely more fortean than I had ever suspected (even for anything already as fortean as a cat with wings!).

The original *MEN* article had been published on 11 August 1972, was entitled 'The cat with wings', and told a truly remarkable story. Its unnamed reporter had apparently been contacted by a lady called Mrs Beatrice Glancey, of Perryman Close, Hulme, Manchester, who had written in asking if the newspaper would be interested in a cat with wings. Naturally intrigued, the reporter confirmed that it would, and Mrs Glancey duly turned up, with two eyecatching photos of what is nowadays known as the Manchester winged cat, stuck on thick card with a typewritten caption providing the now-familiar information concerning this specimen. Namely, that it had wandered into the builder's yard (unnamed in the caption) as a kitten, had been adopted and fed by the foreman, and from an unusual growth of fur on its back had developed its wings after 12 months, growing from its shoulder blades and measuring about 11 inches from the shoulder bone to tip, including a kind of 'wing joint' (probably a lump of gristle). The cat's unusually broad, flat tail was also mentioned.

Photo #1 of Manchester winged cat, in *Living Wonders*

Photo #2 of Manchester winged cat, in *Modern Mysteries of Britain*

Above the *MEN*'s report, two photos of the cat were published. The left-hand photo was the now-famous picture (Photo #2) featured in the Bords' book and numerous other publications since then. However, the right-hand photo (hereafter Photo #3) was one that I had never seen before, showing the cat looking to the right but squatting down, with its body pointing straight ahead and its front paws splayed widely apart. This was a totally different pose from the version captured in the *Living Wonders* photo (Photo #1), in which the cat was not only looking to the right but was sitting (not squatting) with its body pointing to the right as well, and with its front paws close together. The 'squatting' photo showed two men holding its wings up (but was heavily cropped - see later).

Photo #3 of Manchester winged cat, in *Manchester Evening News* report of 11 August 1972

Suddenly, therefore, there were three different photos of the Manchester winged cat on record, not just two. Moreover, as was noted in *Living Wonders*, in response to Mr H. Bootles's letter of 1975 the *MEN* editor had published the winged cat photo (Photo #1) that subsequently appeared in *Living Wonders* (with the cat's head and body pointing to the right) but which had neither appeared nor been mentioned in the original *MEN* report of 1972 – so how had this third photo reached the *MEN* offices? But the strangest aspect of all was yet to come, and included some additional still-unresolved mysteries.

The *MEN*'s report stated that Mrs Glancey claimed to have found these two photographs (Photos #2 and #3) and caption near an area of Manchester called The Shambles, lying in the street! In other words, the history of what is now the world's best known winged cat came to light by the most random and unlikely of chances – the serendipitous discovery of some dropped or discarded photos. Yet this still was not the end of the story – on the contrary, it turned out to be just the beginning of yet another

Alien Zoo

new chapter.

In both of the photographic prints kindly sent to me by the *MEN*, it was clear that some retouching had been made to those used in its article. For the most part, this consisted of nothing more than delineation of the cat's limb edges, whiskers, and cheek fur using a white pen, and increased definition of the fingers of the man holding up the cat's left-hand wing. However, there was also a rather more puzzling, opaque 'ghost' effect in both photos, present between the two outstretched wings, sharpening the edge of one wing, and (in the 'squatting', right-hand photo, Photo #3) increasing their symmetry by obscuring what initially seemed to be the distal portion of the right-hand wing. (In fact, as I was to discover later – see below – it wasn't the wingtip that had been obscured, but the cat's tail, which the cat had lifted up in this particular photo.) It also blanked out a portion of rope visible in the other, left-hand photo (Photo #2, the one published in the Bords' book and elsewhere since). Consequently, on 4 July 1998 I emailed a high-resolution TIF file of each of the two photographic prints to Janet and Colin Bord at the Fortean Picture Library (FPL) to ask their opinion as to the significance, if any, of these modifications. Replying in a fax of 4 July and email of 6 July, Janet informed me that she and Colin agreed that the photos had indeed been heavily 'doctored', but did not feel that there was anything to be concerned about regarding this, as before modern technology arrived it was commonplace to improve the contrast of photographs intended for newspaper reproduction by means of retouching by hand, deleting or toning down superfluous details (such as the rope) and enhancing features that weren't clear. However, the fact that these photos had been found lying in a street remained perplexing, unless their original owner had simply dropped them? But who was their original owner?

On the reverse of the two prints sent to me by the *MEN* was a name and address, given as their copyright holder. The name was a Mr L.G. Morris, of Cheadle Heath, Stockport, but Janet informed me that she was already aware of this, and in her fax to me of 4 July she explained how she had first come in contact with the Manchester winged cat photos' current owner. In January 1986, while she and Colin were preparing *Modern Mysteries of Britain*, Janet had written to the *MEN* to obtain a print of one of the two Manchester winged cat photos that had appeared in its article, for inclusion in their book. In reply, the newspaper had informed her that the photo's copyright was owned by Mr L.G. Morris, and supplied her with his address. Janet had duly written to him, but did not receive any response, so she had then contacted Bob Rickard to ask from where he had obtained the photo (Photo #1) that had appeared in *Living Wonders*.

Bob had told her that he had been sent a print of it by a Mr J.M. Anderson who lived in the same area as Mr Morris. Janet had then written to Mr Anderson, but had presumably received no reply, as she made no further mention of him in her fax to me, telling me that she had gone ahead and used Bob's print of Photo #1 (which he had passed on to her) in their own book. She had then written to Mr Morris again, asking for his invoice for use of the photo in it. Mr Morris had replied this time, stating that he did not want payment but would like a copy of their book, and would also like the print returned to him.

Oddly, his letter had been signed 'J. Morris', not 'L.G. Morris'. However, after Janet had written back to him, asking if he would be prepared to place the photo in the FPL, in his reply Mr J. Morris (the 'J' standing for John) had revealed that L.G. Morris was his father, who had died some time ago, leaving "the picture" (NB - singular, not plural) to him. John Morris stated that he was happy for the FPL to act as agents for it, since when Janet had been sending him payments for its use but had not heard from him again. Now that the 'squatting' photo (Photo #3) had come to her attention via my procurement of the *MEN* article, however, she decided to write to him once more and ask if he had a print of this latter picture, and also whether he had ever seen the cat himself. Before moving on to this next stage in the Manchester winged cat photos' history, however, there is another intriguing mystery needing to be

Alien Zoo

noted here. In her fax to me, Janet mentioned that the photo she had used in *Modern Mysteries of Britain* (Photo #2) was the one sent to her by Bob Rickard that had appeared in *Living Wonders* (Photo #1). However, as I have already pointed out here, these two photos are visibly different – the former (Photo #2) shows the cat looking slightly to the left, the latter (Photo #1) shows it looking to the right (and no, they are not horizontally-flipped, mirror images of one another). So I assume that there must have been two photos (Photos #1 and #2) around at the time of the preparation of *Living Wonders*, and that Janet had received the one that hadn't been used in it (Photo #2).

In any event, both were obviously owned by John Morris, because in the picture credits section of *Living Wonders* he is named as the owner of the photo featured in this book (Photo #1), even though Bob had obtained it from the elusive Mr J.M. Anderson. I wrote to Bob asking for any additional details as to the sources of his book's winged cat photos, but he wasn't able to supply me with any details not already known to me, so this particular mystery remains unsolved. Following her fax and emails to me regarding the retouching aspects, on 13 July 1998 Janet wrote to John Morris requesting a print or negative of the 'squatting' photo (Photo #3) to borrow in order to facilitate the photo's inclusion in the FPL, and received a very pleasant if unexpected surprise. In his reply of 10 August 1998, in which he mentioned that as far as he could recall the cat had been owned by a night watchman on the Trafford Park Industrial Estate in Salford, he enclosed no less than FIVE photos of the Manchester winged cat! These were: the *Living Wonders* picture (Photo #1), the *Modern Mysteries of Britain* picture (Photo #2), the 'squatting' picture (Photo #3) from the 1972 *MEN* article, and two others (hereafter Photos #4 and #5 respectively) that neither Janet nor I had ever seen before. In his letter, John Morris stated that he holds the negatives to all of them, and noted that other photos of the cat may have been taken at the time when his five were snapped, but, if so, no such others are in his possession.

Now, presented exclusively here together for the first time ever in any book are all five of John Morris's photographs – i.e. every photo of the Manchester winged cat known to exist, with the FPL acting as their sole agent for reproduction. As can be seen, unlike those prints that I had received from the *MEN*, these originals are untouched, and in the 'squatting' photo (Photo #3) what was obscured in the retouched *MEN* version of it is clearly revealed to be the cat's thick, upraised tail, not a wingtip as originally assumed by me when I had only the *MEN* version to view. Moreover, in the FPL's uncropped version of this photo, three men are present (the one on the left had been lost in what was a heavily-cropped version published by the *MEN*). As for the two previously-obscure photos: as can be seen here, one of them (Photo #4) shows the cat again with tail upraised but with its head turned to the left and in a standing pose (and which as far as I am aware was published for the first time anywhere as recently as October 2006, within a previous *FT* winged cat article of mine).

And the other one (Photo #5, never published anywhere prior to my *FT* article of January 2009 on which this present book's coverage here is based) shows the cat with its head turned so far to the right as to be almost completely back upon itself and with three men around it, with the faces of two of them largely visible (in three of the other four photos, only two are present, with no faces visible; the 'squatting' photo, Photo #3, includes three men but no faces are seen). It would be interesting to know if anyone recognises either or both of these men, in case they are still alive and able to provide firsthand eyewitness accounts and other details regarding the Manchester winged cat. In the meantime, it would be equally interesting to know why two of these five photos (Photos #2 and #3) came to be lying on a Manchester street. Who dropped them, and why were they mounted on card complete with an accompanying caption? Clearly, they must have been intended for use (or had already been used?) in some presentation or even for reproduction in a publication – but where, or which one? Also, how did the photo that later became the *Living Wonders* picture (Photo #1) reach the *MEN*, and how did the mysterious Mr Anderson come by this same photo, which he supplied to that book's authors? As can be seen, the

Photo #4 of Manchester winged cat, first published in my *FT* winged cat article of October 2006

Photo #5 of Manchester winged cat, first published in my *FT* winged cat article of January 2009 (on which this present book's coverage here is based)

Alien Zoo

history of the Manchester winged cat may now be well known, but the history of some of its photographs is still shrouded in mystery – unless the good readers of *FT* can shed some much-needed light upon it? I hope to hear from you!

I wish to thank Janet and Colin Bord most sincerely for their kindness in making so much of this complex, hitherto-unpublished history known and available to me, for facilitating my viewing of all five of the Manchester winged cat photographs, and for providing me with their much-valued comments concerning photographic retouching; and also the *Manchester Evening News* for so generously supplying me with a copy of their 1972 article and photographic prints of the images used in it.

A WINGED CAT ADDENDUM

On 14 September 2010, Paul Sieveking of *FT* forwarded to me the following details of a feral winged cat that was being regularly fed by Scottish journalist Derek Uchman of Montifieth in Angus:

> We have had a very timid stray cat visit our door for food for a couple of years now. As its fur is extremely long and, and we are unable to groom it let alone entice it indoors, it gets very matted. This fur then slowly peels back to reveal a pair of "wings". Purely composed of hair they cannot be "flapped" like the cats featured recently in *FT*. After a period of several months, the whole lot drops off, and the process begins again.

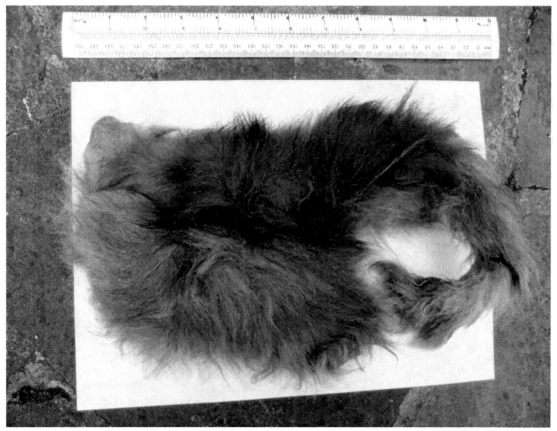

The recently-shed wings of Derek Uchman's feral winged cat (Dr Karl Shuker)

Alien Zoo

From his description, it is clear that in this particular case the wings do not derive from FCA but are merely mats of matted fur, but a winged cat is still a winged cat, regardless of the mechanism by which its wings developed.

After contacting Derek directly, I learnt from him that he had retained the wings that the cat had shed a few months ago. Moreover, he was happy to give them to me if I would like them! Needless to say, my reply was such that, just three days later, a Special Delivery package arrived to my home, containing this most remarkable of unnatural history relics. It consists of a thick, matted, yet clean mass of grey-streaked brown fur comprising two extensions (one longer than the other), and is remarkably resilient in texture, even though there is no skin or connective tissue attached. So now, among my many other zoological curios is a sealed transparent case containing the shed wings of a bona fide winged cat.

Back view (showing wings) of Derek Uchman's feral winged cat (Derek Uchman)

255

Bigfoot family (William Rebsamen)

ALIEN ZOO

2007

BIGFOOT DE-FOOTED! A famous and very imposing 2.5-m-tall, 180-kg wooden statue of a bigfoot, which has attracted plenty of interest and attention from onlookers since it was placed by Washington State chiropractor Tom Payne outside his surgery as a landmark for patients and sightseers over 5 years ago, has been bizarrely mutilated. On 22 January, it was abducted from its familiar site, and 3 days later police discovered it abandoned beneath a pile of debris in a nearby backyard...minus its feet, which had been sawn off, reducing its height by 0.5 m. Payne plans to have the feet replaced, and two suspects have been apprehended, but have offered no explanation for the statue's anomalous abduction and desecration. Hence the mystery remains as to the motive for this strange vandalism – unless, perhaps, there had been plans afoot(!) to make fake bigfoot footprints using those hacked from the statue? **ABC News, 26 January 2007.**

SEEKING THE SEL'AWAA. During late January 2007, Belbis, a district in the Sharkia area of Lower Egypt, was visited by a dreaded canine mystery beast known as a sel'awaa. Superficially similar to a domestic dog but distinguished by its short hind legs and brown or dark yellow skin, it is further differentiated by its considerable savagery. Several residents claim to have been attacked by the unwanted visitor. Yet even though the police later stated that they had shot it (but if so, what happened to the carcase?), sightings are still being reported, though their accounts are not consistent. Could some have been based upon a vagrant hyaena, as suggested by its short hind legs? *Middle East Times*, **6 February 2007.**

LEGALISE BIGFOOT! Canada's Species At Risk Act protects a number of endangered species, including the whooping crane, blue whale, and red mulberry. Now, at least if Canadian MP Mike Lake has anything to do with it, the Act will soon provide protection for another, even more elusive creature

– the bigfoot! This famous cryptozoological man-beast has not received formal scientific recognition, but Lake considers that fact to be of far less significance than the preservation of its existence (should it eventually prove to have one!). Accordingly, in a petition presented to the Canadian parliament, Lake states: "The debate over their (Bigfoot's) existence is moot in the circumstance of their tenuous hold on merely existing. Therefore, the petitioners request the House of Commons to establish immediate, comprehensive legislation to effect immediate protection of Bigfoot". Almost 500 of Lake's constituents in Edmonton, Alberta, have signed the petition, but whether Lake's fellow MPs will prove as enthusiastic when they get around to debating it this year is another matter entirely. *Edmonton Sun*, **5 April 2007.**

My close encounter of the bigfoot kind (Dr Karl Shuker)

STAY AWHILE, CROCODILE.
Crocodiles are something of a novelty in the Cayman Islands. So when one was spotted, and duly captured alive, at Old Man Bay on 30 December 2006, it created quite a stir zoologically, with much speculation as to where it had come from – a mystery that remains unsolved. It appeared to be a freshwater American crocodile *Crocodylus acutus*, albeit one that was far from home, and routine DNA samples were taken to confirm its identity, and thus help determine its population of origin, in order to send it back to its natural environment. When the test results were made public in May 2007, however, they provided just as big a shock as that

which had accompanied the creature's original discovery - because according to its DNA, it was actually a saltwater Cuban crocodile *C. rhombifer*, which is a totally separate species from the freshwater American version. Yet outwardly it looked far more like *C. acutus* than *C. rhombifer*. Not surprisingly, therefore, this enigmatic individual is now the subject of renewed speculation, this time regarding whether it may be a hybrid of the two species, or even a totally new species, as well as where best to send it back, given its mystifying identity. Until this riddle is resolved, therefore, it looks as if this strange crocodile will indeed be staying awhile in the Caymans. ***Caymanian Compass*, 17 May 2007.**

Seeking trolls on Cannock Chase (Dr Karl Shuker)

THE TROLLS OF CANNOCK. Cannock Chase, a vast protected expanse of wooded and bracken-carpeted countryside near Stafford in Staffordshire, England, has been for many years the scene of alleged encounters with a bizarre hairy entity variously likened by its startled eye-witnesses to a bigfoot or even a werewolf. These reports (some of which, according to one investigator, have come from such credible claimants as scout leaders on patrol, military personnel, police, ex-police, and even a local postman) have attracted considerable interest and attention from local paranormal and cryptozoological researchers. Moreover, a wide variety of suggested identities have been proffered – from paranormal man-beasts, or crazed tramps, to huge stray dogs, or even extraterrestrial aliens. In May 2007, however, a local resident (who has chosen to remain anonymous) hit the headlines in this area with a new and truly extraordinary proposal. He has soberly claimed that a tribe of primitive humanoid beings may be inhabiting the vast honeycomb of subterranean tunnels and passages beneath the surface of Cannock Chase (which was formerly a major mining area), but occasionally coming above-ground to hunt deer and other wildlife for food. And it is rare sightings of these latter-day trolls, he believes, that is responsible for the accounts of hairy bigfoot- or werewolf-type entities – as well as for an unexpectedly high number of local pets going missing here. ***Stafford Post*, 30 May 2007.**

QUAILING BEFORE A POTENTIAL REDISCOVERY. In November 2006, I visited the island bird sanctuary of Tiritiri Matangi, off Auckland, New Zealand, home to some of this country's rarest birds, including the famous takahe *Porphyrio mantelli*, the tieke *Philesturnus carunculatus*, North Island kokako *Callaeas cinerea wilsoni*, and stitchbird *Notiomystis cincta*, all of which I was fortunate enough to espy in the wild state here. On my way back down one of its paths leading to the ferry, I also briefly spied a couple of brown quail-like birds that I couldn't identify, but I didn't think anything more of this – until May 2007. That was when I was startled to read that this tiny island may still be home to a species hitherto thought to have become extinct everywhere as long ago as 1875. And the species in question? The New Zealand quail or koreke *Coturnix novaezealandiae*. Once common throughout New

Alien Zoo

Zealand, by 1840 the koreke had become far less so due to habitat destruction for farming. By 1870, it had been exterminated on North Island, and by 1875 it had also disappeared on South Island, since when the koreke has been regarded as extinct. There is, however, a quail species that is still known to exist in New Zealand – the introduced Australian brown quail *Coturnix ypsilophora*, which not only is smaller than the koreke, but also is smaller, intriguingly, than the previously-unstudied quails present on Tiritiri. Consequently, Massey University researcher Mark Seabrook-Davison is currently conducting comparative DNA analyses on this non-native species and on the Tiritiri mystery quails, in order to determine whether or not these latter birds are actually surviving specimens of the officially-demised native koreke. **Stuff.co.nz 8 May 2007.**

A pair of New Zealand quails, painted by John Gerrard Keulemans (1888)

MONSTERS ON VIDEO. Lake Kanasi (aka Hanas) in northwestern China's Xinjiang Autonomous Region has long attracted cryptozoological interest, due to unconfirmed reports that it contains huge red-coloured water monsters. Now a tourist claims to have videoed these mystery beasts. Aired on Chinese television and also online, the video appears to show several very large animate objects moving rapidly across the lake just beneath the water surface. Their most ardent investigator is Prof. Yuan Guoying of the Xinjiang Institute of Environmental Protection, who experienced his first sighting in 1985. The creatures, which have been likened to enormous taimen *Hucho taimen* (a salmon relative), are reddish in colour, and are said to be 10-15 m long - which is far bigger than any known fish of this type. *Daily Telegraph*, **18 July 2007.**

NITTAEWO-SEEKING IN SRI LANKA. A well-known field cryptozoologist planning a major new search for mystery beasts is Czech explorer Ivan Mackerle. He is presently focusing upon the cryptids of Sri Lanka, most notably the nittaewo - a short red-furred type of ape-man reputedly exterminated by

Alien Zoo

the Veddah (Wanniyala-Aetto) people at the turn of the 18th Century, but which may still persist in small numbers in Sri Lanka's least accessible regions. Ivan is planning a late-autumn search of the island, hoping in particular to trace the footsteps of an expedition led by Army Captain A.T. Rambukwelle in 1968. Captain Rambukwelle was seeking a large cave inside which the last nittaewos were supposedly trapped and burnt to death by the Veddahs. Other Sri Lankan mystery beasts that Ivan will be investigating are the horned jackal and the devil-bird, the latter named after its hideous, blood-curdling shriek. **Ivan Mackerle, pers. comm., 23 July 2007.**

DROPPINGS OF DELIGHT FOR THYLACINE SEEKERS? One of Australia's most enigmatic creatures is the thylacine *Thylacinus cynocephalus*, also called the Tasmanian wolf or Tasmanian tiger, which closely resembled true wolves in outward form but was striped, and was a marsupial. Confined in historic times to the island of Tasmania, the last known wild specimen was shot around 1930, and the last captive specimen died in 1936. Since then, despite numerous claimed sightings, no conclusive evidence for thylacine survival has been confirmed. However, zoologist Dr Jeremy Austin from the Australian Center for Ancient DNA and colleagues are now investigating some mysterious animal droppings collected in the Tasmanian bush during the late 1950s and 1960s by thylacine expert Dr Eric Guiler, and preserved ever since at the Tasmanian Museum. Interest in these droppings stems from their apparent resemblance to those of thylacines, and now, thanks to advances in DNA analysis techniques, they can finally be tested to discover whether they contain any thylacine DNA. If they do, this will be the very first proof that this spectacular mammal did indeed survive beyond the early 1930s in the wild, and will increase hope that it may still exist there today. **http://www.sciencedaily.com/ releases/2007/06/070626214417.htm 27 July 2007.**

FINS ON FILM AT LAKE TIANCHI. A veritable flotilla of finned mystery beasts has been videoed swimming across Lake Tianchi, a famous 'monster lake' in Jilin province, northeastern China, close to the border with North Korea. TV reporter Zhuo Yongsheng filmed no less than six animate objects for 20 minutes and also took some still photographs of them during the early morning of 6 September. One of the stills shows the black dot-like objects swimming parallel in three pairs, the others show them grouped closely together and leaving deep ripples on the surface of the lake. Tianchi is a volcanic lake, 373 m deep, and critics have claimed that it is too cold to sustain large animals. However, reports of lake monsters here have been circulating for over a century. Zhuo alleges that the objects that he filmed were seal-like beasts but with huge fins that were longer than their body length, and that they could swim as fast as yachts but sometimes disappeared completely beneath the water surface. He has sent photos of them to Jilin's provincial bureau for examination and evaluation. **http://www.china.org.cn/ english/China/223790.htm 9 September 2007.**

IN SEARCH OF TROPICAL APE-MEN. Two different man-beasts of the tropics are, or will be, the subject of searches during 2007. On 2 September, British cryptozoological explorer Adam Davies and a History Channel camera crew set off for Sumatra, renewing his quest for the elusive orang pendek or 'short man'. Adam has already achieved considerable success in his search for this entity. On previous expeditions, he has obtained a cast of an alleged orang pendek footprint that displays distinct differences from the tracks of all known primates and other mammals, and he has also obtained hair samples that again cannot be identified with any known species. He has now returned to the UK, having collected a number of new tracks that await formal study. And in November, courtesy of funding from a videogame company, Richard Freeman will be leading a CFZ team of investigators to Guyana in pursuit of the didi - claimed by the locals to be a large species of ape-like entity that walks on its hind legs and lacks a tail. Perhaps the most infamous, controversial evidence for such a creature in northern South America is the photo snapped in 1920 by geologist Dr François de Loys of what he claimed was a bipedal tailless ape shot dead on the Venezuela-Colombia border by his expedition, and which was

Alien Zoo

later christened *Ameranthropoides loysi*. The creature's carcase had been propped upright with a pole beneath its chin for the photo, but it has since been widely denounced as nothing more than a spider monkey modified deliberately to look like an ape. However, irrespective of the photo, there are many eyewitness accounts and descriptions of ape-like creatures from much of South America. And although no true ape is known from this continent, palaeontologists have in recent years unearthed the remains of some officially extinct species of giant spider monkey, leading to speculation that perhaps the didi and other Neotropical man-beasts are surviving representatives of these extra-large monkeys. The CFZ team, conversely, believe the didi to be a living species of ground sloth – an identity previously proposed by zoologist Dr David Oren for another South American mystery ape-man, the Brazilian mapinguary. **Adam Davies, pers. comms; http://www.theregister.co.uk/2007/09/10/guyana_apeman/ 10 September 2007.**

MYSTERY DEER IN IRAN. Three specimens of an unusual, unidentified species of deer have recently been spotted by provincial environmental guards on patrol in northeastern Khorasan Razavi province, Iran, near the border with Turkmenistan. Characterised by a distinctive stripe on its back, and weighing an estimated 35-40 kg, apparently this elusive plains-dwelling deer form has not been reported before, and Iranian officials are now checking the sightings to determine its zoological identity. **http://www.iran-daily.com/1385/2750/pdf/i5.pdf 14 September 2007 [no longer online].**

DINOSAUR-HUNTING IN CAMEROON... Since the early 1980s, Scottish-born but Canada-based cryptozoological field researcher Bill Gibbons has led several expeditions to the People's Republic of the Congo (formerly the French Congo) and Cameroon in search of elusive long-necked dinosaur-like mystery beasts, most notably the Congolese mokele-mbembe. Bill now informs me that in early 2008 he will be returning to Cameroon to conduct the most extensive river exploration ever conducted in search of these and other exotic cryptids. He is also preparing his long-awaited book, *Mokele-Mbembe: Mystery Beast of the Congo Basin*, documenting the history of the mokele-mbembe and similar Central African 'neo-dinosaurs'. We await both his new expedition and his new book with great interest. **Bill Gibbons, pers. comm., 30 September 2007.**

The mokele-mbembe (Richard Svensson)

A FLYING WOMBLE? Until recently, Wimbledon Common has been most famous zoologically as the home of those furry, fictitious litter-pickers the wombles. However, since late September 2007 a cryptozoological interloper of the factual variety has been spasmodically reported here. According to one eyewitness: "It was squirrel-like but its face looked more mouse-like, with long whiskers, black eyes, and small ears. As it jumped between trees, flaps of skin stretched between its front and back legs, and it gilded to the next tree. I couldn't believe my eyes". This description suggests an escapee European (or American) flying squirrel, or, less plausibly, an escapee Australian flying phalanger (such as a sugar glider *Petaurus breviceps*) – always assuming, of course, that evolution has not engineered a flying womble! *Guardian*, **12 October 2007.**

A COW-CLUTCHING TREE? I've always had a soft spot for mystery carnivorous plants, albeit from a safe distance, so recent news of a cow-snatching tree in India made irresistible reading. According to the report, the bovine victim in question was owned by Anand Gowda and had been left to graze in Uppinangady forest range by a cowherd on 18 October 2007. Suddenly, a tree grabbed the unfortunate animal in its branches and hauled it off the ground. The terrified cowherd raced away to the nearby village of Padrame, where the cow's owner, Godwa, lives, and returned with him and some other villagers, who succeeded in freeing the cow by hacking at those branches of the tree that had seized it until the branches became limp, releasing the cow from their deadly clutch. According to Subramanya Rao, an Uppinangady range forest officer (RFO), the tree was referred to in the villagers' local language as the pili mara – 'tiger tree', and he had received many complaints in the past about cattle returning home from this forest with their tails missing. Furthermore, other field staff later confirmed that a similar tree had been discovered partially felled. Like so many previous reports of mystery carnivorous trees, it is difficult to know how seriously such stories be taken, but as this is such a recent one, surely it would be possible for local scientists to investigate the trees in question, if only to determine whether their species is known to science. **Express News Service, New IndPress, 23 October 2007.**

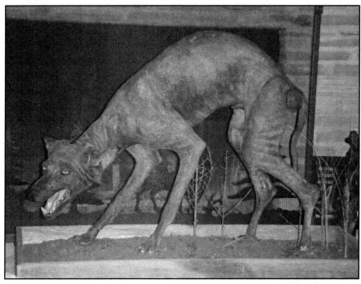

The Cuero blue dog carcase (CFZ)

COYOTE, NOT CHUPACABRAS. In July 2007, Phylis Canion found the dead bodies of two large but very strange-looking animals outside her ranch at Cuero, southeast of San Antonio, Texas. Suspecting that they may explain the bizarre deaths of several of her chickens, found apparently sucked dry of blood, she preserved the head of one of these animals, and speculated that it may be a chupacabras.

Intrigued, KENS-TV, a San Antonio-based affiliate of CBS, arranged for tissue samples from the head to be tested by genetic experts at Texas State University, in order to discover the creature's zoological identity.

The results have now been released – Phylis's mystery beast was not a chupacabras. Instead, it was a coyote *Canis latrans* (later, more extensive tests specified a coyote x wolf hybrid) – but a grotesque, perplexing one, nonetheless, because it was totally hairless and had blue-grey skin. Skin tests are now being undertaken to find out why. *San Antonio* **(TX)** *Express-News*, **2 November 2007.**

Alien Zoo

A DEVIL OF A MYSTERY FROM SMETHWICK. Much closer to home: There is a pub in Smethwick, just outside Birmingham, in the West Midlands, England, called the Blue Gate that may lay claim to erstwhile cryptozoological fame. Mysteries researcher and author Nick Redfern now resides in the USA, but he formerly lived only a few miles from Smethwick, and recently he gave me the following fascinating information - derived from the great-uncle of a friend, Eddie, of Nick's father. Around the end of the 19th Century, Eddie's great-uncle Ned was driving a pony and trap on Rolfe Street, Smethwick, late one night when he heard some strange noises behind him. Suddenly, a weird-looking animal leapt out at him, but he supposedly fought it off with his horse-whip. The creature was killed, placed in a glass case, and displayed in the Blue Gate pub on Rolfe Street for some time, where the locals dubbed it 'Old Ned's Devil'. Sadly, however, this mystifying specimen, for which no morphological description exists, apparently vanished years ago, and nothing more is known of it. Living not too far from this area, I have made some enquiries myself, but no-one has been able to add to the above details given to me by Nick. So at present this is where the matter rests – except for the intriguing fact that the general storyline of the Smethwick mystery beast's attack upon Eddie's great-uncle is very similar to that for another bizarre beast from that same bygone period, the Shropshire Union Canal man-monkey. As Nick has actually written a book about this latter beast, he is naturally particularly curious about the Smethwick equivalent, as indeed am I. Could the two stories stem from a single source, and, if so, how? Alternatively, were there truly two separate albeit very comparable incidents and, if so, what exactly was Old Ned's Devil? An escapee/released exotic pet, possibly? **Nick Redfern, pers. comm., 5 November 2007.**

The Shropshire Union Canal man-monkey (Richard Svensson)

SAHARA'S SECRET CROCODILES. The arid desert lands of the Sahara may not be the likeliest place to expect water-loving crocodiles to thrive. Nevertheless, a relict, hardy population of Nile crocodiles *Crocodylus niloticus* was known to exist in the Tagant region of Mauritania until 1996, when,

Alien Zoo

following years of depletion, the last-known specimen was killed and eaten by locals; its surviving head was preserved for scientific research. In 2007, however, Ursula Steiner photographed at least three living crocodiles here, confirming that the Tagant's most unlikely reptiles do still survive. Unless there is permanent water, desert crocodiles aestivate in deep burrows, just like lungfishes, during the hot season – which may explain how they have been overlooked by scientists during the past decade. **http:// www.wildlifeextra.com/desert-crocodiles893.htm 9 November 2007 [no longer online].**

A SHUNKA WARAK'IN REFOUND? In 1886, a large, decidedly odd-looking canine beast was shot and killed by Israel Ammon Hutchins, the grandfather of present-day zoologist Dr Ross E. Hutchins, on his ranch in the Madison River Valley north of Ennis, Montana. Moreover, this strange creature was actually preserved, and for many years afterwards it was exhibited in a glass case by taxidermist-entrepreneur Joseph Sherwood (who had received it in trade from Israel Hutchins) at his store-cum-museum near Henry's Lake, Idaho. In addition, a decent b/w photo of it was taken and published in Dr Hutchins's autobiography *Trails to Nature's Mysteries: The Life of a Working Naturalist* (1977). What is especially interesting about this creature as subsequently recognised and publicised by American cryptozoologists Loren Coleman and Jerome Clark is that it bears a close resemblance to a longstanding mystery beast known to the Ioway people and other native Americans living along the USA-Canada border as the shunka warak'in - described as a dark-furred beast with a lupine head and high shoulders but sloping back and short hind legs, thus resembling a cross between a wolf and a hyaena. Naturally, to have a preserved specimen of such an animal to hand for scientific examination, particularly in these technologically advanced times of DNA analysis, is a great boon – or would be, were it not for the regrettable fact that several years ago it vanished, having been moved to some unspecified location in the West Yellowstone area. Happily – and very unexpectedly – however, this unique specimen has just been rediscovered. After reading a story about it in late October 2007, Jack Kirby, another grandson of Israel Hutchins, tracked down the elusive exhibit to the Idaho Museum of Natural History in Pocatello. Moreover, the museum has agreed to loan it to Kirkby in order for it to be displayed at the Madison Valley History Museum. A new examination of this famous specimen has revealed some previously-undocumented details. It measures 48 inches from the tip of its snout to its rump, not including its tail, and stands 27-28 in high at the shoulder. Its snout is noticeably narrow, and its coat is dark-brown, almost black, in colour, with lighter tan areas, and includes the faint impression of stripes on its flanks. Despite its age and travels around America, this potentially significant taxiderm specimen is in remarkably good condition, with no signs of wear or tear or even any fading of coat colouration. Could it truly be a shunka warak'in? And, if so, what in taxonomic terms *is* the shunka warak'in? Now that the lost has been found, DNA analyses of hair and tissue from the long-preserved exhibit may at last provide some answers. ***Bozeman (MT) Daily Chronicle*, 15 November 2007.**

LOOK OUT, IT'S BIRDZILLA! For years, there have been many reports and sightings across San Antonio and South Texas of a truly monstrous thing with wings - spanning as much as 6 m, and recently dubbed Birdzilla by the American press. One eyewitness, Frank Ramirez, likened what he saw to an enormous man-faced bird, another to a giant prehistoric-looking bird. Now, Ken Gerhard, author of a book on these and other cryptozoological "big birds", has launched a formal investigation of such sightings here. In November 2007, he appeared in a History Channel documentary on Birdzilla, and he has been installing cameras in Harlingen, where one eyewitness, Guadalupe Cantu, wants his big bird sighting documented and validated. Ramirez, meanwhile, keeps a light outside his house to ward off any possible return of Birdzilla, and with good reason. "I know what I saw. It took me more than a week to step out of this house. I wouldn't step foot out of this house," said Ramirez. "It had this very, very horrible demeanour-look on its face. Like I was lunch". **KENS-TV, Channel 5 (San Antonio, TX), 16 November 2007.**

Alien Zoo

IN THE WAKE OF THE WATER TIGER. One of the least-documented South American cryptids is the water tiger, which had scarcely been referred to even in cryptozoological publications until my book *Mystery Cats of the World* (1989) appeared, and included reports of it from various countries. Now, a series of detailed eyewitness accounts and descriptions of this mystifying aquatic animal has been collected by the recent CFZ expedition to Guyana, led by Richard Freeman. According to this new data, Guyana's water tiger is spotted, bearing black markings resembling those of a jaguar upon a white background, and has a striped head like a tiger. A pelt of one such beast, matching this description and roughly 10 ft in length including a long tail, which had been shot, was seen back in the 1970s by Joseph, one of the team's guides. However, the water tiger can also apparently exhibit a range of other colours. Intriguingly, a local man named Elmo from a township called Point Ranch claimed that the water tiger hunts in packs, each pack led by an alpha individual that Elmo termed 'the master', which organises the hunting, carried out by younger members of the pack. Half a century ago, Dr Bernard Heuvelmans briefly referred to South American water tigers in his book *On the Track of Unknown Animals* (1958), and suggested that a giant otter, possibly discrete from the famous saro *Pteronura brasiliensis*, may be the explanation. However, I have never been convinced by this identity, and the new data emanating from Guyana further suggests to me that the water tiger is not an otter. In my mystery cats book, I pointed out the similarities between water tigers reported in South America and certain aquatic cryptids from tropical Africa variously termed water lions, water leopards, etc. Heuvelmans argued that these latter might be surviving sabre-tooths that had become secondarily aquatic. Could the same explanation be tenable for South America's water tigers? Intriguingly, an enigmatic river-dwelling walrus-like cryptid called the maipolina has already been reported from Guyana (within an area called Maripasoula), as discussed in my book, but is seemingly distinct from the water tiger reported to the CFZ team, because the maipolina is said to be unspotted. The plot thickens! If only that 1970s pelt had been preserved! **http://cfzguyana.blogspot.com/ 22 November 2007.**

The Guyanan maipolina (Markus Bühler)

The Lost Ark

A BLENNY FOR YOUR THOUGHTS!
FISHING FOR A BITE IN THE MIDDLE EAST.

From Tehran comes a report of a diminutive black fish found in the Shatt al Arab River. It reputedly has killed twenty-eight people with a venomous bite. *Death is said to be swift. No other information is presently available. (No other fish is known to have a venomous bite, and this report is at least suspect.)*

Roger Caras – *Dangerous To Man*

Size isn't everything - and mystery beasts are no exception to this rule. The above-quoted snippet, from Roger Caras's definitive book on hostile creatures, *Dangerous to Man* (1975), is the only published item that I have seen concerning a hitherto uninvestigated mini-monster with a serious attitude problem. Assuming that such a fish really does exist, what could it be? Various candidates can be selected from the many thousands of fish species already known to science, but none can offer a comprehensively satisfying solution.

Take, for instance, the Asian stinging catfish *Heteropneustes* [formerly *Saccobranchus*] *fossilis*. Just like our mystery fish, this species inhabits the Shatt al Arab River, is blackish in colour, is often only around 4 in long (though it can measure up to 27 in), and is venomous. A likely identity? Not quite, for whereas our mystery fish has a venomous *bite*, this species' source of poison is the spines in its pectoral fins. Furthermore, it is not native to the Shatt al Arab, but was imported here from Indochina and India (despite its dangerous fins, it is an edible species). In any case, although its spines can inflict painful wounds, they are not lethal. Exit *H. fossilis* from further consideration.

Alien Zoo

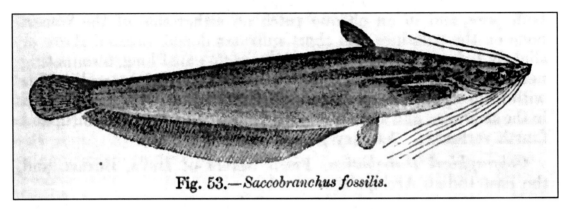

Fig. 53.—*Saccobranchus fossilis.*

The Asian stinging catfish

Rather more hopeful on first sight is the long-tailed or slender giant moray eel *Strophidon sathete* (aka *Thyrsoidea macrura*). This is a widespread marine species in the tropical Indo-Pacific, but frequently enters estuaries, and can be found quite a distance up rivers - including the Shatt al Arab. Moreover, it does have a toxic bite - after a fashion. There is still controversy as to whether moray bites are venomous in the true sense, but particles of rotting flesh adhering to the moray's teeth from past meals could well poison a wound caused by such a bite (this can also occur with crocodiles and mammalian meat-eaters such as lion and tigers). However, no human fatalities from moray bites are on file, and bearing in mind that adult long-tailed morays are up to 12 ft in length, this can hardly be called a 'diminutive' species. Don't call us, *S. sathete*.

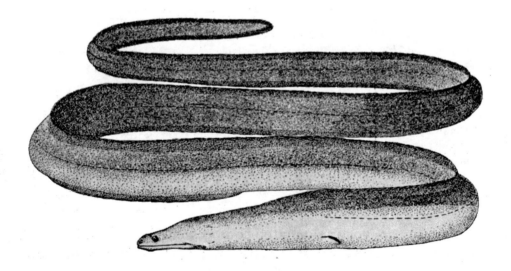

The long-tailed (slender giant) moray eel

Alien Zoo

The latter discrepancy also eliminates the bull shark *Carcharhinus leucas* - quite apart from the fact that its bite, though indisputably inimical, is not of the venomous variety. The vast majority of sharks are exclusively marine, but this particular species is well known for entering freshwater lakes and rivers in many parts of the world, including the Shatt al Arab, where it does indeed kill people from time to time. That unpleasant claim to fame, however, would appear to be its only shared similarity to the incognito fish in Caras's account.

Whereas the above three candidates are relatively familiar species, the final example on offer is far less so. Nevertheless, it is the one that seems to offer the best hope for solving this mystery. Blennies comprise a large group of mostly small and mainly marine species, distributed worldwide in temperate and tropical waters, but with some freshwater representatives too. Among their most noticeable features are their scaleless skin, and their numerous teeth. These are often very small, but some species have pairs of much larger, curved teeth resembling daggers in the sides of their jaws.

The Red Sea and the Gulf of Suez and Aqaba are home to a small but sufficient blenny called *Meiacanthus nigrolineatus*, measuring 2.5 in long. Its head and foreparts are blue-grey in colour, the remainder of its body is pale yellow, and there is often a thin black stripe running lengthwise just beneath its dorsal fin. What makes this species so significant in relation to the Shatt al Arab mystery fish is that it does indeed boast a venomous bite - thereby disproving Caras's earlier claim. Its lower jaw has sizeable canine teeth, with grooved sides and venom-producing tissue at their base, and with the aid of these it can inflict a bite sufficiently painful to deter all forms of piscean predator, irrespective of their size.

No human fatalities have been recorded with this particular blenny, but what if the Shatt al Arab River is harbouring a still-undescribed, related species of darker colour and capable of producing a more potent venom? If such a creature is captured one day, then our mystery fish will surely have been unmasked at last - turning up like the bad blenny that it is.

My special thanks to French ichthyologist Dr François de Sarre for very kindly sharing his own thoughts and comments with me regarding this fishy affair.

REFERENCES

CARAS, Roger (1976). *Dangerous To Man* (rev. edit.). Holt, Rinehart & Winston (New York).
SARRE, François de (1992). Pers. comm., 22 November.
SHUKER, Karl P.N. (1999). *Mysteries of Planet Earth: An Encyclopedia of the Inexplicable.* Carlton (London).

The Tahina palm tree (John Dransfield/Wikipedia)

2008

A NOTEWORTHY SEARCH, MOA OR LESS. Veteran cryptozoologist Rex Gilroy and his wife Heather have among their many items of zoological interest some recently-obtained plaster casts, and also evidence of what they consider may be a nesting area, that in their view indicates the survival of bush moas. These were (are?) relatively small members of New Zealand's famous clan of vanished flightless birds that also included the huge ostrich-like *Dinornis* moas, but all moas officially became extinct several centuries ago. However, there is some tantalising anecdotal evidence on record suggesting that a small species of bush moa may have lingered on until much more recently. Now, encouraged by their finds, during the end of February 2008 the Gilroys plan to scour their source - North Island's Urewera Forest - in search of whatever was responsible for the prints and the apparent nesting area. Not everyone shares their enthusiasm and optimism, however, with cryptozoological sceptics opining that the real culprits are much more likely to be escapee Australian emus *Dromaius novaehollandiae*. They also note that most moa reports emanate not from North Island but instead from the larger and much less densely populated South Island. ***Hawkes Bay Today*, 8 January 2006.**

MAN-EATER NOT SO UNLIKELY AFTER ALL? The legendary Madagascan man-eating tree (see my book *The Beasts That Hide From Man*, 2003, for full details) has been dismissed as unlikely not just because of its extraordinary morphology and behaviour but also because it seemed highly improbable that even in Madagascar, a giant tree could remain undetected by science. In fact, this is no longer improbable – due the remarkable discovery here in 2007 of a hitherto-unknown species (and genus) of amazing 70-ft-tall tree! Scientists have just revealed that 92 specimens of an extraordinary new palm tree, formally christened the Tahina palm tree *Tahina spectabilis* in early 2008, have been revealed in a small area of Madagascan swampland accessible only by air. Equally as surprising as its discovery, moreover, is its unique, suicidal style of reproduction. Just once in its lifetime, it sprouts a huge spire of tiny nectar-dripping flowers, attracting swarms of pollinating insects, which is essentially

Alien Zoo

an act of suicide, because as soon as the tree has bloomed and fruited, its nutrient resources are totally exhausted, causing it to collapse and die. If such a huge, astounding species of tree can remain undetected by science until the 21st Century on Madagascar, who can still categorically deny on the grounds of being too conspicuous the existence here of the man-eating tree? *BBC News*, **17 January 2008.**

MYSTERIES OF MINDANAO. According to recent news stories, the Philippine island of Mindanao may be home to two very different but equally intriguing mystery creatures. One of these is said to be a large monkey with a very distinctive red face, known to the local people as the uma-ay. A group of these creatures was recently spied in high trees on the sacred mountain of Sal-dab, in northern Mindanao. They are notable not only for lacking hair on their faces and therefore looking surprisingly human, but also for their alleged ability to mimic the human voice, which they supposedly use to scare away the enormous, predatory Philippine (aka monkey-eating) eagle *Pithecophaga jefferyi*. Perhaps this latter, uncanny talent explains the widespread native belief that seeing one of these animals is unlucky, causing the unfortunate eyewitness to become lost in the forest or to suffer misfortune, possibly even death. Even hearing its laugh is said to bring bad luck. The only species of monkey known to inhabit this island is the crab-eating macaque *Macaca fascicularis*, which does not exhibit the uma-ay's very eyecatching red-coloured face. Even stranger, however, is this island's second putative cryptid – a bizarre-sounding nocturnal bird called the ukang (owl) or gulus (ghost) that sports mammalian hair, and, like the uma-ay, mimics the human voice, as well as emitting a cry resembling that of a newborn baby. The description of hair is unusual for a bird to say the least, unless it refers to fine, hair-like feathers of the type possessed by birds such as the kiwis and the tiny Inaccessible Island rail *Atlantisia rogersi*, for instance. However, its baby-like cry recalls that of Sri Lanka's infamous mystery bird, the ulama or devil-bird, believed to be either a species of owl or a species of nightjar, and in view of its local name of ukang this Mindanao 'hairy owl' may be well be much the same. *Philippine Star*, **12 February 2008**; http://www.abs-cbnnews.com/storypage.aspx? Storyid=108846 **12 February 2008** [no longer online].

The ukang (Andy Paciorek)

Alien Zoo

REDISCOVERING A DODO? One of the most infamous losses in zoological museum history is that of the world's only stuffed specimen of the dodo *Raphus cucullatus* – which was discarded and burnt on 8 January 1755 on the orders of an over-zealous committee of trustees at Oxford's Ashmolean Museum, because they considered it looked tatty. Happily, a less finicky assistant had the foresight to rescue its head and one of its feet before the flames reached them.

These precious relics were later transferred to the University Museum of Zoology, where they are now among its most prized specimens. Being well aware of this dark episode, I was nothing if not startled by an email published on 16 March 2008 in the *Sunday Mirror* newspaper's 'Treasure Hunters' column, compiled by TV antiques and collectables expert James Breese. The email in question was from a Ray Holmes (no address or location details given), in which he asked Breese to tell him the likely worth of what he described as a stuffed dodo in near-perfect condition, owned by him and preserved inside a domed glass case, which had originally been acquired by an ancestor of his sometime during the mid-1600s. In reply, Breese rightly pointed out that if a genuine stuffed dodo did indeed exist anywhere, it would be priceless both in value and in scientific importance.

However, he also revealed that fake dodos have been produced by taxidermists (notably Rowland Ward of London), created by using feathers and tissues from other birds, and opined that this is very probably the explanation for Holmes's dodo. Nevertheless, the fact that the latter specimen apparently dated from the mid-1600s – a period when the dodo was still alive (this species' official extinction date is traditionally given as 1681, though recently a slightly later date has also been proposed) – was significant enough in Breese's view for him to suggest that Holmes should take his dodo to a museum for a closer inspection.

I certainly agree with this, because, although highly unlikely, if by any chance it actually is the real thing, a complete, near-perfect, genuine stuffed dodo would be one of the greatest zoological finds of modern times – almost as amazing as discovering a living, breathing dodo!

If Ray Holmes is reading this column and would like to get in touch with me or with *FT*, we'd be delighted to hear from him. ***Sunday Mirror*, 16 March 2008.**

A 19th-Century engraving of the dodo

LOOF LIRPA? YOU'RE HAVING A LAUGH! On 1 April 2008, the nowpublic.com website published a news report by Bill Adler announcing the escape from Washington DC's National Zoo of a particularly rare creature known as the loof lirpa. Its photo revealed it to be a llama-like beast but sporting a pair of long slender gazelle-like horns.

Native to Patagonia, it is further distinguished by its distinctive odour, likened to cheap cologne, and although usually a mild-mannered species, this particular individual, an adult male, had been ready to

Alien Zoo

The loof lirpa (Tim Morris)

mate when it escaped and so may well be feeling aggressive. Needless to say, however, there was no need to be alarmed - as long as you took the trouble to notice the date of this report and the phrase that this exotic animal's name spelt out if read backwards! **Nowpublic.com 1 April 2008.**

Another equally delightful zoological April Fool hoax for 2008 featured a recently-discovered, unique colony of flying penguins, as filmed in a BBC trailer for the BBC iPlayer, and first screened on 1 April 2008 (but which can still be viewed today on YouTube at http://www.youtube.com/watch?v=23qDl1aH9l4). Presented by TV star and film maker Terry Jones, the trailer revealed how these extraordinary penguins utilise their aerial ability to migrate away, far from their chilly Antarctic abode during the winter months, spending them instead ensconced snugly amid the warm, humid rainforests of South America, basking in the tropical sun.

AN UNRECORDED RAT KING FROM NEW ZEALAND. One of forteana's favourite animal anomalies is the rat king – a bundle of rat specimens inextricably bound to one another by virtue of their tails, tortuously intertwined in a knot of Gordian complexity. Quite a number of these grotesque but highly intriguing specimens have been recorded down through the centuries, and even today they constitute prized museum specimens. Few new examples, however, come to light nowadays, which makes the following news particularly noteworthy. Courtesy of Janet Bord, I recently learned that according to New Zealander Martin Phillipps, a rat king has actually been on public display for several decades in the Otago Museum, preserved in a jar of what he presumes is formaldehyde. Yet as far as I am aware, this important specimen has never before been documented in the fortean literature. Consequently, I contacted Martin, who duly informed me that the rat king fell down onto the ground, alive, from the rafters of the company shed of Keith Ramsey Ltd on Birch Street, Otago, sometime during the 1930s – followed swiftly by a parent rat that defended them vigorously. The king consists of eight rats, which are black (ship) rats *Rattus rattus*, whose tails are bound together with horse hair (used as nesting material by rats). **Janet Bord, pers. comm., 6 April 2008; Martin Phillipps, pers. comms, 24 & 29 April, 5 May 2008.**

The Otago rat king (Andy Paciorek)

274

Alien Zoo

HORSE-EELS A-PLENTY! I was delighted recently to hear from Gary Cunningham, an indefatigable investigator of Irish mystery beasts, whose work will already be familiar to *FT* readers by virtue of his extensive article on the dobhar-chú or master otter (*FT*168).

I am pleased to announce that Gary is continuing with his in-depth cryptozoological researches, having achieved particular success of late in relation to Ireland's horse-eels. After collecting a number of hitherto-unreported sightings of these sinuous water monsters, including early references and observations in Counties Galway, Mayo, and Kerry, Gary and his fiancée Deidre are now planning to visit Clifden in Connemara, Co. Galway, to pursue them in further detail, and he is also preparing a report on various much more recent sightings, i.e. ones made since 1990. One particularly memorable example featured a head, neck, and several humps, seen by the eyewitness in question while swimming about 50 m away! Other Irish cryptids under investigation by Gary include the Irish wildcat, and the little-known Achill Island dwarf wolf (which I documented in my book, *Extraordinary Animals Revisited*, 2007), plus more master otter researches. **Gary Cunningham, pers. comm., 7 April 2008.**

THE NAMING OF THE SHREW. During research for his PhD on the diet of barn owls *Tyto alba*, Dave Tosh from Queen's University, Belfast, was dissecting some owl pellets obtained in Tipperary when he found a number of shrew skulls within them that he was unable to put a name to, taxonomically speaking, because they seemed too big to be from Ireland's native pygmy shrew *Sorex minutus*. Consequently, some humane traps were set at four different Tipperary locations in the hope of capturing living specimens of this mysterious mega-shrew.

The traps proved successful, because in March 2008 no less than seven specimens were obtained, and to everyone's surprise they were found to be greater white-toothed shrews *Crocidula russula* - a widespread continental species, but never previously recorded from the Emerald Isle, thus indicating that it had probably been recently introduced here by man rather than constituting a hitherto-unobserved native species. *Express and Star* **(Wolverhampton), 29 April 2008.**

LESSER-KNOWN LAKE MONSTERS. Two previously little-documented lake monsters have recently hit the news headlines. Croatia has witnessed a spate of reports concerning a mysterious 6.5-m-long creature said to inhabit Lake Peruca. Some investigators have speculated that it is a huge snake, but this identity is discounted by local eyewitness Marija Duynjak, who described what she saw as having a wide, fat body and a man-sized head. Meanwhile, in Minnesota, USA, a 50,000-dollar reward has been offered to anyone who can hook, capture, or record on film the serpentine monster known as Pepie, said to frequent the depths of Lake Pepin. The reward has been put forward by Pepie eyewitness Larry Nielson of Lake City, who owns the 125-passenger paddle wheeler 'Pearl of the Lake' and is also a member of the Lake City Tourism Bureau.

Nielson believes that his Pepie reward may help to attract tourists to the lake, which has hosted many alleged sightings of the monster down through the years, dating back at least as far as 1871. A sturgeon is one identity mooted for Pepie, but the entity seen four years ago by two fishermen was greenish-yellow and undulating out of the water. *Daily Star*, **24 June 2008;** *St Paul Pioneer Press*, **2 June 2008.**

THEY'RE JUST WILD ABOUT WILD MEN. The Centre for Fortean Zoology (CFZ) expedition to Russia's Caucasus Mountains, led by Richard Freeman, in search of its elusive wild man the almas (=almasty), returned in triumph to England in July 2008, armed with skull fragments, hair, and dung samples alleged to be from this mystery man-beast.

These are to be examined for putative almas DNA - which if found will then be analysed by human

Alien Zoo

The almas (Richard Svensson)

genetics expert Prof. Bryan Sykes from Oxford University to ascertain this cryptid's taxonomic identity. Many eyewitness accounts were also obtained, and Richard with fellow investigator Adam Davies came within a few yards of a suspected almas. **CFZ press release, July 2008.**

MUSEUM MUSINGS OVER A MYSTERY INSECT. With over 28 million insects in its vast collection, it is eminently fortean that London's Natural History Museum in South Kensington should be perplexed over the identity of a tiny insect lately discovered in its own garden. Only the size of a rice grain, red and black in colour, and roughly almond-shaped, this entomological enigma was first seen in March 2007 on some of the museum garden's plane trees, but within three months it had become the commonest insect in the entire garden and could also be seen in other central London parks. Yet, remarkably, it does not seem to correspond with any of the museum's countless insect specimens. The closest match is with *Arocatus roeselii*, a species of hemipteran bug normally found in central Europe, but brighter red and living on alder trees. Consequently, museum staff have speculated that the mystery hemipteran may comprise a variety of *A. roeselii*, or even a new, related species. Whatever it is, however, it has shown itself to be decidedly hardy, by surviving the 2007 British winter and continuing to thrive in London. ***Wolverhampton Express and Star*, 15 July 2008.**

HERE'S THE HAETAE – KOREA'S FOUR-LEGGED FIRE-EATER. Who says that the internet isn't educational? While idly browsing recently through the online treasure trove of information that is Wikipedia, I was delighted to stumble upon a mythological beast that was totally new to me – the Korean haetae (also spelt 'haechi'). Although it superficially resembles a lion, this formidable but sacred creature of traditional Korean and Chinese legend is actually a huge fire-eating dog. Generally deemed to be a symbol of justice and a guardian against prejudice and disaster, it is occasionally portrayed bearing a unicorn horn. Very large, imposing statues of the haetae can be

A haetae statue at a Korean temple (Carty239/Wikipedia)

Alien Zoo

found throughout Korea, and in the future this impressive beast may well acquire international renown - because on 13 May 2008, the haetae was formally adopted as an official symbol of Seoul, South Korea's capital, by this city's citizens. **Wikipedia, accessed 23 August 2008; http://www.visitseoul.net/ en/article/article.do?_method=view&art_id=32903&lang=en&m=0004007001004&p=07 accessed 22 April 2010.**

OUT OF THE BLUE. It's not every day that you look out to sea and not only spy but also successfully photograph a colossal marine creature the size of an oil tanker, but as reported in September 2008, this is precisely what happened to amateur whale-watcher Ivan O'Kelly, who spied a veritable leviathan off the coast of County Kerry. Although he was not sure of its species, its immense size was sufficiently spectacular for him to send his photos to the Irish Whale and Dolphin Group for formal identification. To his amazement, he discovered that it was a blue whale *Balaenoptera musculus* – the largest species of animal on the planet, which can grow to more than 30 m long. Moreover, this was the first time that one had ever been photographed in Irish waters. *Wolverhampton Express and Star*, **19 September 2008.**

DODO IS A NO-NO. In a previous Alien Zoo (*FT241*), I revealed how *Sunday Mirror* antiques columnist James Breese had lately received an email from reader Ray Holmes possessing what he believed to be a genuine stuffed dodo *Raphus cucullatus* dating from the 17th Century (the century when this famously demised species became extinct), but that both Breese and I considered that it was most probably a later, fake specimen, many of which have been produced. I have since received several communications from dodo author-historian Anthony Cheke, supporting my suggestion and providing further insights into this intriguing subject, which he has kindly permitted me to document here. Most significantly, the methods of preservation available during the 17th Century were so poor that there are no stuffed birds from that time period still in existence anywhere – they have all long since rotted away or been eaten by insects. As stated by Anthony to me:

> In the 17thC (i.e. 1600s) skins were just cleaned and dried (rarely stuffed) without preservative - or any they used was short-lived. So the specimens simply over time succumbed to moths and mites, only the hard bits generally surviving (i.e. beaks, bones, antlers, carapaces etc.). Very few survived as far as the mid-18thC, the decaying dodo [at Oxford's Ashmolean Museum] was a late survivor, and the Ashmolean dumped ALL their remaining 17thC specimens at the same time as the dodo [1755]... - [museums/ cabinets of curiosities] didn't even invent pickling specimens in alcohol until the mid-1660s ! As far as I know there are NO 17thC specimens intact anywhere, and very few that pre-date the mid-1700s - most of the huge (former) royal collection in Paris (including almost all of Brisson and Buffon's types) was destroyed around 1800 when they tried to save it by fumigation, but the sulphur fumes damaged the specimens more than the bugs they were trying to kill !...The stuffed dodo in Prague was at some point reduced to just a skull and some bones that still survive. Two more in Oxford's Anatomy School disappeared around 1750.

Further details can be found in Anthony's authoritative book *Lost Land of the Dodo: The Ecological History of Mauritius, Reunion and Rodrigues* (2007), co-authored by Julian Hume. In short, as suspected from the beginning, the dodo owned by Ray Holmes is undoubtedly a later, facsimile dodo, constructed from the skin and feathers of other, non-dodo species of bird. Thanks once again to Anthony Cheke for his valuable information, and also to Dr Darren Naish for independently mentioning Anthony's researches to me. **Anthony Cheke, pers. comms, 16, 17, 18 & 24 September, 9 October 2008.**

Dodo statue at Chester Zoo, England (Dr Karl Shuker)

SIGHTING A SEA SERPENT. Kent-based cryptozoological researcher Neil Arnold recently passed on to me, with kind permission from its source to publish it here, a remarkable, previously-undocumented sea serpent sighting from the late 1950s. Neil's source was a correspondent called Nick (whose email address I have on file), and one of the eyewitnesses in question was his father. Nick is anxious that the sighting be preserved, so here is his account of it:

> I am 39 years old with three children and my father is now 72. He was one of the last people to have been conscripted to national service, serving as navigator on RAF Shackleton aircraft, in 1957, flying missions over Europe and much of South America. The official job of the Shackleton was as a spotting aircraft, and as such the pilot and crew were highly trained in this respect; their job was to search for submarines and relay the positions back to base. One of the tell-tale signs they used to look for high above was a dark shape underwater, which could often be a shark or whale or other creature but occasionally they would get lucky and find a submarine near to the surface. On one occasion, having seen such a dark shape near the surface, they

Alien Zoo

flew down to investigate, and as they flew by this shape, every crew member on board all saw the same thing - the neck and head of a large sea creature protruding from the surface. The nearest thing they could think of at the time was a plesiosaur. They were all in agreement but of course the pilot on the way home forbade them absolutely from talking about it; if they had done so they would all be immediately grounded on returning home and accused of being drunk on duty. So anyway, my father decided to research what the creature might have been, and found that it was almost definitely an 'elasmosaurus' [long-necked plesiosaur], as the one thing that was very obvious was the huge length of the animal's neck. My father was forced to conclude, and of course he'd read all about the coelacanth, that there were indeed much larger 'extinct' creatures out there, undiscovered to science. When I reached adulthood he was to tell me of the incident. Twenty years on, he still will swear to his grave that this is what they all saw on the flight...The year was 1957, the location was "about 500 miles north of the Canary Islands, in open sea". The RAF Squadron was 228 Squadron, flying out of St Eval in Cornwall. (This air base has since been closed but I believe remnants of the site remain.) I don't know who the other crew members were and I have been thinking about making an official enquiry to the RAF archives. However what I hope is important is the location (vague, sorry), species and date. I am more than aware that most scientists consider the possibility of any large dinosaur [sic] still alive as astronomically small, and my father himself also thinks now with sadness that he might have seen one of the last ones; his own reasoning is that at this point, not ever so long after the war, there really was not much shipping around; indeed, the area where they had the sighting was entirely devoid of shipping and maybe this was the reason the creature had considered it safe to surface, far from the throb of ship's propellers.

One zoologist who read Nick's account suggested that perhaps they had seen a tentacle from a giant squid rise above the water surface, but Nick's father totally discounted any such possibility. **Nick, via Neil Arnold, 28 November 2008.**

FOR CROWING OUT LOUD. My book *Extraordinary Animals Revisited* (2007) documents many reports of mysterious snakes that can allegedly crow like cockerels. These have been reported widely from Africa, as well as various Caribbean islands, Samoa, and China. However, I was very interested to discover recently that a Filipino version has also been reported. Known locally as the banakon, and familiar to a number of tribes (in particular the Matigsalug), it reputedly inhabits Mindanao's rice paddy fields. According to native accounts, the banakon is very large, venomous, aggressive, and highly territorial. They describe it as black or very dark in colour, with a diamond-shaped mark on its mullet-shaped head ('banakon' apparently translates as 'mullet-head'), plus two small horn-like head projections, and the distinctly un-snake-like ability to crow, albeit with a deep pitch. As I have discussed in my book, there are various basic physiological reasons why we should not expect snakes to hear, want to make, or indeed be able to make, loud noises, but that in spite of these, certain species of snake can definitely produce vocal sounds beyond those reptiles' traditional sibilant hisses. One such snake is the king cobra *Ophiophagus hannah*, which is known to emit unusually deep growling sounds

Alien Zoo

that its species can indeed hear, and it is interesting to note that some investigators have suggested that the banakon may be this selfsame species. Having said that, if descriptions of its morphology are accurate, it is quite different in appearance from the king cobra. Moreover, some descriptions of the banakon are wholly different from the version given earlier, stating instead that it is of variable size, silver in colour with iridescent scales, and, most remarkable of all, a pair of short legs near its head. This last-mentioned feature readily recalls Mexico's legged amphisbaenians (worm lizards) known as ajolotes, and also certain species of skink lizard. Could there be two entirely different but equally cryptic reptiles involved here, one a large snake, the other some form of amphisbaenian or true lizard? Or perhaps one or both are merely fantasy creatures of local legend. **http://www.cryptozoology.com accessed 4 December 2008.**

The Mexican ajolote, a legged amphisbaenian

The Lost Ark

A NEW EDEN IN NEW GUINEA.

There ought to be a similar zoological principle, based on the difficulty of making field observations of large animals, man's natural rivals, which tend to disappear as soon as man arrives on the scene. The degree of difficulty depends on their habitat: obviously it is greater in the water than on land, and greater on land than in the air; greater in mountains than in plains; greater in the forest than in the savannah or the bush, and greater in the bush than in a sandy desert. This is why we know so little about the creatures of the sea and so much about birds – except those species that cannot fly. The difficulties are almost insuperable when two coincide, as they do in mountainous country with thick vegetation. That is why most unknown animals live in marshy or mountainous forests. It is a mere matter of logic.

Bernard Heuvelmans – *On the Track of Unknown Animals*

Judging from the rampant hyperbole filling the headlines of worldwide news reports covering this expedition story in 2006, one might have assumed that at the very least a herd of sauropod dinosaurs had raised their heads above the leafy canopy or a phalanx of pterodactyls had swooped over the intrepid explorers in this "Lost World" or "Jurassic Park". Inevitably, the truth was less dramatic, but still no less significant, or interesting, as revealed here.

UNVEILING THE HIDDEN HONEYEATER OF FOJA

The geographical region in question is an undeniably remote, little-traversed area of the world - Irian Jaya, or Indonesian New Guinea, occupying the western half of the island of New Guinea. And the precise location is notably inaccessible - the mist-shrouded forested slopes of the 7,000-ft-high Foja Moun-

Alien Zoo

tains, uncolonised by local tribes, and never previously visited by a major modern-day scientific expedition. During November and December 2005, however, its exile from human scrutiny came to an end when a global team, co-led by Dr Bruce Beehler of Conservation International, spent a month here - exploring, examining, and identifying its rich diversity of wildlife, which included some noteworthy surprises. Indeed, so significant were the team's findings that in 2006 the expedition was commemorated philatelically, by Indonesia's issuing of a set of postage stamps, miniature sheets, and first day covers depicting some of its finds.

The four-stamp set of Indonesian stamps from 2006, featuring the wattled smoky honeyeater (bottom left) and the golden-fronted gardener (bottom right)

The first of these surprises also turned out, taxonomically-speaking, to be the best, because the first bird to attract the scientists' attention after being dropped into the area by helicopter proved to be a totally new species. Readily distinguished from all related forms by its wattled, orange-hued face, it was a previously unknown species of honeyeater, which in 2007 Dr Beehler formally dubbed the wattled smoky honeyeater *Melipotes carolae* (after his wife, Carol), and is the first completely new species of bird discovered in New Guinea for several decades.

A LOST GARDENER AND A GROWLING PARADISE BIRD

Two very important avian rediscoveries were also made here during this expedition. One was a famously elusive species of bowerbird called the golden-fronted gardener *Amblyornis flavifrons*. Until 1981, it was known only from a handful of museum skins of unknown provenance (they had been purchased at plume markets in Europe), and over a dozen searches of potential native locations in New Guinea had failed to uncover its homeland. In 1981, however, a breakthrough happened when American zoologist Dr Jared Diamond, visiting the Foja Mountains, encountered quite a number of living specimens. He even filmed the eyecatching male's bower-building, blue berry-luring courtship of the dowdy female - only to lose his precious evidence later when all of his films and camera equipment were ruined during the capsizing of his boat in a river. Now, however, this new expedition has repeated his success by spotting, and filming, this near-legendary bird, thus confirming its provenance as Foja.

Equally evanescent was - until now - a beautiful type of six-wired bird of paradise known as Berlepsch's parotia, usually classed as a subspecies of Queen Carola's parotia *Parotia carolae* but sometimes accorded distinct specific status as *Parotia berlepschi*. Formally described in 1897 and named after German ornithologist Hans von Berlepsch who had chanced upon some taxiderm museum specimens, all male, it was another 'homeless' New Guinea enigma - its stuffed specimens had definitely originated somewhere in this huge island, but where? Once again, Diamond had helped to fill in a gap, when in 1979 he had encountered what he considered to be some female specimens of this rarity, alive and well and living in the Foja Mountains. In December 2005, however, Beehler's team achieved a notable first. When a male and female flew into their camp, their startled observers duly became the first scientists to observe a living male Berlepsch's parotia - its throat adorned in gleaming metallic plumage,

Alien Zoo

Miniature sheet in the 2006 Indonesian series of Foja-commemorating stamps, featuring a male golden-fronted gardener

its head sporting a striking crest of six disc-tipped, wire-like plumes, and giving voice to an unsettlingly raptorial growl, far removed indeed from the dulcet tones that one might expect to emerge from the throat of a paradise bird!

GIANT RATS AND GIGANTIC RHODODENDRONS

Other Foja discoveries included 20 new species of frog, four new species and several new subspecies of butterfly, some new small species of mammal (including a pygmy possum), plus a giant rat measuring approximately 55 cm long and weighing in at a hefty 3 lb (1.4 kg), five new species of palm tree, and, most spectacular of all, a currently unidentified form of gigantic rhododendron - whose white, scented, six-inch-wide flowers are as large as the biggest rhododendron flower ever previously recorded.

Wandering fearlessly into the scientists' camp in search of food, the giant rat caused quite a stir when it was first spotted – due to its extraordinary size, it was initially mistaken for a cat! Laying claim to the title of the world's biggest known form of rat, this very rangy rodent is further distinguished by its long grey hair and very wide underside. Belonging to the *Mallomys* genus of woolly rats, it may well constitute a totally new species.

Moreover, although not a new species, another very important zoological find in the Fojas was a sizeable marsupial called the golden-mantled tree kangaroo *Dendrolagus pulcherrimus*, because this mammal had hitherto been documented only from Papua New Guinea, not Irian Jaya. So too, until now, had

Xenorhina arboricola, a little-known species of frog.

PINOCCHIO FROGS, GARGOYLE GECKOS, AND DWARF WALLABIES

In November 2008, Dr Beehler led a second Conservation International expedition to the Fojas, lasting a month. And as the official press release announcing their findings subsequently revealed when it was made public in May 2010, the team had once again proved to be highly successful at discovering some truly remarkable animals. One of their finds this time was a seemingly new species of *Litoria* tree frog instantly distinguished by the long pointed protuberance on its nose that raises up when the male is calling out, but deflates and drops back down when he becomes less active. Like so many interesting zoological discoveries, this curious amphibian, aptly dubbed the Pinocchio frog, was chanced upon completely by accident, when herpetologist Paul Oliver happened to notice it sitting on a bag of rice in the campsite.

Another extraordinary find was a grotesque, gargoylesque gecko with bent toes and huge bright-yellow eyes. Nor should we overlook – though it may well be easy to do so – the minute *Dorcopsulus* forest wallaby encountered here that is not just a new species but may well be the smallest known species of wallaby anywhere in the world. Other noteworthy finds included a new *Syconycteris* blossom bat, and an unexpected tricolour-plumaged form of *Ducula* imperial pigeon (sporting rust, white, and grey feathers) that once again could prove to be a hitherto-undescribed species.

True, they may not be on the Sir Arthur Conan Doyle scale of lost world discoveries (media headlines notwithstanding), but the fauna and flora of Foja have proven to be serious eye-openers, especially to the armchair cynics who routinely disparage suggestions that our world has any little-explored, virgin territory still awaiting scientific investigation.

EDEN REGAINED?

To my mind, however, perhaps the most inspirational aspect of these expeditions' revelations is not even the discovery and rediscovery of novel species. What amazed me, in a modern-day world in which the animal kingdom inevitably flees both metaphorically and literally at the coming of humankind, was to learn that the fauna of Foja showed no fear whatsoever of the expedition members. Even species as notoriously shy and reclusive as long-nosed spiny anteaters, rarely before seen by scientists in the wild, showed no inclination to hide from the team - on the contrary, they actually allowed themselves to be picked up and casually examined. Just as unafraid and seemingly content to be held was the aforementioned tree kangaroo, squatting affably like a pet dog in the laps of researchers. Living in an area unpopulated - and hence not hunted by - native tribes, and previously unexplored by scientists, these and other animals here have simply never learnt to be afraid of humans, a state of existence in this 21st-Century planet that is both exceptional and uplifting.

Indeed, whereas I must agree to differ with headlines that have labelled the Foja Mountains as a Lost World or a Jurassic Park, there is one description that has been applied to this astonishing region which I consider to be singularly apt - a New Eden. Truly, the Foja Mountains, whose animals continue to live without fearing humans and show nothing but gentle curiosity when encountering our species, must be the closest vision of that blessed Garden remaining on earth today. Let us pray, therefore, that now its glorious domain has been entered by modern-day humanity and its treasures revealed to the world outside, we will preserve its sanctity for all time - and not become the serpent of downfall and destruction for this New Eden.

SELECTED REFERENCES

ANON. (2007). Giant rat found in 'lost world'. BBC News, 18 December.

ANON. (2007). Giant rat discovered in Indonesia [Irian Jaya]. Independent Television News (ITN), 18 December.

BEEHLER, Bruce M., *et al.* (2007). A new species of smoky honeyeater (Meliphagidae: Melipotes) from western New Guinea. *Auk*, 124 (July): 1000-9.

DERBYSHIRE, David (2010). Found in the wilds, a frog like Pinocchio. *Daily Mail* (London), 18 May.

FOGARTY, David (2010). Scientists find tiny wallaby, spiky nosed frog in Asia [Irian Jaya]. http://www.newsdaily.com/stories/tre64g41d-us-indonesia-species/ 17 May.

HANLON, Michael & SHEARS, Richard (2006). Paradise found. *Daily Mail* (London), 8 February, pp. 22-3.

KIRBY, Terry (2006). Paradise found! Scientists hail discovery of hundreds of new species in remote New Guinea. *The Independent* (London), 7 February, pp. 1-2.

Alongside thylacine picture in my study (Dr Karl Shuker).

2009

TWO TASSIE EXPEDITIONS. Two separate expeditions seeking the 'officially' extinct thylacine or Tasmanian wolf *Thylacinus cynocephalus* (or Tassie for short!) in its island homeland have set out in February 2009. One features longstanding Tassie enthusiast Chris Rehberg, Sydney-based webmaster of the in-depth Where Light Meets Dark website devoted to Australian cryptozoology, who has been placing a number of trail cameras in strategic locations during the previous weeks in partnership with the Tasmanian Parks and Wildlife Service, and will now be investigating whether any of them have recorded any evidence of thylacine presence. The other expedition is a one-man quest launched by fellow cryptozoological researcher Matthew J. Coefer of New Jersey, his third in search of evidence for Tassie survival in Tasmania. In 2002, Matt founded a company called ESIR Inc (Extinct Species Investigation & Rediscovery), and his first Tassie expedition, which took place that same year, was a preliminary assessment of the areas that he had been researching as potential thylacine hotspots. Expedition #2 occurred in 2005, during which time he met up with veteran Tassie seeker James Malley. Fully equipped with infrared cameras, military night vision goggles, and a night-vision camcorder, Matt's baited camera areas acquired superb photos of every major bush species – except for the thylacine. So now he hopes for a third-time-lucky result in February, as indeed we all do for him. **http://www.wherelightmeetsdark.com accessed 4 February 2009; Matthew J. Coefer, pers. comms., 29 January 2009.**

UNPUBLISHED TASSIE PHOTO? Matthew J. Coefer has also emailed to me what would appear to be a hitherto-unpublished b/w photo of two thylacines in captivity, and which is now reproduced here with Matthew's kind permission. He mentioned that on the reverse of the photo was a pencilled title "Tigers Hobart. Wolves", and I learnt from him that the photo had been taken by a Mr George H. Judd, who travelled extensively around the world during the early 20th Century. The person who had provided

Alien Zoo

Matthew with these details believed that the photo had been taken around 1920, but was under the misapprehension that the thylacine in front was suckling some young pups along its abdomen. However, when the photo is examined more closely, it can be discerned that these 'pups' are actually the folded up legs and paws of the front thylacine and also one leg and paw of the back thylacine, which are projecting out from beneath the front thylacine's chin. In any event, the thylacine was not a true (placental) wolf but a marsupial, in which the female's teats (four in number, arranged as two pairs) were present not along her abdomen but were within her rearward pouch instead. So thylacine pups suckled inside her pouch, not along her abdomen.

I showed this photo to Scottish cryptozoologist Alan Pringle, who has an unrivalled knowledge and collection of thylacine material, and although he was immediately certain that he had seen it before, he could not find any trace of it in any published source of thylacine pics known to him. What Alan did find in his collection, however, was an unlabelled print of this selfsame photo on Kodak copier paper. He suspects that it was brought back for him by fellow cryptozoological enthusiast Dennis Vrettos during his first visit to Tasmania back in 2001, as Dennis brought back a number of Tassie photos that he had purchased from a market stall in Hobart – and whose owner had told him that some of them had never been published. If any *FT* reader can find a published version of this enigmatic picture, we'd love to hear from you. Otherwise, it may indeed be that its reproduction here is an *FT* world exclusive! [No-one has since come forward with news, so it does indeed appear to be previously unpublished.] **Matthew J. Coefer, pers. comms, 29 January 2009; Alan Pringle, pers. comm., 2 February 2009.**

ATTENBOROUGH AND APE-MEN. Since his fondly-remembered 1970s BBC series 'Fabulous Animals', which surveyed a wide range of mystery and mythological creatures, TV naturalist Sir David Attenborough has seldom commented on cryptozoological matters. During a recent interview with Jonathan Ross on the latter's 'Friday Night With...' show, however, Attenborough confessed that he was "baffled" by reports of the Himalayan yeti or abominable snowman, stating: "Very convincing footprints have been found at 19,000 ft. No one does that for a joke. I think it's unanswered". **Press Association National Newswire, 28 February 2009.**

THE TIGER THAT THINKS IT'S AN ANT! In the current issue of *BBC Wildlife Magazine*, TV explorer-presenter Steve Backshall reveals his close encounter with what appears to have been a very remarkable yet previously-undocumented and still scientifically-undescribed insect. While filming the documentary 'Expedition Borneo' in 2006, at one stage Steve found himself hanging precariously by his fingers from the edge of a steep vertical cliff face rising out of a remote hitherto-unclimbed canyon called Imbak. And while there, he was startled to see what looked on first sight to be a huge bright red ant, the size of a baby's finger and the colour of molten lava, scuttling speedily along the rocks around his fingers.

Closer observation, however, revealed to his amazement that this 'ant' was in fact an extraordinary, totally novel ant-mimic. Namely, a species of tiger beetle whose abdomen was greatly attenuated in an hour-glass shape, thus deceptively imitating the familiar 'waist' of genuine ants. Moreover, this imposter was actually hunting the smaller black ants milling around, seemingly tricked into complacency by their persecutor's ant-like disguise. Because of his predicament, however, Steve was unable to capture this singular specimen, but when he eventually managed to extricate himself (by which time of course his veritable beetle in ant's cuticle was long gone) and described it back at camp to expedition entomologist Dr. George McGavin, George duly confirmed that nothing like it had been described before, and that it was therefore certainly a new species! [Since writing this article, I've also found an earlier account of this beetle by Steve, in *Exotic Pets*, autumn 2007, published by CFZ Press] ***BBC Wildlife*, vol 27, March 2009.**

The previously-unpublished thylacine photograph (Matthew J. Coefer/Alan Pringle)

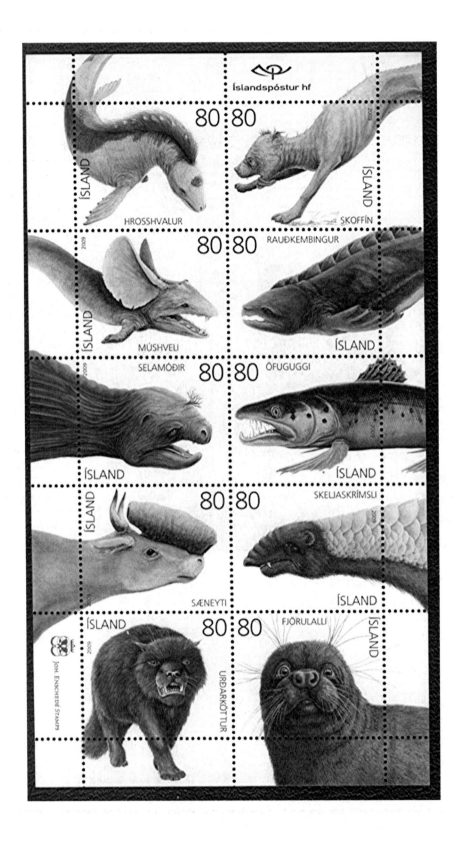

Alien Zoo

FACING PAGE: The full 10-stamp sheet of legendary Icelandic beasts from my collection of cryptozoological stamps (Dr Karl Shuker)

ICELAND'S STAMP(S) OF CRYPTOZOOLOGICAL APPROVAL. One of my more recent books, *Dinosaurs and Other Prehistoric Animals on Stamps* (2008), contains a special appendix devoted to stamps depicting creatures of cryptozoology, including such famous examples as the yeti, bigfoot, Nessie, thylacine, Ogopogo, and all manner of sea monsters. On 19 March 2009, moreover, a new set of postage stamps with a cryptozoological and zoomythological theme was issued by Iceland, and the creatures portrayed on this mini-sheet of ten extremely attractive stamps are so unusual and hitherto-obscure that they definitely deserve a mention here. Perhaps the best-known is the skoffin, a basilisk-like entity depicted on its stamp as mammalian in nature, which I have documented in the past. Far less familiar, however, are bizarre creatures such as the coast-inhabiting skeljaskaimsli (shell monster), a rusty-brown, multi-limbed animal with a tapir-like trunk and pangolin scales; a grave-robbing mystery felid called the urdarkottur or ghoul cat; and the highly poisonous ofuguggi or reverse-fin trout, whose fins all turn forwards instead of back. Completing this interesting set are three different varieties of mystery whale - the red-maned hrosshvalur (horse-whale), the raudkembingur or red-crest (both inspired perhaps by oarfishes *Regalecus glesne*?), and the massive-eared mushveli (mouse-whale); the sheep-molesting, shore-dwelling fjorulalli or beachwalker; the huge selamodir (seal mother), which protects normal-sized seals if they are threatened; and the amphibious saeneyti (sea cattle) that sometimes mingle with true cattle. A useful new book, *Meeting With Monsters*, which documents many of these curious Icelandic crypto-beasts, and is available in English and Icelandic, can be ordered from Postphil Iceland. **https://www.postur.is/cgi-bin/hsrun.exe/Distributed/Postphil/Postphil.htx;start=DetectLanguage? Language=LANG_English accessed 19 March 2009.**

IN SEARCH OF SIBERIA'S SASQUATCH. During late March 2009, a two-day scientific expedition led by veteran man-beast seeker Igor Burtsev (Director of the International Centre for Hominology) journeyed to a remote Siberian cave, where there has been a recent spate of sightings of entities referred to colloquially by Burtsev as bigfoots, but by locals as 'the dark people'. Azasskaya Cave, which is several km long and passes beneath a riverbed, is situated 120 km off the town of Tashtagol in Kemerovo Region. Here, eyewitnesses have lately reported seeing thickset man-beasts, standing upright, approximately 2 m tall, covered in black and red fur, very reminiscent of bears, and able to climb trees. The team includes several of the local hunters who reported sightings. A second expedition to this region in search of its 'dark people' is planned for June 2009. **http://www.mosnews.com/ weird/2009/03/23/bigfoot/ 23 March 2009.**

THE SEAL(S) OF APPROVAL. The publication in a peer-reviewed academic journal of a paper dealing with cryptozoological animals is probably just as rare as the beasts documented in it – which is why the recent appearance in *Historical Biology* of a paper dealing with putative pinnipeds (mystery eared and earless seals) is of particular note, and, one hopes, an indication of increasing mainstream approval for serious cryptozoological research. Authored by Royal Holloway, University of London postgraduate biology student Michael Woodley and Portsmouth University palaeontologist Dr Darren Naish, both with well known cryptozoological interests, together with Royal Holloway computer scientist Dr Hugh Shanahan, it is entitled 'How Many Extant Pinniped Species Remain To Be Described?'. In it, the authors examine the description record of the pinnipeds using non-linear and logistic regression models in an attempt to ascertain the number of still-undescribed forms, and they combine that work with an evaluation of cryptozoological data, featuring such alleged pinniped cryptids as the longneck sea serpent, the merhorse, Vancouver's Caddy (=Cadborosaurus), and the tizhurek. From the results obtained, they reveal that three possibly new, currently undescribed species of pinniped match their statistical expectations, but even these, the authors feel, would need to possess some exceptional characteristics if they do indeed exist. ***Historical Biology*** **press release, 24 March 2009, doi:10.1080/08912960902830210**

DEALING WITH THE DUENDE. Traditionally reported from Mexico and several Central American countries, the duende or dwendi are said to be short, hairy, goblinesque entities, mainly nocturnal, and sometimes wearing rags for clothes. Certain reports claim that they wear large straw hats, others that they hold the big leaves of the banana palm over their heads as makeshift sun-shielding sombreros (behaviour also documented, incidentally, from chimpanzees). Although the duende have attracted some cryptozoological attention, they are usually discounted as folklore and fairytale rather than factual, with no objective reality. However, in view of a remarkable first-hand eyewitness report recently emailed to me, that notion can no longer be deemed tenable.

In May 2009, I was contacted by Don Campbell, who had read a recent ShukerNature blog of mine at www.karlshuker.blogspot.com on the subject of New World littlefeet (mini-humanoids). Don had experienced a littlefoot encounter in Chiapas, Mexico, during 1976, while working as a nature photographer for a now-defunct adventures travel group, taking tourists to exotic locations worldwide. Quoted here with his kind permission, this is how Don described to me his experience with the duende:

Duende-seeker Ken Gerhard in Belize, 2004 (Ken Gerhard)

Alien Zoo

While in the Chiapas jungles and in the highlands we were warned by the locals about the bigfoot type creatures (Guymas) and the little Duendes del Bosque or forest goblins. The bigfoot creatures were aggressive and we told (by the local police) to shoot them if we encountered them. We did encounter the duendes and they were a small hirsute human about four feet tall. The males were four feet tall and the females were about four to five inches shorter. The hair was black mostly but I did see some with reddish brown hair. The hair was about two to three inches long. These duendes would run through our camps and steal food or bits of clothing - usually hats or shirts. We would later find the clothing hanging on a bush. I tried to have my camera available to take pictures but they came at irregular intervals and it was impossible. I asked the locals about the the duendes and were told that they were harmless tricksters who raided peoples huts and homes looking for food and other useful things likes pots.

Judging from this fascinating and very sober, matter-of-fact account, the duende are real, corporeal entities, as opposed to mythical Little People. And far from being diminutive man-beasts either, as has also been suggested by some authors in the past, they clearly constitute a tribe of small, pygmy-like, forest-dwelling people. Indeed, Ivan Sanderson suggested long ago that the duende may be relics of a pygmy Mayan race. Of particular interest is Don's mention of female duende – only very rarely are female representatives reported in relation to any of Latin America's supposed Little People, and yet obviously they must exist for any real-life tribe to be viable. Also of great worth here are the apparent duende tracks obtained by field primatologist Ken Gerhard during a duende-seeking expedition to Belize in 2004, as previously documented in Alien Zoo. **Don Campbell, pers. comm., 6 May 2009.**

AND WHATEVER HAPPENED TO THE KUIL KAAXS? Returning to Mexican mini-men: Those mysterious dwarf-sized humanoids known as the alux, long reported from the Yucatan Peninsula, are, like the duende, well documented in the cryptozoological and mythological literature. In stark contrast,

a third Mexican mini-human entity rates scarcely a mention. I first learnt of the kuil kaax many years ago, when reading a traditional Mexican folktale concerning the basilisk lizard. According to the tale, the kuil kaaxs existed when the world was still young, were nature spirits of the forests that resembled little old gnome-like or dwarfish men, and – recalling duende descriptions - wore huge sombreros made out of leaves from the banana palm. Kind and gentle, the kuil kaaxs were revered almost as demi-gods by the animals of the forest. This description (sombreros excepted!) closely corresponds with reports of diminutive humanoids reported widely across the Americas in modern times, yet, bizarrely, the kuil kaax has received so little formal attention that even a Google search by me failed to elicit more than a brief note within a Russian account (see url below) of Mesoamerican dwarf deities. So if anyone out there has any additional information concerning the kuil kaax, I'd love to hear from you. **http://www.mesoamerica.ru/indians/maya/dwarfs.html accessed June 2009.**

A kuil kaax (Tim Morris)

Alien Zoo

TRACK ME KANGAROO DOWN SPORT. One of the most memorable cryptozoological discoveries of modern times was the dingiso *Dendrolagus mbaiso* – a sizeable, dramatically new species of tree kangaroo, distinguished by its panda-like black-and-white markings and its whistling cry, which was successfully tracked down in Irian Jaya, western New Guinea, by Australian zoologist Dr Tim Flannery in 1994. However, there may be a similarly noteworthy sequel still to take place on the neighbouring eastern island of New Britain.

In 2007, John Lane, principal scientist at Chico Environmental Science and Planning in Chico, California, led a six-week expedition to New Britain, and while there his team not only reported a number of new species of butterfly, frog, snake, and fish, but also sighted some tree kangaroos. They duly filed a report of their findings to the Nature Conservancy, and earlier this year Lane was informed that the presence of tree kangaroos on New Britain was unexpected and may involve a new species. Aware of the conservation publicity value of discovering a major new mammal here (this area is at risk from loggers and palm-oil encroachment), in July 2009 Lane returned to New Britain with a team of explorers, to track and possibly even capture a specimen of this intriguing creature in order to obtain DNA samples.

They have now returned home to the USA, so we await the release of Lane's latest findings report with great interest – not least of all because Lane is also aware (and has not discounted the possible reality) of mysterious creatures resembling living pterodactyls reported from New Britain's sparsely-explored rainforests. **http://www.newsreview.com/chico/content?oid=1004993 4 June 2009.**

LOOK OUT, LOOK OUT, THERE'S A DINOSAUR ABOUT! Over the years, there have been a number of so-called 'living dinosaur' reports filed from the USA, describing a mystery reptile seen running erect on its hind legs and likened by its startled eyewitnesses to a small bipedal dinosaur.

I personally consider it likely that these sightings are based upon encounters with escapee monitor lizards (varanids), which can indeed run for short bursts on their hind legs. However, this very morning I received a letter from a longstanding colleague of mine, English zoologist Prof. John L. Cloudsley-Thompson, that contains a very different, and hitherto-unreported, American mystery dinosaur report:

> During 1969, I was a National Science Foundation Senior Research Fellow at the University of New Mexico, Albuquerque, U.S.A. One evening, my wife Anne and I were invited to dinner by an anthropologist we knew. He told us how the driver of the school bus was driving to collect the children when he saw a large animal step over the fence at the side of the road. Some weeks later, he saw our friend's children with a dinosaur book. He claimed he had never seen a picture of a dinosaur before and the animal he had seen was exactly like the picture of *Diplodocus*. So our friend advised him to keep a camera in the bus. He saw the animal once again, but he had forgotten the camera!

Apart from the extraordinary thought that there could be anyone out there in the States who had reached adulthood without ever having seen the image of a dinosaur before, this is a very intriguing account, to say the least! If anyone can shed any light on it, or know of similar reports of American long-necked quadrupedal dinosaur-lookalikes - veritable mokele-mbembes in miniature! - we'd love to hear from you! **Prof. John L. Cloudsley-Thompson, pers. comm., 7 July 2009.**

Alien Zoo

OOPS, IT'S THE UPAH – SUMATRA'S GIANT SHRIEKING CENTIPEDE! Never heard of this cryptid before? Neither had I until I read explorer-ecologist Jeremy Holden's illuminating account of it in this month's *BBC Wildlife Magazine*. Several years ago, while he had been visiting a small village in western Sumatra, the locals had solemnly revealed that the jungle creature they feared the most was a giant 1-ft-long tree-dwelling centipede with a thick ghostly-green body, an agonisingly painful bite, and – most extraordinary of all - the unique ability (for a centipede) to shriek or yowl like a cat. Although initially sceptical of such a bizarre beast, which the locals termed the upah, several weeks later while walking through the forest Jeremy suddenly heard a loud cat-like cry coming from high up in the canopy, followed by a rattling 'churr'.

He tracked the precise location of this eerie sound, using his binoculars, to a hollow branch, and his native guides excitedly confirmed that this was indeed the upah calling, but the creature itself remained unseen. More recently, however, while walking through Kerinci Seblat National Park with the eminent birdwatcher Frank Lambert during a return trip to Sumatra, Jeremy heard the very same cry – the cat-like yowl, then the rattling churr – once again emanating from the canopy. Sadly, however, the true originator of these sounds, although certainly elusive, proved to be something rather less exotic than a vocalising centipede.

Instead, as instantly recognised by his birder companion, the species responsible was none other than the Malaysian honeyguide *Indicator archipelagicus*, a bird famous in ornithological circles for being much easier to hear than observe. Faced with such an anomaly – a creature often heard but seldom seen - it is understandable, perhaps, that the villagers' imagination would, down through the ages, ultimately conjure up a truly dramatic yet wholly fictitious monster to explain it. There's a lesson in here for cryptozoology somewhere! *BBC Wildlife Magazine*, **vol 27, August 2009.**

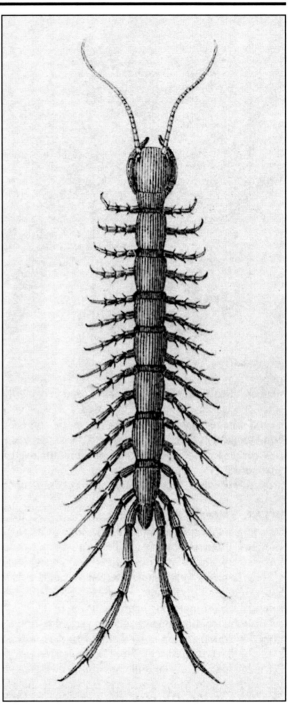

When is a centipede not a centipede?
When it's an upah

Alien Zoo

New Caledonian lorikeets, painted by 19th-Century artist John Gerrard Keulemans

LOOKING FOR THE LOST. In August 2009, BirdLife International announced that it is launching a global bid to seek 47 so-called 'lost' species of bird, some of which have not been officially recorded for more than 180 years.

The species named include some 'classic' extinct species, like the pink-headed duck *Rhodonessa caryophyllacea*, Himalayan mountain quail *Ophrysia superciliosa*, crested shelduck *Tadorna cristata*, New Caledonia's endemic lorikeet *Charmosyna diadema* plus its owlet-nightjar *Aegotheles savesi*, and the glaucous macaw *Anodorhynchus glaucus*, as well as such infamously elusive examples as the Australian night parrot *Pezoporus* (=*Geopsittacus*) *occidentalis*, eskimo curlew *Numenius borealis*, slender-billed curlew *N. tenuirostris*, and ivory-billed woodpecker *Campephilus principalis* – these latter species having all been reported in recent years, only to vanish again as swiftly as they reappeared. The rediscovery of any of the species on BirdLife's list of the lost would constitute significant ornithological events, and provide real hope that even after many years without verified sightings, species can and do persist. **http://www.birdlife.org/news/news/2009/08/lost_and_found.html 21 August 2009;**
http://www.wildlifeextra.com/go/news/extinct-birds.html#cr 25 August 2009.

DOWN AMONG THE DEATH WORMS. At the end of August 2009, after braving two weeks in Mongolia's inhospitable Gobi Desert seeking its elusive (and potentially lethal) death worm or allghoi khorkhoi – claimed by the local nomads to spit venom and kill by a mystifying process suspiciously similar to electrocution – journalist David Farrier and cameraman Christie Douglas have returned home to New Zealand. Although they haven't brought a death worm back with them, they recorded around 30 hours of expedition footage, in which they interviewed a number of alleged eyewitnesses, and hope to compile a 90-minute documentary from it. This will be a particularly valuable endeavour, because death worm sightings appear to have peaked during the 1950s, which means that many of the witnesses from that period spoken to by the NZ team are now elderly, and if their testimony had not been filmed, it may well have been lost forever in the not-too-distant future. David believes that the death worm does exist, and has not ruled out the possibility of a second expedition in search of it, which would also pursue various other Mongolian cryptids, such as the almas, a reputed man-beast. **http://www.odt.co.nz/entertainment/film/72038/deathworm-discovered-documentary-definite 31 August 2009.**

A MUDDLE OF MERMAIDS. On 29 September 2009, I was emailed by American coast guard Christopher M. Younger, who kindly brought to my attention an extraordinary report filed a couple of months earlier by the online Fanoos Encyclopedia. It reads as follows:

Alien Zoo

Real mermaid discovered in Batroun

For six years now, rumours of an alleged mermaid visiting the shores of Batroun [in Lebanon] around early spring have been received with much scepticism. The reported encounters were so numerous that many residents of Batroun swore mermaids are real and not mere fantasy creatures from fictitious legends. Some were even convinced, like Emme Aoun, who claims to have seen one in 2004, that they might be trying to warn us about something such as an upcoming Tsunami.

But recently, the astonishing discovery of what appears to be a dead "mermaid" that had perished on the shores of Batroun, near The Phoenician Wall, has left even the most pragmatic speechless! Toufic Shebtini and his girlfriend Mona Chedid were the first to come across the decaying carcass, while on a romantic stroll. They have been officially warned not to give out any interviews or answer to inquiries even by their closest relatives. This amazing incident was quickly contained and intentionally smothered! It is said that two foreign men alongside a woman (most probably German) working aboard a German vessel present in the Lebanese waters at that time, were seen on the premises within minutes. The corpse was immediately shipped to Germany under very tight security, before heading to the University of Birmingham where it was put in the care of Dr. Karl P. N. Shuker.

British marine biologist Dr. Shuker is actually a zoologist specialized in cryptozoology, which is the study of "hidden" animals. Cryptozoologists look either to find creatures like the Loch Ness Monster and the Yeti, or "re-find" seemingly extinct species like the Tasmanian wolf (Thylacine) and New Zealand's giant moas etc...

However, the found creature is far less attractive than the common mermaids in folktales usually represented with the body of a young, beautiful woman. (See the two-tailed siren illustration of "Hortis Sanitati" from 1491). Actually, this one has the withered body of a monkey and the dried tail of a fish. On first consideration, it is believed to most likely be an unknown form of primate adapted to sea-life! The photographs depicting the "mermaid" are absolutely authentic and are neither computer generated nor digitally retouched.

Awaiting the final verdict of authenticity by Dr. Shuker expected very soon. They have all refused to give any clarifications regarding the matter, complying with the imposed discretion.

Alien Zoo

Fascinating though it may be, this report (accompanied on the website by a photo of a 'stuffed' Feejee mermaid specimen, clearly composed of monkey and fish remains sewn together) has no basis in reality. This is because I can categorically confirm that no mermaid remains, alleged or otherwise, have ever been sent to me to examine! [Obviously I am not including here the Feejee mermaid that I received several months later as a gift from Alan Friswell – see below – because that specimen is wholly unrelated to this incident and in any case was not even made public by us until March 2010.] Hence I have no idea whatsoever how or where this claim originated.

Equally odd is the statement that I am based at the University of Birmingham (England), bearing in mind that although I did indeed obtain my PhD in zoology and comparative physiology there, I subsequently left to pursue a freelance writing and consultancy career over 20 years ago! All in all, a very baffling little vignette, and it isn't even 1 April yet! **Christopher M. Younger, pers. comm., 29 September 2009;** http://www.fanoos.com/special/real_mermaid_discovered_in_batroun.html accessed by me 29 September 2009.

MY VERY OWN MERMAID! In November 2009, after several months of skilful, painstaking work, expert movie model-maker Alan Friswell of London emailed me with the exciting news that I'd been eagerly awaiting since the beginning of the year when he had very kindly promised to manufacture one

My Feejee mermaid (Dr Karl Shuker)

Alien Zoo

Me and my Feejee mermaid (Dr Karl Shuker)

for me. Yes, my very own Feejee mermaid was finally complete!

During the 19th Century and earlier, these fabulous frauds were artfully constructed from the dried remains of small monkeys and large fishes by Eastern vendors, to sell at exorbitant sums to gullible Western travellers believing them to be bona fide mermaids, and later exhibited in sideshows by canny showmen to ingenuous paying visitors.

As mine is a modern-day Feejee equivalent, I'm very happy to say that although, thanks to Alan's diligence and expertise, it looks every bit as realistic and antiquarian as its infamous predecessors, no monkeys or fishes were utilised in any way in its construction.

In any case, as Alan pointed out when emailing me a detailed breakdown of its vital constituents, even the original Feejee mermaids owed a fair amount of their physical composition to simulated substances rather than genuine biological tissues:

> The figure is built up onto a reinforced steel skeleton, made from parts welded together.
>
> The muscles and bones are wood and resin, cut and sculpted into shape. The skin is formed from a type of gelatine glue, much the same as would have been used for the original mermaids in the old sideshows. The glue takes ages to 'cure', but in the drying process takes on a desiccated look, rather like dry skin. The teeth and fingernails are sculpted and then cast in a tough resin, and varnished with a shellac mixture, as were the teeth for the old mermaids - real monkey teeth were often too small, and replaced with handmade ones. The hair is from a section of hair that I bought from a suppliers years ago.

And so it was that after I informed my friends and colleagues that on 5 March 2010 I would be going out for the day to collect a mermaid, that is precisely what I did - travelling to Hornchurch that day to meet Alan and bring back home with me my latest and most extraordinary cryptozoological exhibit. It is, without question, spectacular testament to the survival of a unique and truly marvellous artform - creating with infinite care and ingenuity the mortal remains of exotic beasts that have never existed. My sincere thanks to Alan Friswell for so very kindly and generously donating so much of his precious time and very considerable skills to fulfilling a longstanding dream of mine of one day owning a Feejee mermaid. Who says that wishes can't come true? **Alan Friswell, pers. comm., 16 November 2009.**

THE APE-MAN COMETH? Over the years, I have received and documented a number of accounts from correspondents claiming to have seen an unknown, second photograph of Loys's infamous South American 'ape' (dubbed *Ameranthropoides loysi*) allegedly shot during a 1920s scientific expedition on

Alien Zoo

the Colombia-Venezuela border led by geologist Dr François de Loys, and officially known only from a single iconic photo in which its corpse is depicted by itself, propped upright on a crate, by way of a stick placed beneath its chin. Apart from the crate, there is nothing in that photo that can be used for scale purposes to ascertain the creature's size, leading to much speculation as to whether, despite its spider monkey appearance, it was much bigger than any known species of spider monkey.

However, the second, mystery photo is said by its eyewitnesses to have people standing alongside the alleged ape, thereby giving a good estimate of its size. In December 2009, after a long lull in reports of this alleged second photo, I received a most interesting email from Canadian correspondent Dorian Gallant, who believes that he saw it in 2002-2004 in a damaged book held at a library in Ottawa. This is his description of the book and its contents:

> It had a dark cover with no picture on it. The title may have had the word unexplained in it. The book seemed to talk a lot about mysterious animals, at least the parts of the book I remember did, although there may have been a few pictures of UFOs, however, I'm not 100% sure, as I was not interested in UFOs and probably skipped that part, if so, then there was most likely some stuff about Ghosts. Now what I remember most about it was the photo of Ameranthropoides. I saw two men standing beside it, one on each side, and it was clear to me that there was no way that was any <u>known</u> type of Monkey, it was too big, however it did look like one.

Alien Zoo

The men standing beside it didn't have a weapon as far as I can remember, and they had serious expressions on their faces. There was no way it was a chimp or a siamang, as it didn't look like either of those. Here's some other details about the book:

- The book talked about living pterosaurs too, and had a drawing of one, beside a picture of a rainforest.

- The book was not very long.

- The book had a picture of a Yeti Footprint

- The book talked a bit about the Gazeka

- Oddly enough, there was very little about Bigfoot in the book, and nothing about Nessie.

- I found the book in the Adult section of the library.

I have also been looking for the odd picture, with no success at all. I suspect that the book was discarded.

Dorian wondered whether I could identify the book from the above description, but unfortunately I have not been able to do so.

The gazeka (aka the Papuan devil pig) is a cryptid rarely alluded to in cryptozoology books other than various of my own, so this is probably the most significant detail in Dorian's account when attempting to uncover the book's identity. Is there any *FT* reader out there who knows which book this is? If so, please get in touch! **Dorian Gallant, pers. comm., 14 December 2009.**

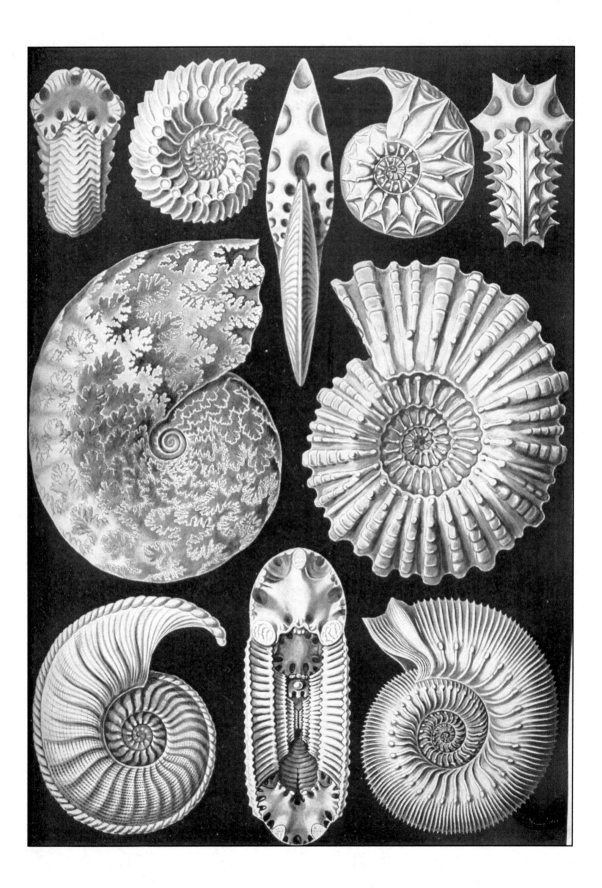

The Lost Ark

A SERPENT IN STONE
- FROM ADDER TO AMMONITE?

*When Whitby's nuns exalting told,
Of thousand snakes each one
Was changed into a coil of stone,
When holy Hilda pray'd;
Themselves, within their holy ground,
Their stony folds had often found.*

Sir Walter Scott - *Marmion*

One of the most baffling zoological mysteries is the spasmodic discovery of solid slabs of stone or rock that, when split asunder, are found to contain a toad or (less commonly) a frog, very much alive (if rarely well!), and often adult. The official explanation for any immured amphibian like this is that it originally entered the slab when young and small, through some inconspicuous hole, and, as it grew older, became too large to escape, but survived by feeding upon tiny insects and suchlike attracted inside the rock by the amphibian's odour. Moreover, as toads and frogs are able to absorb air directly through their skin as well as by inhaling it into their lungs, as long as the hole remained open or the slab itself was composed of a porous material (e.g. limestone), the creature would not suffocate.

FACING PAGE: Ammonites portrayed by Ernst Haeckel in 1904.

Alien Zoo

AN INCARCERATED ADDER?

Although a plausible explanation for certain instances of entombed toad or frog on record, it certainly cannot explain all cases, including the example considered here, which does not feature an amphibian, but rather a reptile. While browsing through the 8 March 1823 issue of a bygone English journal comprehensively entitled *The Mirror of Literature, Amusement, and Instruction*, I noticed the following account of a very remarkable reptile:

> The newspapers of June 1772, state, that a living adder [*Vipera berus*] was found in a block of stone of 30 French feet diameter, the centre of which it occupied. It was twisted nine times around itself in a spiral line; it could not support the weight of the atmosphere, but died in a few minutes after it was taken from the stone. On examining the stone, not the least crevice could be discovered through which it might have crept, nor the minutest opening through which it could have received fresh air, or inhale any sort of sustenance.

To explain the adder's incarcerated existence via traditional belief, we must postulate that the stone initially possessed an opening sufficiently large to permit the entry of a 5-8-in snake (the size of a young adder) before eventually becoming sealed in some way and thereby entombing the hapless animal indefinitely. And if the snake were already an adult, an opening large enough to facilitate the passage of a 20-24-in-long snake would have been required.

Even more perplexing is how any snake, young or adult, could respire whilst so interred, as scaly reptiles like this cannot respire through their skin. And how could it feed? Perhaps it didn't - perhaps it underwent a prolonged period of torpor, comparable to the enforced rest embarked upon by snakes during the cold winter months in temperate climates?

Only one thing, in fact, *is* certain. Assuming the report to be genuine, more than 200 years after the finding of this serpent in stone, science is still unable to provide a conclusive explanation for it. In contrast, an answer may be forthcoming from folklore.

...OR AN UNRECOGNISED AMMONITE?

Long-extinct relatives of today's squids, octopuses, and nautiluses, ammonites were characterised by their tightly-coiled shells, which have been discovered as well-preserved fossils within Mesozoic strata worldwide. Ammonite-containing rocks are common in many parts of Britain, and have inspired a great deal of unusual folklore, some of which is of particular relevance to early reports of entombed snakes.

For example, in his *Britannia* (1586), scholar William Camden

An ammonite (Dr Karl Shuker)

Alien Zoo

spoke of strange stones discovered near Whitby in Yorkshire: "...if you break them you find within stony serpents, wreathed up in circles, but generally without heads". These are obviously ammonites, which do look very like decapitated snakes on first sight. Indeed, there are even legends to explain why these ammonite-identifiable 'stony serpents' or 'snakestones' were headless.

In *'Formed Stones', Folklore and Fossils* (1982), Dr Michael Bassett, the National Museum of Wales's Curator of Geology, notes that according to early local belief, snakes were very common in the Whitby vicinity - until the 7th Century AD, and the coming of the Saxon abbess St Hilda. She successfully cleared a site for the building of her convent by transforming all the snakes in this area into stone, whereupon their heads dropped off too.

(Interestingly, there is many an ammonite specimen in existence, including the example depicted on the front cover of Dr Bassett's very informative booklet, that now sports a carefully-sculpted snake's head - carved by local sculptors from the rock encompassing the aperture of the ammonite's shell, in order to revive the ancient legends, and enhance the sale of fossils in the area!)

Dr Michael Bassett's fascinating booklet, alongside a model of a living ammonite and some fossil ammonite exhibits (Dr Karl Shuker)

There is also, of course, a similar and even more famous legend in the rich lore of Ireland - whereby as a reward for converting the Irish nation to Christianity, St Patrick was permitted to banish all serpent life from the Emerald Isle. In the words of some versions of the story, he achieved this by touching each snake with his staff, which instantly changed its sinuous form into the stone coils that we now

Alien Zoo

A very large, unmodified ammonite - and not a giant, headless serpent in stone as once believed (Dr Karl Shuker)

know to be fossil ammonite shells.

In view of the widespread occurrence of snake-related ammonite fables like these, it is quite possible that the case of the alleged adder imprisoned in stone was based upon nothing more substantial than a distorted account of an ammonite discovery. Well worth noting in support of this proposal is the close morphological correspondence between the flat, tightly-coiled spiral of many ammonites' shells and the report's description of the entombed adder as being "...twisted nine times around itself in a spiral line". Moreover, at the time of this report the true zoological identity of ammonites had not been widely disseminated in non-scientific publications, thus increasing the likelihood for confusion with snakes.

Like so many others before it and since, this intriguing little tale of the adder and the ammonite readily confirms that a mystery is merely the answer to a question that we have yet to formulate.

REFERENCES

ANON. (1823). [Living adder discovered encased in block of stone.] *The Mirror of Literature, Amusement, and Instruction*, 1 (no. 19; 8 March): 303.

BASSETT, Michael G. (1982). *'Formed Stones', Folklore and Fossils*. The National Museum of Wales (Cardiff).

SHUKER, Karl P.N. (1989). Reptiles in stone. *Fate*, 42 (March): 83-5.

SKINNER, Bob (1985. *Toad in the Hole* (*Fortean Times Occasional Paper*, no. 2). Fortean Times (London).

SHUKER, Karl P.N. (2010). Sea-dragons, fairy loaves and serpents of stone. Fables and fossils of Lyme Regis. http://www.darkdorset.co.uk/the_dorsetarian/0/fossil_folklore posted 23 July.

2010
- LATEST NEWS AND UPDATES.

SHORT TALE OF A SHORT TAIL. A disquieting yet recurring theme in cryptozoology is the capture or shooting of a strange creature and the sending of its body to a scientific research body or institution for examination and identification, only for it to be subsequently lost or forgotten about afterwards, and never referred to again. One such example, briefly reported in a local American newspaper as recently unearthed by cryptozoological researcher Richard Muirhead but previously undocumented in the fortean literature, features a very unusual wolf. Shot in Idaho's Boise National Forest in early 1909 by the forest's supervisor, E. Grandjean, it was said to be of huge size, with a black shaggy coat – and, unlike any previously-reported wolf, sporting an incongruous bobtail, instead of the lengthy tail typical for wolves. As no-one who observed its body had ever see anything like it before, Grandjean sent it for formal investigation to the Biological Survey at Washington DC, but, true to tradition, nothing more has ever been heard of it. ***The Standard*** **(Ogden, Utah), 8 April 1909; Richard Muirhead, pers. comm., 1 January 2010.**

LOCKING HORNS WITH THE GIANTS OF SPANISH HILL. I have always been greatly intrigued by claims that during the 1880s, several horned skulls of giant men (average height 7 ft) had been found inside a huge native American mound called Spanish Hill at Tioga Point, near Sayre, in Bradford County, Pennsylvania, and I briefly documented this subject in my book *The Unexplained* (1996). I had assumed that if such objects had truly existed, they must surely have been either malformed human skulls or even misidentified fossilised skulls of some exotic-looking prehistoric ungulate. In fact, as I have now learnt, courtesy of American correspondent Andrew D. Gable, the truth may well be very different, and rather more prosaic, albeit amusing. In an email of 17 January 2010, Andrew

Alien Zoo

kindly provided me with the following information:

> Louise W. Murray, a resident of the area on whose property many Indian artefacts had been found, wrote an article, "Aboriginal Sites in and Near "Teaoga", Now Athens, Pennsylvania" in 1921 that clarified the story of the skulls' discovery. According to Mrs Murray, the story of the horned men grew from a misinterpretation by the media of a quote spoken by a workman. Apparently one of the skeletons had a deer-antler headdress on, indicating that he was a chieftain, and the man's comment that "He has horns on his head!" was interpreted as meaning that he literally had horns sprouting from his head as opposed to merely wearing them on his head.
>
> It is true, however, that the Susquehannock Indians that were native to Carantouan (the old name for the Spanish Hill site) were unusually tall. Many were apparently 6-7 feet tall. One of the central figures in a local ghost story also was apparently part Susquehannock, and he was reputed to have been of great size as well, suggesting that it was not just that population that was tall but quite possibly the Susquehannocks in general.
>
> This is interesting because in a book making reference to the history of the Lenape (Delaware) Indians, reference is made to a neighbouring tribe of "giants". A reference to the Carantouan Susquehannock? The Lenape lived all along the Delaware River, especially in the area of Monroe and Pike Counties.

If this statement of Louise W. Murray is correct, and the skulls' horns were non-existent, this may explain another claim, made in Jim Brandon's classic book *Weird America* (1978). Namely, that although some of these skulls were later sent to the American Investigating Museum, the museum apparently has no knowledge of them. **Andrew D. Gable, pers. comm., 17 January 2010; http://www.spanishhill.com/skeletons/horns.shtml accessed 17 January 2010.**

NO YETI YET. I'm always pleased to receive an update of an ostensibly long-forgotten cryptozoological story, especially when it's a personal favourite of mine, like this one. As reported by Heuvelmans, myself, and others, back in 1953 a Tibetan lama called Chemed Rigdzin Dorje Lopu announced that he had personally examined the mummified bodies of two yetis – one at the monastery at Riwoche in the Tibetan province of Kham, the other in the monastery at Sakya, southern Tibet. According to Heuvelmans's account of this lama's very interesting claim:

> They were enormous monkeys about 2.40 m high. They had thick flat skulls and their bodies were covered with dark brown hair about 3 to 5 cm long. Their tails were extremely short.

The thought that such extraordinarily significant cryptozoological relics (if genuine) may still survive today has long intrigued me. Consequently, I was delighted when in February 2010 I was contacted by

Alien Zoo

The yeti (Richard Svensson)

correspondent Peter Pesavento who informed me that he had been actively pursuing this mystery himself, and had emailed both monasteries. Unfortunately, he did not receive a reply from Sakya, but Samten O'Sullivan had very recently replied to him on behalf of Riwoche. Samten informed Peter that, tragically, the monastery had been razed to the ground following China's annexing of Tibet and all of its precious contents had been looted or burnt. Consequently, although the monastery was subsequently rebuilt (and as an exact replica of the original), any yeti mummy that may have been in the original building is certainly not present in the new one.

Whether it was removed and taken elsewhere or simply destroyed, however, is another matter, which seems unlikely ever to be resolved. Nevertheless, even knowing where something is not (namely, in the new Riwoche monastery) is still better than knowing nothing about it at all. **Peter Pesavento, pers. comm., 13 February 2010.**

THE LITTLEST CHUPACABRAS. When a small but supposedly bona fide chupacabras was captured alive – a rare event indeed! – on a man's back porch in Dry Gulch, Oklahoma, in February 2010, and was found to be hairless, it seemed likely to be yet another of those odd fur-lacking canine creatures that have been seen and occasionally found dead in Texas during the past few years (but shown merely to be coyote or coyote-wolf hybrids rather than the dread chupa itself). However, when examined by a vet at a local animal rescue centre, Kojak (as the little critter was dubbed, despite being female!) proved to be something quite different. Namely, a completely bald specimen of the raccoon *Procyon lotor*, whose odd condition has been attributed to an advanced case of mange - the usual (if not always necessarily correct) explanation also offered for the hairless Texas wild dogs. Intriguingly, Kojak isn't the first such animal on record – only a month earlier, a hairless raccoon was found dead on the Runaway golf course in Wise County, Texas, where it had quite likely frozen to death in the wintry weather. Moreover, this specimen was confirmed by biologists *not* to be suffering from mange, hence its hairless condition is most probably congenital, as is true with the famous Mexican hairless breeds of dog and cat. **Kens5.com, 3 March 2010 (Oklahoma raccoon) & 28 February 2010 (Texas raccoon).**

A PLATINUM PUZZLE. On the evening of 13 March 2010, Shaun Histed-Todd was driving a bus along a Dartmoor road when he saw a most unusual creature run down the edge of the moor and stand at the road side, where the bus's headlights afforded him an excellent view of it for roughly half a min-

Alien Zoo

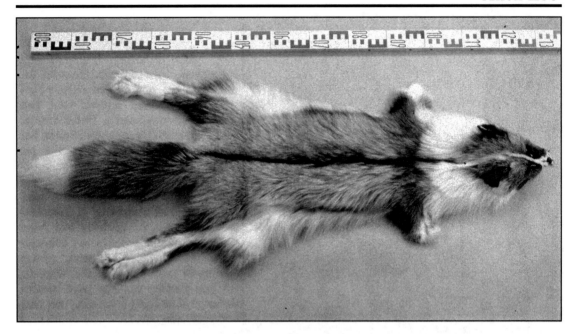

Pelt of a platinum fox (Mickey Bohnacker/Wikipedia)

ute before it ran back up onto the moors (Shaun has asked me not to make public the precise location, to protect the animal). Shaun contacted me a few days later, as he was unable to identify it, and provided me with a detailed description, whose most notable features were as follows. It resembled a young fox and had a bushy white-tipped tail, but its coat was dark silvery-grey, it had noticeably large ears, white paws, and a black raccoon-like facial mask. Reading this, I was startled to realise that Shaun's description was an exact verbal portrait of a most unusual yet highly distinctive animal – a young platinum fox. After checking photos of platinum foxes online, Shaun confirmed that this is indeed what he had seen. Arising in 1933 as a mutant form of the silver fox (itself a mutant form of the red fox *Vulpes vulpes*), its extraordinarily beautiful and luxuriant fur meant that platinum foxes were soon being bred in quantity on fur farms as their pelts became highly prized. But what was a platinum fox doing on Dartmoor, where, as far as I know, there are no fur farms? The platinum condition results from a dominant mutant allele (gene form), and as it has arisen spontaneously in many unrelated, geographically-scattered fox litters since 1933, perhaps it has done so again, recently, in a litter of Dartmoor foxes. Shaun has since learned of other sightings of this animal, with one made only 2 miles away from the site of his own observation. He now plans to continue investigating this intriguing case by looking out for the animal and making discreet enquiries relating to its movements, so we await further news with interest. **Shaun Histed-Todd, pers. comms, 17 March 2010.**

A PRECIOUSSSSSS SIGHTING? It's not every day, or night, that a mysterious creature is specifically likened by its startled eyewitnesses to the sibilant Gollum from Tolkien's classic fantasy trilogy *The Lord of the Rings*, but this is precisely how a husband and wife described to the Big Cats in Britain research group the snarling entity encountered by them at Fairhaven Lake, Lytham St Annes, in Lancashire, just before midnight on 19 March. As big as a medium-sized dog but lacking a tail, it was also distinguished by its disproportionately lengthy, twisted limbs, its brownish fur-less body, its "almost human-like but deformed in a way" face, and apparent absence of ears. Could this be yet another of those bizarre-looking mange-ridden hairless specimens of various carnivorous mammalian species that

Alien Zoo

seem to be cropping up globally lately? **Big Cats in Britain, April 2010.**

ENDING THE TALE OF TWO QUAILS. In a previous Alien Zoo, I mentioned seeing two unusual quails while visiting the island bird sanctuary of Tiritiri Matangi, off Auckland, New Zealand, in November 2006 and learning several months later that some scientists believed it possible that instead of merely being introduced Australian brown quails *Coturnix ypsilophora*, the quails on this small island might actually be surviving New Zealand quails *C. novaezelandiae* - a native species also known as the koreke and officially believed to have become extinct in 1875. The principal scientist investigating this exciting prospect, PhD researcher Mark Seabrook-Davison, from the Institute of Natural Sciences at Albany, planned to pursue the matter by conducting comparative DNA analyses, using samples from known Australian brown quails, from museum specimens of the New Zealand quail, and from living quails on Tiritiri. Sadly, these analyses subsequently revealed that the Tiritiri birds, although larger, were genetically identical to the Australian brown quail, and so New Zealand's own long-lost species has not been resurrected after all. **http://www.massey.ac.nz/?aaadb1846e?mnarticle=tiri-quails-found-to-be-aussie-imports-23-10-2009 updated 7 April 2010.**

SOLVING A PERFECT PICTURE OF MYSTERY. It was in my Alien Zoo for **FT*165*** (December 2002) that I first documented the mystifying painting created by an equally mysterious artist known only as Canzanella that depicted two very strange, seemingly unidentifiable beasts. The painting had

The colour-reversed ('positive') version of the Canzanella painting – a mystery no longer
(Stephanie Sparkman, née Canzanella)

lately been sold on eBay, but all of my attempts to trace the artist and discover the beasts' identity had ended in failure, although it engendered much interesting speculation from readers when I revived the subject on my ShukerNature blog in October 2009. One reader made the intriguing suggestion that the painting looked rather like a photographic negative and wondered what a colour-reversed ('positive') image of it would look like, so I duly created one on my computer (the result is shown below), but sadly it did not help in identifying the creatures. Over the years, I had sent a number of enquiries via eBay's communication system to the painting's seller, but I had never received a response. In early April 2010, however, I succeeded in tracing the seller's email address (not visible on eBay), so I then tried emailing this person directly - and, to my great delight, on 10 April I finally received a reply! The seller proved to be a lady artist from Florida called Stephanie Sparkman, and in just a few lines she successfully cleared up all of the longstanding mysteries surrounding this painting and its subjects:

> I painted the picture over 28 years ago myself. It is a representation of what I felt the porpoise of today used to look like as a land animal. First the fur stage, then to smooth skin as shown in the beast behind the one in the foreground. I have NO recall of how much I got for the picture (for sure not enough as it was exceedingly detailed with hairs on the hairs [and] with several different colours on each hair) and didn't even recall what had become of it as I was looking for it not too long ago and forgot what I had done!!! Thanks to you I know what happened to it LOL.... [Canzanella] was my maiden name as I kept the name of my late husband which is Sparkman but always painted under my maiden name to give credit where credit was due.

So the seller and the artist were one and the same person, and the animals were her own concept of what the porpoise's erstwhile land ancestors may have looked like. It had taken almost eight years but Canzanella's perfect picture of mystery had finally been solved, thanks to none other than the hitherto-elusive Canzanella herself! **Stephanie Sparkman (née Canzanella), pers. comm., 10 April 2010.**

PYGMY PACHYDERMS ON PARADE! Although never scientifically confirmed, there have been many reports over the years of mysterious pint-sized elephants inhabiting various forests within south-western India's Kerala state. The latest of these reports comes from two wildlife photographers, Sali Palode and Jain Angadikkal. They encountered at Marakappara a very diminutive but otherwise adult-resembling elephant at a distance of just 100 m while trekking with a local tribal chief through south-ernmost India's Peppara Wildlife Sanctuary in March 2010. Stating that it had a long tail, a wrinkled face, and a trunk that touched the ground, but resembling in overall appearance resembled a miniature tusked adult, they also photographed it and released their pictures to the region's media. Unfortunately, these did not contain any features that could be used as a scale for estimating the elephant's size. Al-though familiar to this area's native Kani tribes, who refer to them as kallana or kallaana, these Indian pygmy elephants are generally dismissed by zoologists merely as dwarf specimens of the Asian ele-phant *Elephas maximus* rather than deeming them to be a taxonomically-discrete form in their own right. *The Hindu*, **12 April 2010; The Press Trust of India Limited, 12 April 2010.**

CALLING OUT FOR KAWEKAWEAUS. Known from just a single taxiderm specimen in Mar-seilles Natural History Museum that is believed to have originated in New Zealand but which for well over a century was not recognised as representing a hitherto-undescribed species, Delcourt's giant gecko *Hoplodactylus delcourti* is not only the world's biggest gecko species but also the most mysteri-ous. Its lone specimen bears a remarkable resemblance to a supposedly mythical New Zealand reptile known in Maori folklore as the kawekaweau, and strange lizards whose eyewitness descriptions readily recall the latter creature have long been reported from North Island. Consequently, it is exciting to learn that during this year's Christmas period, New Zealand cryptozoologist Tony Lucas will be visiting

North Island's Rotorua region to investigate claims of a remnant kawekaweau population there, and collect any eyewitness reports. If *H. delcourti* were to be discovered alive and confirmed to be one and the same as the kawekaweau, this would certainly be one of the greatest cryptozoological success stories of modern times, so I wish Tony all the very best! **Tony Lucas, pers. comm., 26 April 2010.**

Delcourt's giant gecko (Markus Bühler)

FACING UP TO "THE UGLY ONE". In late May 2010, news media worldwide carried stories of an extraordinary-looking creature that had allegedly been discovered dead in Big Trout Lake, at the heart of a remote Oji-Cree Native American community south of Hudson Bay in northern Ontario, Canada. The animal had been found by two nurses walking by the lake earlier in May who photographed it but then left it on the lake's rocky shoreline; when searched for a few days later, it had vanished. Based upon eyewitness descriptions and the photos, the creature was roughly 30 cm long with brown fur and a rat-like tail, but its facial skin had been entirely eroded away, leaving only white flesh behind, and the same had happened to its right forefoot. According to the community, it is a rare local beast of ill omen known as the omajinaakoos or 'the ugly one', seen here only very occasionally, with the last previous sighting made around 50 years ago. Suggestions as to its zoological identity have varied dramatically, from a muskrat, beaver, mink, or otter, to an opossum, bear cub, chupacabras, or even a wholly unknown species. Based upon the photographic evidence, which clearly shows the presence of canine teeth (thereby eliminating a rodent identity), this animal was most likely either a young otter or an American mink *Neovison vison* (though in the photos it seemed too sturdy to be the latter). ***Globe and Mail* (Ontario), 21 May 2010.**

Alien Zoo

FINDING OUT – **THE END OF THE QUEST.** After posting details on my ShukerNature blog for 18 September 2009 regarding the elusive 1960s magazine partwork *Finding Out* and its spectacular back cover illustrations by the late, highly-acclaimed artist Angus McBride (1931-2007) of mystery and mythological entities (as documented by me in previous Alien Zoo reports), on 26 January 2010 I received the following response from ShukerNature reader Anne Craig:

> I have just bought 12 issues of Volume 17 and they all have the legendary beasts you are interested in. There is the White Buffalo, the Little People, the Lamassu, Sphinxes, Cyclopses, Gnomes, the Cailleach-bheur, the Midgard Serpent, the Morrigan, Werewolves, Thunderbird and Centaurs.

Needless to say, I was delighted to receive this news and swiftly replied to Anne's post, asking whether there was any possibility that she could scan at least some of these pictures and email them to me. Unfortunately, I did not receive any reply. Happily, however, my longstanding quest for sight of any of these near-legendary *Finding Out* illustrations finally came to an end on 5 June 2010, when I discovered that the original McBride artwork for several of them (namely, the naga king, white buffalo, leshy, cyclops, and Peruvian sun birds) had lately been sold on the fine arts website 'Books Illustrated Ltd' at http://www.booksillustrated.com - and I can readily confirm from the images of these works included there that they are indeed truly breathtaking!
ShukerNature, http://karlshuker.blogspot.com/2009/09/finding-out-about-finding-out.html?showComment=1264531407441#c9125608861411975070 26 January 2010; http://www.booksillustrated.com/en-UK/angus-mcbride accessed 5 June 2010.

SNAPPING AUSTRALIA'S CAMERA-SHY MOOLYEWONK. During the past 45 years, Aussie cryptozoologists Rex and Heather Gilroy have amassed a very sizeable dossier containing hundreds of eyewitness reports concerning New South Wales's Hawkesbury River Monster, also known as the moolyewonk. However, they are now seeking to obtain something never before achieved – an actual photograph of this notoriously camera-shy cryptid. According to traditional Aboriginal folklore, the moolyewonk would lurk along the river's banks, waiting to attack women and children. The Gilroys state that it also features in ancient rock art on the banks of the river. Moreover, it is still being seen today, especially around Mooney Mooney and Long Island, and there are even stories of houseboats on the river being lifted up at one end when something very large underneath tried to surface. The most recent sighting was near Wisemen's Ferry in March 2010, when a local fisherman spied a serpentine head and a long 2-m neck rise briefly above the water surface before submerging

What lurks in Australia's Hawkesbury River? (William Rebsamen)

Alien Zoo

again. Rex Gilroy and his assistant Greg Foster claimed a similar sighting near here in August 2009. Based upon observers' testimony, the moolyewonk is roughly 24 m long and seemingly plesiosaurian in overall morphology, with a large bulky body that is grey and mottled in appearance, sporting two pairs of paddle-like flippers, a thick eel-like tail, a long neck, and a snake-like head.
http://hornsby-advocate.whereilive.com.au/news/story/does-the-hawkesbury-have-a-loch-ness-style-monster-of-its-own/ 9 June 2010.

DID DOUGLAS DEVOUR THE LAST HARPAGORNIS? Weighing 22-33 lb in adult female specimens, Haast's eagle *Harpagornis moorei*, formerly native to New Zealand's South Island, was the largest eagle of all time, and is widely accepted as the origin of traditional Maori legends referring to a monstrous human-killing raptorial bird known variously as the pouakai or hokioi. As it is believed to have preyed upon the very large flightless moas that also once inhabited South Island, it should come as no surprise to learn that this colossal bird of prey's disappearance seemingly coincided with the demise of the moas (caused in turn by hunting and habitat destruction by humans), with its official extinction date estimated at approximately 1400 AD. Very recently, however, while browsing the Wikipedia entry for this species, I discovered a remarkable snippet of information suggesting that either Harpagornis itself or else another now-extinct but very sizeable bird of prey endemic to New Zealand may have lingered into much more recent times, only to be extinguished in the most demeaning manner:

> A noted explorer, Charles Douglas, claims in his journals that he had an encounter with two raptors of immense size in Landsborough River valley (probably during the 1870s), and that he shot and ate them. These birds might have been a last remnant of the species, but some might argue that there had not been suitable prey for a population of Haast's eagle to maintain itself for about five hundred years before that date, and nineteenth century Maori lore was adamant that the pouakai was a bird not seen in living memory. Still, Douglas'[s] observations on wildlife generally are trustworthy; a more probable explanation, given that the alleged three-metre wingspan described by Douglas is likely to have been a rough estimate, is that the birds were Eyles'[s] harriers [*Circus eylesi*]. This was the largest known harrier (the size of a small eagle) — and a generalist predator — and although it is also assumed to have become extinct in prehistoric times, its dietary habits alone make it a more likely candidate for late survival.

If, however, these were the last Haast's eagles, what an ignominious finale for such a spectacular species – slaughtered and served up as dinner by a hungry Westerner, for whom the chilling age-old Maori legends of murderous winged marauders from the ancient skies were nothing more than quaint fables, despite the very substantial evidence to the contrary that his twin kills had presented not only to his eyes but also to his stomach! **http://en.wikipedia.org/wiki/Harpagornis accessed 11 June 2010.**

Reconstruction of Haast's eagle (Markus Bühler)

Alien Zoo

A GIANT MYSTERY OF GIANT HELMETS. In early June 2010, while browsing at the Tuesday bric-a-brac market in Wednesbury, West Midlands, I purchased the old sepia-tinted picture postcard reproduced here, as I was – and remain – very intrigued but perplexed by its picture, which depicts a pair of extremely ornate and exceptionally large helmets on display in what appears to be a museum or art gallery. Whereas the right-hand helmet takes the form of a huge feathered eagle or similar bird of prey, the left-hand helmet constitutes a winged curly-tailed dragon-like beast gripping a spherical object. But what is the history of these extraordinary helmets, and where is that museum or gallery? The postcard's reverse bears only the following simple, if somewhat clumsily-worded, caption: "View on the helmets of the Giants", which is also given in French (first) and in German (second). No other detail, not even the location of these remarkable-looking exhibits or any postcard manufacturer/printer information, is present. The postcard has not been used, so there is no writing or stamp to offer any clues as to its age or origin either.

After spending some time vainly searching for any information regarding the helmets, on 14 June I included a short account and photo of this postcard on my ShukerNature blog, and I subsequently received a number of interesting suggestions from readers concerning it. The most popular idea was that the helmets were theatrical props for some early 20th-Century stage production, which is certainly possible, though to my mind they seem too sturdy in appearance to be lightweight constructions.

In contrast, Danish cryptozoologist Lars Thomas opined that they may be genuine, revealing a most interesting snippet of historical information that could be very relevant. Namely, that according to legend, Alexander the Great always left behind giant helmets and weapons following one of his conquests, in order to scare off any challengers and to keep those conquered by him in a subdued state, and Lars

My picture postcard depicting "the helmets of the Giants" (Dr Karl Shuker)

Alien Zoo

vaguely recalls reading somewhere that some of these giant Alexandrian helmets were displayed in various early museums. If true, this could indeed explain the picture on my baffling little postcard, but my major qualm with this notion is that if such spectacular exhibits do exist, surely they would be very famous. Yet no-one who has seen my postcard recognises them at all, nor the locality in which they are housed. For now, therefore, my quest for information regarding these enigmatic "helmets of the giants" continues. Just what, precisely, are they, and where were they being exhibited? Any suggestions or information would be very greatly welcomed. Answers on a postcard - but not this one! **http://karlshuker.blogspot.com/2010/06/giant-mystery.html 14 June 2010.**

LAIR OF THE WHITE EELS. The CFZ's daily blog carried a very intriguing crypto-snippet recently, which, it stated, had originally been discovered on an angling website. I subsequently tracked this down; it was posted by 'peakstroller' on the fishing section of an internet forum called Mombu under the heading 'luninous [sic] white albino eels' on 25 September 2004, and read as follows:

> I was diving in Wast Water, Wasdale, Cumbria and found thousands of weird bright white eels about a foot to a foot and a half long. Twas a night dive and our approaching torches made the eels stick their head into the mud to avoid our light. They didn't have any fin on their back as I remember. Any one any idea what they could be? I thought these could be some ancient fish trapped along with the Arctic Char in Cumbria's deep waters or are they just another weird Lamprey type creature? I didn't notice if they had teeth or even a jaw. Any feedback most welcome. Dave

Yet despite the passing of almost 6 years since then, no replies had ever been posted, so clearly these fishes had mystified other readers of that site too. Assuming that the report is genuine, if they truly lacked any dorsally-sited fin, a typical anguillid eel could be ruled out, and their pallid colouration is also perplexing. Without knowing whether or not they were jawed makes any speculation of lampreys futile at this point (and, in any case, lampreys also possess dorsal fins). There are a number of elongate fish species that routinely stick their heads into mud or sand, so that particular behavioural attribute is not novel, but it would be most interesting to receive any thoughts as to their possible taxonomic identity from angling-enthused *FT* readers. **http://forteanzoology.blogspot.com/2010/07/lair-of-white-worms.html 10 July 2010; http://www.mombu.com/fishing/fishing/t-luninous-white-albino-eels-2043217.html 25 September 2004.**

MINI-WOLVES AND MEGA-OTTERS. Irish cryptozoologist Gary Cunningham is best-known for his researches into the Emerald Isle's lough-frequenting horse-eels, the controversial Irish wildcat, and the legendary master otter or dobhar-chú of Glenade Lake, County Leitrim. However, as I recently learnt from him, he is also currently pursuing a longstanding investigation of the little-known dwarf wolves allegedly native to Achill Island, off County Mayo, as well as reports of giant otters on Omey Island, in Connemara, County Galway. Interestingly, back in October and November 2009, I was contacted independently by Waterford artist Sean Corcoran with details of a sighting made by himself and his wife back in 2003 on this very same island of what may have been one such creature, sporting orange-red flipper-like feet, which gave voice to a haunting screech and swam away at high speed. Full details of Gary's plentiful Irish crypto-searches are contained within his newly-published book, *Mystery Animals of Ireland*, co-authored by Ronan Coghlan. **Gary Cunningham, pers. comm., 19 July 2010; Sean Corcoran, pers. comms, 18 October, 2 & 24 November 2009.**

Alien Zoo

TRUNKO REVEALED! One of the most bizarre cryptids ever reported was the huge white-furred trunked sea monster allegedly observed from the shores of Margate, South Africa, by several eyewitnesses one day during the early 1920s (the precise date differs between accounts) as it battled two whales out to sea for several hours before its lifeless 47-ft-long carcase was later washed up onto the beach, where it lay for several days before the sea took it back out, never to be seen again. During those several days, however, not a single scientist came to examine or even observe this extraordinary creature's remains. Nor had its remains ever been photographed...or so I had always assumed - until now!

On 4 September 2010, I learnt from German correspondent Markus Hemmler that he had discovered a website containing an allegedly genuine photograph of Trunko, snapped by none other than Mr A.K. Jones of Johannesburg - the correspondent and photographer for an article on this entity that had been published in the *Rand Daily Mail* and also in *Wide World Magazine* way back in August 1925, thus providing a promising air of authenticity to the image. The photo is currently included by the Margate Business Association's website in a page devoted to the creature's history, at:

http://www.margatebusiness.co.za/index.php?option=com_content&view=article&id=64:the-legend-of-trunco&catid=1:mba-news&Itemid=2

Moreover, after tracking down A.K. Jones's account in the above-noted *Wide World Magazine* issue, Markus and I were delighted to discover that it included two more of his photographs of the beached Trunko carcase! The fibrous, amorphous appearance of Trunko's remains in these three images vehemently reaffirms my abiding opinion – discussed fully in my book *Extraordinary Animals Revisited* (2007) - that Trunko was not a bona fide cryptid at all, but merely a globster. Namely, a gelatinous mass of collagen encased in a fibrous skin-sac of rotting blubber from a dead whale, and that it was the

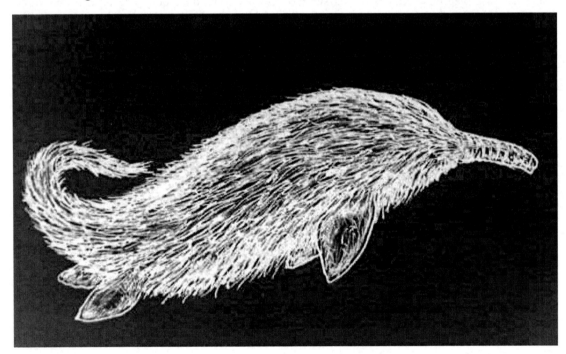

Reconstruction of Trunko as living animal, based upon eyewitness claims (Markus Bühler)

Alien Zoo

The Trunko carcase, washed ashore on Margate beach (A.K.Jones)

sight of two whales out to sea tossing this mass into the air that fooled eyewitnesses on the beach into thinking that they were witnessing the whales battling a living furry sea monster. My grateful thanks to Markus Hemmler (http://www.kryptozoologie-online.de/) for bringing these cryptozoologically-significant photographs to my attention, and also for alerting me to another Trunko-related discovery.

In November 1930, a second, smaller trunked mystery carcase, a veritable 'Son of Trunko', was discovered floating off Alaska's Glacier Island by two fox farmers. This time, a team of scientific investigators (led by W.J. McDonald, district forest supervisor of Alaska's Chugach National Forest) did come out and examine it, but there did not appear to be any public announcement of their findings. However, Markus has discovered that on 2 May 2008, an article in Alaska's *Cordova Times* newspaper revealed that the carcase had been found to be the skeleton of a minke whale *Balaenoptera acutorostrata*.

In January 1931, this spectacular specimen was purchased for $600 by taxicab owner Tom Vevig who toured widely across the USA, Canada, and Mexico exhibiting it, before donating it later that same year to the National Museum of Natural History in Washington DC, where it remains to this day, catalogued as specimen USNM 256498. **Markus Hemmler, pers. comms, 4-18 September 2010; http://www.margatebusiness.co.za accessed by me 4 September 2010; ShukerNature reports posted on http://www.karlshuker.blogspot.com 4, 9, and 16 September 2010.**

A male great crested newt in mating finery exhibiting its distinctive dorsal crest
(Wikipedia/public domain)

The Lost Ark

THE MONSTER OF LAKE KHAIYR – AFTER FOUR DECADES, THE TRUTH AT LAST...?

Fantasy abandoned by reason produces impossible monsters; united with her, she is the mother of the arts and the origin of their marvels.

Francisco Goya – text caption from the Prado etching version of Goya's 'The Sleep of Reason Produced Monsters' fantasy artwork plate

It is always sad when an icon is toppled from its pedestal. In cryptozoology, certain iconic photographs and other illustrations that were once deemed to be of great scientific significance have subsequently been tainted by controversy and claims of fraudulence - as with, for instance, the Surgeon's Photo of Nessie, and Dr François de Loys's memorable photograph of the alleged South American ape shot by his expedition team. Now, it is my sorrowful duty to act as official iconoclast to another well-known cryptozoological image - the much-published sketch of Siberia's Lake Khaiyr monster, as prepared by its principal eyewitness back in 1964 during Moscow State University's Northeastern Expedition (surveying mineral deposits), led by Dr G. Rukosuyev, and published later that same year in a *Komsomolskaya Pravda* report that also contained an interview with Rukosuyev.

Since then, this very striking picture has appeared in numerous cryptozoological works, my own included. Due to an extraordinary revelation, however, published once again by *Pravda* (6 August 2007), the whole episode featuring the creature so depicted has now been exposed in a new and very shocking

Alien Zoo

Nikolai Gladkikh's illustration of the Lake Khaiyr monster

light. But to begin at the beginning...

UNMASKING THE GIANT NEWTS OF SIBERIA

Within my book *In Search of Prehistoric Survivors* (1995), I concisely documented as follows what was assumed at that time to be the truth concerning the Lake Khaiyr (aka Khainyr) monster sighting:

> One of the most distinctive long-necks ever reported was spied in 1964 by Soviet biologist Dr Nikolai Gladkikh, while visiting Lake Khaiyr, a remote body of water in the east Siberian province of Yakutia and largely ignored by science until then. Standing at the lakeside one early morning, Gladkikh caught sight of a very big and extraordinary beast on the opposite shore [apparently there to eat the grass, according to some versions of this sighting]. Its huge body was bluish-black in colour, with two pairs of ill-defined limbs, a long sturdy tail and an elongate gleaming neck carrying a small head at its tip. Its most eye-catching feature, however, was the low, triangular dorsal fin, containing vertical rays, that ran from the base of its neck to the beginning of its haunches.

> By the time Gladkikh had alerted his colleagues to come and see it, this unidentified mystery beast had vanished, but he was able to prepare a good sketch. Before the end of their visit the expedition's leader and two other members were able to confirm the sketch's accuracy, when they briefly saw the head and dorsal fin of the monster appear above the surface at the centre of the lake, and watched its long tail furiously lashing the water. Its dorsal fin sets this animal well apart from any species currently known to science, but it is not the only time that such a feature has been reported for long-necks.

> In 1942, two Soviet pilots surveying this lake reported seeing two bizarre animals that were later likened to giant newts. As some newts, exemplified by males of the familiar [great-crested newt] *Triturus cristatus*, bear a conspicuous dorsal crest, this lends support to Gladkikh's story.

Unlike the recent *Pravda* report, conversely, which tersely strips his story of any support whatsoever.

To begin with, this radical reassessment of the Lake Khaiyr monster episode claims that far from being a Soviet biologist, Gladkikh was nothing more than a migrant worker hired by the expedition. Moreover, it alleges that in addition to confirming Gladkikh's account of the monster by way of his own sighting, team leader Rukosuyev had made some unusual extra allegations - asserting that there were no fishes in the lake, that birds never landed on its surface, and that the locals often heard "muffled sounds and splashes of water" coming from the lake.

An expedition was duly dispatched to Lake Khaiyr to investigate all of these claims (a fact not generally mentioned in Western reports of this case), and it swiftly refuted all of them. Yes, there were fishes in the lake, and yes, birds did land upon it. Most significant of all, however, was this new expedition's discovery that none of the locals had ever seen any strange creatures in the lake. Moreover, despite combing the bottom of this lake (which, incidentally, is actually a very shallow body of water, only about 7 m deep), scuba divers dispatched by the expedition found nothing unusual.

Consequently, the expedition had what the *Pravda* report describes as "a heart-to-heart talk" with Gladkikh, who confessed that he had invented the entire story but was unable to explain clearly why he had done so. Accordingly, *Pravda* concludes: "He concocted it either to entertain himself and his friends or as an excuse for shirking his duties at work".

Yet in spite of *Pravda*'s ostensibly decisive dismissal of Gladkikh's testimony, and, in turn, the whole monster encounter, there are some glaring omissions to consider. What about Gladkikh's colleagues who also saw the creature – if any of them are still alive, why weren't they interviewed by the new expedition, especially Rukosuyev? And if they are dead, surely that should have been noted in the *Pravda* report, if only to avoid the very questions being posed here? Ditto for the two Soviet pilots from 1942.

Nevertheless, the damage has been done - sufficient doubt has been cast upon this whole episode by *Pravda* for Gladkikh's Lake Khaiyr monster sketch to be viewed forever more with scepticism by all but the most trusting of investigators. Exit another item of tantalising mystique from the panoply of mystery beast memorabilia previously assumed to constitute valid cryptozoological evidence untainted by the shameful shadow of fraud.

SERPENT-HEADS AND FLEECY-SKINS

Yet even if the Lake Khaiyr monster has been dishonoured, there may yet be other water beasts of more reputable, less questionable status still lake-lurking amid the former Soviet Union. In the selfsame *Pravda* article, there is, for instance, mention of a huge serpent-headed, crocodile-bodied monster said to conceal itself within underwater karst caves at the bottom of 56-m-deep Lake Somin in western Ukraine, except when it surfaces to prey upon any unwary domestic animals straying too near to its shores – and even the occasional human, if local stories are to be believed.

In addition, *Pravda* recalls the collecting by late Russian cryptozoologist Maya Bykova of several eye-witness reports concerning large mystery beasts with fleecy dark-brown skin that rise quickly to the surface, force air violently out through their noses, and then sink back beneath the water surface in several lakes within Siberia's Irtysh basin. If such beasts truly exist, perhaps they are landlocked seals, comparable with the famous species peculiar to Siberia's Lake Baikal.

WATER GOBLINS AND BIRD SNATCHERS

Also documented by *Pravda* and in need of an explanation are the 'water goblins' that reputedly inhabit the startlingly blue, fish-filled waters of Lake Vedlozero and others in Russian Karelia. According to local descriptions, these goblins, first reported during the mid-1990s and since sought unsuccessfully by cryptozoologist Viktor Sapunov, resemble hairy midget men with rounded heads. Could these be wandering seals from other inland lakes? Certainly, seals have been sighted in the Vedlozero area, and one was even captured and photographed there by a local inhabitant several years ago. The reptilian monster of Lake Somin, conversely, is apparently very different.

Nor should we overlook the long-necked, snake-headed, bird-snatching cryptid allegedly dwelling in Yakutia's Lake Labynkyr - which was supposedly observed back in the 1950s by a visiting team of geologists and several local reindeer hunters, and is documented in my *Prehistoric Survivors* book.

In short, it is evident that even if the 'giant newts' of Lake Khaiyr truly are bogus beasts, there may still be more than enough creatures of cryptozoological interest in other ex-Soviet bodies of freshwater to intrigue and inspire mystery beast seekers for a long time to come.

SELECTED REFERENCES

GRACHEV, Guerman (2007). Mysterious water mammoths inhabit Siberian lakes. *Komsomolskaya Pravda*, 6 August, pp. 1-4.
SHUKER, Karl P.N. (1995). *In Search of Prehistoric Survivors*. Blandford Press (London).

The Lost Ark

THE GIANT MOLE OF MESTY CROFT.

*Strong-shouldered mole,
That so much lived below the ground,
Dug, fought and loved, hunted and fed,
For you to raise a mound
Was as for us to make a hole;
What wonder now that being dead
Your body lies here stout and square
Buried within the blue vault of the air?*

Andrew Young – 'A Dead Mole'

[This item was originally intended as an Alien Zoo report, but was too long; equally, however, it was too short for a Lost Ark article. Consequently, except for an appearance in my online ShukerNature blog on 28 April 2010, its subject became the forgotten Fortean – until now.]

People often claim that objects, buildings, animals, etc seemed much bigger when viewed as a child than they do now when viewed as an adult. Yet I can't say that I've ever noticed this, personally speaking. Everything that I can recall from childhood and which still exists in unchanged form today looks just the same to me now as it did then – which is why I cannot use this notion as an explanation for the giant mole of Mesty Croft. Between the ages of seven and eleven (i.e. during the late 1960s), I attended Mesty Croft Junior School in Wednesbury, West Midlands, and whereas for many other erstwhile pupils of this fine establishment their abiding memories may well be of childhood classmates or even of teachers whom they liked or disliked, my most tenacious recollection of Mesty Croft is an extraordinary exhibit – one that fascinated me throughout my time there, and which I secretly coveted for what I now realise to be reasons of slowly-awakening cryptozoological awareness. The exhibit was a taxiderm mole, mounted on a wooden plinth, and placed on a shelf beneath a window in the main assembly hall. For a child who hoarded dead insects in matchboxes, enjoyed birdwatching in the countryside at weekends, collected frogspawn and sticklebacks in jam jars, gathered up seashells off the shore on seaside holidays, and visited every major zoo in Britain, it is hardly surprising that I entertained daydreams of owning this wonderful item and displaying it in pride of place in my bedroom. However, thanks to a loving but responsible upbringing by parents who left me in no doubt about the evils of theft and other misdemeanours, a wistful daydream is all it ever remained.

It wasn't until some time after leaving Mesty Croft that I saw other taxiderm moles, and was astonished at how small they were (no more than around 5 inches long at most) in comparison with the example that I had known so well at junior school, and which I now realise was almost as big as a rat in terms of body size (but with only a very short tail). Not only that, their fur was always velvety smooth, whereas the Mesty Croft giant mole's was decidedly curly and much harsher, as I well remembered from having stroked it on a couple of occasions when walking by its shelf in the assembly hall.

Could it be that the latter specimen's fur had deteriorated over the years since its pelt had originally been prepared by the taxidermist? Yet if so, a more likely outcome would surely have been fur loss, not increased curliness? And how can this megamole's size be explained? Is it possible that it was not a specimen of the common mole *Talpa europaea*, but of some much larger, exotic species from outside Europe? Its fur was jet black in colour, whereas that of the common mole is more often dark grey, thus providing further support for the prospect that this was no ordinary British mole. Indeed, I had speculated about Townsend's mole *Scapanus townsendii* as a possible candidate, bearing in mind that it is North America's largest species of mole, but having checked it out I can confirm that not only is it too small at 'only' 8 inches long (including its tail) but also its shiny, velvety fur does not correspond at all with that of the Mesty Croft specimen - which was matt rather than shiny in appearance, and definitely coarse or harsh to the touch, with those peculiar curls or ringlets all over its body. Ditto for various suggestions made in response to my blog's coverage of this specimen that perhaps it was either an Australian marsupial mole *Notoryctes* spp. or one of southern Africa's golden moles (family Chrysochloridae); both of these mammalian groups are superficially mole-like but are unrelated to true moles. Very odd indeed. Today, the trail of Mesty Croft's giant mole is so cold that there seems little hope that it can ever be

The common mole *Talpa europaea*

Alien Zoo

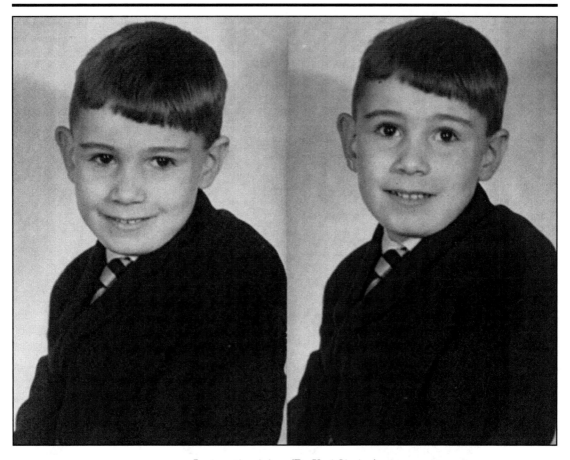

Junior school days (Dr Karl Shuker)

thawed out. As far as I am aware, none of the teachers who were at Mesty Croft during my time there are still alive; and less than a decade after I left, the entire school was razed to the ground and later completely rebuilt in a totally new layout. So whatever happened to its mole of mystery? Was it simply discarded during this major upheaval, or did someone rescue it from such a cruel fate, so that somewhere out there it may still exist today - preserved in scientific anonymity, perhaps, on someone's windowsill, bookshelf, or mantelpiece?

I readily appreciate that it is a very remote possibility, but if there is anyone reading this who can shed any light on the very curious case of the long-lost giant mole that once resided proudly on a shelf in Mesty Croft Junior School, Wednesbury, West Midlands, I'd very much like to hear from you.

REFERENCE

SHUKER, Karl P.N. (2010). The giant mole of Mesty Croft. ShukerNature (http://karlshuker.blogspot.com/2010/04/giant-mole-of-mesty-croft.html), 28 April.

Do bat-men exist on Seram and elsewhere in the world? (Richard Svensson).

The Lost Ark

DRIVEN BATTY BY THE ORANG BATI.

Lightless, unholy, eldritch thing,
Whose murky and erratic wing
Swoops so sickeningly, and whose
Aspect to the female Muse
Is a demon's...

Ruth Pitter – 'The Bat'

There is a wonderfully wry poem entitled 'The Blind Men and the Elephant', written by John Godfrey Saxe, in which six blind people each touch a different part of an elephant (a flank, a tusk, its trunk, a knee, an ear, its tail), and thus voice a completely different, but equally erroneous, opinion of what the elephant must be like (a wall, a spear, a snake, a tree, a fan, a rope). I am reminded of this situation when attempting to reconcile the likely – and unlikely – identity of Indonesia's mystifying orang bati, based upon the wildly disparate descriptions given of it over the years.

TYSON HUGHES'S TESTIMONY

It was veteran field cryptozoologist Bill Gibbons who first brought to my attention this bizarre entity (undocumented at that time in the cryptozoological or fortean literature). In a letter to me of 29 July 1993, Bill included the following information – which had been collected by one of his colleagues, a tropical agriculturalist called Tyson Hughes, who in June 1986 had arrived on the island of Seram (=Ceram) in Indonesia's Moluccas group to act on behalf of the VSO (Voluntary Service Overseas) as a project manager of a model farm, spending 18 months there:

Alien Zoo

During his free time, Tyson became interested in native accounts of a bizarre creature known as the orang bati, meaning 'flying men' in Indonesian. The villagers of the coastal regions live in literal terror of the orang bati, which, they claim, live in the interior of long-dead volcanic mountains. At night the orang bati will leave their mountain lairs and fly across the jungles to the coast, where they seize human victims from the coastal villages and carry them back to the mountains.

Although the local police are aware of the native accounts of the orang bati, they dismiss them out of hand, mocking any reports that filter their way into civilisation. The majority of serving police officers are actually Javan, so their scepticism is, perhaps, not unexpected. However, the fear of the orang bati is considerable among almost all the indigenous inhabitants of Seram island. Those abducted by the orang bati are never seen again. These mysterious creatures are described as human in form, with red coloured skin, black wings, and long, thin tails. Tyson was able to persuade several village hunters to lead him through a dense area of jungle towards the alleged lair of the orang bati. Although he became the first white man to explore part of Seram island, the orang bati remained elusive.

In a follow-up letter to me of 5 August 1993, after having chatted with Tyson, informing him of my interest in the orang bati, Bill included some additional details, received from Tyson. The coastal vil-

lage whose inhabitants especially fear this entity is Uraur, whose inhabitants claim that it is children and infants that the orang bati abduct, presumably to eat. Their home is a system of caves inside a dormant volcano at the centre of Seram, they utter a long mournful wail, stand 4-5 ft tall, sport large bat-like wings and a long thin tail, and are black in colour with a red tinge. Interestingly, Tyson also noted that this island does not contain any monkeys.

Based upon this information, I suggested to Bill and in various subsequent publications that perhaps the orang bati is an unknown species of giant bat, humanised somewhat in native folklore, as I was mindful of reports describing ostensibly comparable cryptids filed from elsewhere in the world, but notably from the much larger Indonesian island of Java. Here a huge bat-like beast called the ahool is said to exist, and whose unique triple cry has even been heard by a western naturalist, Ernst Bartels. Moreover, in the absence of monkeys on Seram, it could be argued that perhaps a giant carnivorous bat would seek out the nearest equivalent – human infants.

JAAP VAN DEN BORN'S TESTIMONY

In April 2003, however, a dramatically different notion regarding the orang bati's identity was aired, this time by *FT* reader Jaap Van Den Born from the Netherlands within a letter published in *FT169*. According to information given to him by inhabitants of Seram, the orang bati are a normal tribe of humans, taller than average for Moluccan people and with strange eyes, but lacking wings or tail, yet which nevertheless possess the ability to fly and also to become invisible...allegedly. Moreover, it is claimed that they do indeed abduct humans – adults as well as children, and especially foreigners – but merely to bolster their own numbers, not to devour them!

Jaap also referred to an incident documented in *Ambon, Island of Spices* (1979), written by Shirley Deane, who taught English on Ambon (a small island off Seram's southwest coast). Deane claimed to have once met one of the leaders (radja) of the orang bati at a Seram police station. Deane described him as:

> ...very tall and thin, with an angular face and high cheekbones...[and] enormous, piercing and very hypnotising eyes, which seemed to look straight through me. 'Laserbeam eyes', I thought, and quickly looked away.

Boldly, Deane asked the radja if she could visit his village, and in reply he said that any such visit would have to be announced well in advance to an even more senior leader of his village, and that in any case she would not see anything as the entire village would be invisible to her eyes. Furthermore, when, feeling even more reckless, Deane asked him if it was true that the orang bati could fly, after some hesitation the radja replied that although many of his people could indeed fly, he personally did not possess this ability, after which he said goodbye and left.

After having listened to this conversation, the Chief of Police stated that the radja probably could fly, but had not admitted this in his, the Chief's, presence because according to tradition an orang bati could not be initiated into the art of flying until he had killed a human. And the radja was hardly going to admit to murder in front of the Chief of Police! Interestingly, Jaap actually encountered some orang bati when visiting Ambon, but far from possessing piercing eyes, to him their eyes seemed very withdrawn,

FACING PAGE: Reconstruction of orang bati as winged humanoid entity (Tim Morris)

Alien Zoo

due to the whites (the sclera) being almost black, like the eyes of aliens!

So now there seemed to be two totally discrete, mutually exclusive versions of the orang bati. One was a winged, tailed, child-abducting bat-man. The other was a slightly odd-looking but nonetheless entirely human being, with a distinct absence of wings and tail. Could it be that the former was merely a greatly exaggerated, mythicised version of the latter, comparable with the mythification of the shy, diffident gorilla as a marauding man-devourer back in the early days of African exploration by westerners before its true nature was exposed by science? Yet according to Tyson Hughes, the winged bat-like orang bati was a frightening reality, not a fanciful folktale.

(In a more recent *FT* article, published in its April 2010 issue, Jaap continued to doubt Tyson's testimony - questioning how well Tyson understood the local Bahasa Indonesia language and how well his informants understood English, and suggesting that he was misled by them. However, this notion ignores the self-evident fact that as Tyson resided and worked in close contact with native Seramese helpers for a very appreciable time, he and these latter people would have been unable to function satisfactorily in their respective VSO-linked roles without possessing a good working knowledge of each other's language, as I noted in a *FT* letter of August 2010.)

RUCHI MEHTA'S TESTIMONY

In May 2009, this tangled tale became even more complex, because that was when I received news of a first-hand eyewitness account of an orang bati – which offered yet another entirely different view of this anomalous entity. The eyewitness in question was 32-year-old Ruchi Mehta from London, who had been on holiday in the Moluccas during March 2009, after living in Indonesia for the previous two years, before returning to London. In her initial email to me, of 17 May, Ruchi stated:

> I have recently come back from Seram, Indonesia, where I saw the 'orang bati'. Unfortunately, I didn't manage to get a photo of it as it happened quite quickly and then I was too scared to go back. I knew nothing of the creature before I went to Seram but upon returning to London, I've been reading up on it...your name came up whilst researching this creature so I thought I'd write.

Needless to say, I lost no time in emailing Ruchi back, with a request for details of her sighting, which later on 17 May she kindly provided as follows:

> I arrived in Seram by boat from Ambon late in the evening, and was making my way by car from the port in the west (Kairatu I think) to Masohi.
>
> After about 4 hours driving, I saw something sitting in the middle of the road (it was around 11 pm). At first I thought it was a monkey – a big monkey – but then realised Seram doesn't have any. The driver (from Ambon, later told us he had heard of orang bati, but had never come across one) was terrified and told us not to do anything.
>
> We drove past it, and I saw it side-on. Initially it reminded me of Helena

Alien Zoo

Bonham Carter in the film 'Planet of the Apes'. Sitting down (on its bum, in a crouching position), it was about the height of an average 4-5 year old. It was eating something out of a box that was, strangely, in the middle of the road. I didn't see how many digits it had in total, but it was 'scraping' the food out of the box using all digits.

It was all so quick I didn't really have a chance to see its features in detail, but it was quite hairy (less than a chimp, more than a human) [and] didn't seem to take any interest in our car (though the driver said this is because by sounding the horn, he is asking for 'permission' to cross by peacefully). I thought it strange that it didn't react in any way to the presence of a car – quite a big car – it didn't seem at all worried that we could actually drive straight into it.

I've trawled through hundreds of pictures and the one that I think best fits the description (though the picture is still too human-like) is at the following link:

http://images.google.co.uk/imgres?imgurlʃhttp://biblioteca.udg.edu/fl/ sahara/gifs/tallan.jpgˈimgrefurlʃhttp://biblioteca.udg.edu/fl/sahara/p- 2.htmˈusgfˍ5JnhoShIwhgsNoYzOBohV85nFgcfˌhʃ687ˈwʃ751ˈszʃ425ˈhlʃenˈ startʃiˈtbnidʃ5odKGN31cf8AVM:ˈtbnhʃi29ˈtbnwʃi41ˈprevʃ/imagesǽ3Fqǽ 3DHomoǽ2Bhabilisǽ26gbvǽ3D2ǽ26hlǽ3Denǽ26saǽ3DG [the picture is of an early African hominid, *Homo habilis*]

There were more of these things in the trees, I didn't see it properly, but at one point as we were driving, from the corner of my eye I saw something jumping down from the top of the tree, but I can't give any more details on that.

In my reply, I asked Ruchi if she had seen any suggestion of wings or a tail, and I also enquired whether, based upon her comparison of the creature with the *Homo habilis* picture, what she had seen could have been a primitive human, perhaps some form of pygmy. On 19 May, Ruchi emailed me with the following additional information:

I don't recall seeing a tail or wings, and from the illustrations I've seen of orang bati, it doesn't really fit the description, but it all happened so quickly and I was focusing more on the face.

At first I thought it was probably one of those 'feral' children you hear about in the Ukraine or something, and then I thought it was probably some sort of lesser-evolved human. I asked the driver (who wasn't really clued up to be honest) and he referred to them as 'mountain people' that come into the villages occasionally and cause havoc. He told us that 'they' would leave 'traps' on the road to cause accidents, and if you came across them, you needed to ask for their 'permission' to pass and they'd leave you alone. He told us a story about how they would hide out in the trees, and would over-turn cars (I didn't understand why they'd want to do that). [This may explain the presence in the middle of the road of the box containing food – fallen from some previous vehicle that had experienced an accident caused by one of their traps?]

Along the way, closer to the communities (trans-migrants), we'd see groups of men walking down the road with rifles strapped to their backs. I assumed they were out hunting at night, but the driver said it was for protection from these creatures. I know it wasn't a 'human' – but I can't seem to find any info about it apart from orang bati.

Even the tribal people weren't too keen on discussing it (it's got to be bad when head-hunter tribes are scared to talk about it!) so I am quite tempted to go back and spend some time there, preferably with some SAS guys, just in case J

So now, performing an extraordinary metamorphosis that would turn even the shape-shifting Greek god Proteus positively emerald with envy, the orang bati has transformed from winged man-bat to alien-eyed human to hirsute man-beast! How can this be?

(In his afore-mentioned *FT* article of April 2010, incidentally, Jaap van den Born proposed, very implausibly, I feel, that the hirsute wildman entity spied by Ruchi may have been an RMS - Republik Maluku Selatan - guerrilla fighter left over from the RMS uprising against the Indonesian government over 30 years ago, and still surviving amid Seram's remote mountain lands as a hairy old man in his 70s.)

AND THEN THERE WERE THREE...?

Is it conceivable, perhaps, that the term 'orang bati' has been applied to three totally different mysterious entities – a winged creature of cryptozoology, a modern but secretive tribe that has inspired various fears and fanciful folk beliefs among other tribes, and a primitive (or at least a feral) mountain-dwelling tribe comparable with (or even directly equivalent to) hairy wildmen reported widely across the globe? Or is the orang bati more than any of these? Could it be, I wonder, that, just like the blind men and the elephant, we are all, in our ignorance of this entity's true nature, touching a different part of it – an entity whose true, complete, but presently unperceived form is something more than we can even begin to imagine?

REFERENCES

DEANE, Shirley (1979). *Ambon, Island of Spices*. John Murray (London).
GIBBONS, Bill (1993). Pers. comms, 29 July & 5 August.
MEHTA, Ruchi (2009). Pers. comms, 17 (two) & 19 May.
SHUKER, Karl P.N. (1996). *The Unexplained:*

An Illustrated Guide to the World's Natural and Paranormal Mysteries. Carlton (London).
SHUKER, Karl P.N. (2010). Orang bati. *Fortean Times*, no. 265 (August): 70.
VAN DEN BORN, Jaap (2003). The bati-men. *Fortean Times*, no. 169 (April): 55.
VAN DEN BORN, Jaap (2010). The bat-man returns. *Fortean Times*, no. 260 (April): 56-7.
VAN DEN BORN, Jaap (2010). Orang bati update. *Fortean Times*, no. 263 (June): 70.

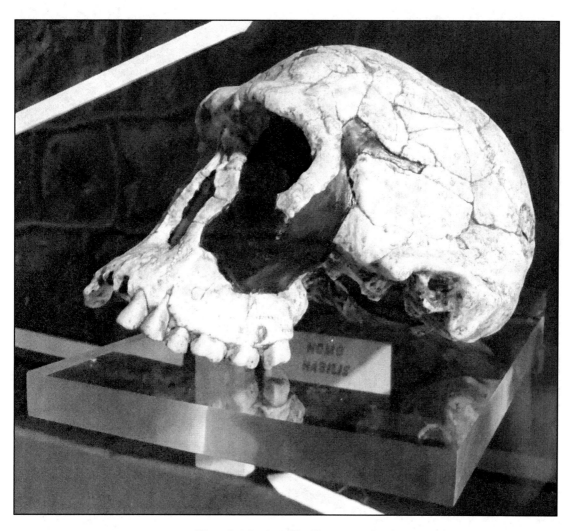

Homo habilis skull (Dr Karl Shuker)

Yes, I had Nessie all along (Dr Karl Shuker)

Reviews:
THE DVD THAT GOT AWAY,
AND A BEVY OF BOOK REVIEWS.

What creature thrives below the waves,
Beneath the surface grim?
What beast appears in photographs –
Obscure, opaque, and dim?
What creature gave the myths and tales
From bygone days new fame?
Of water horses, fierce, malign,
Which from the waters came,
To strike the hearts of every man
With terror of their forms.
The kelpies – dark, malignant ghosts,
And harbingers of storms.

Karl P.N. Shuker – 'The Loch Ness Monster',
Star Steeds And Other Dreams

For reasons of space, the following lengthy review of the DVD 'Loch Ness Discovered', which I submitted to *FT* on 12 July 2006, was never published. On 25 February 2009, I posted it on my Shuker-Nature blog, but this is the very first time that it has been published in hard-copy print format, and I have chosen to do so because, as will be seen, this DVD's content contains information and raises interesting issues that had rarely been publicised before.

For this same reason, I am also including here a selection of book reviews written by me that have been published in *FT* over the years.

'LOCH NESS DISCOVERED'
- REVIEWING THE DVD.

Discovery Channel, 2005, approximately 1 hr 17 min.

The cover of this Discovery Channel DVD prominently displays the familiar image of the Surgeon's Photo, purportedly depicting Nessie, but as its title suggests, the scope of the DVD's contents goes beyond Nessie to encompass Loch Ness as a whole. Indeed, of the four films included on it, the principal one, sharing its title with the DVD itself and lasting for 45 minutes, is primarily concerned with the loch's natural - as opposed to unnatural - history.

Alien Zoo

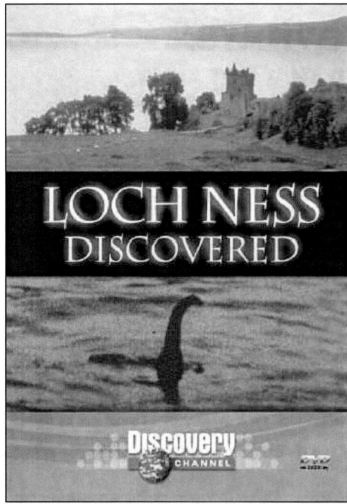

Cover of 'Loch Ness Discovered' DVD

Originally released in 1993, Film #1 follows Project Urquhart, featuring the researches of two scientific teams working at Loch Ness, studying its complex but hitherto little-investigated underwater ecosystem. One team, from the Freshwater Biological Association (FBA), is particularly interested in the intriguing fact that Loch Ness is to all intent and purposes two separate lakes, comprising a warmer upper layer, where little lives, above a colder, wilder, under layer containing fauna and dramatic underwater weather. The second team, from London's Natural History Museum, is surveying global pollution, which it is investigating at Loch Ness by examining its microscopic but pollution-sensitive nematode worms.

Interspersed with coverage of these ongoing mainstream studies are cryptozoologically-interesting segments focusing on various aspects of the Loch Ness monster phenomenon - such as Peter MacNab's 1955 photo of a Nessie-type form close to Urquhart Castle, the underwater flipper photos of Dr Robert Rines, the Surgeon's Photo, Tim Dinsdale's film, assorted eyewitness accounts, and psychologist Dr Susan Blackmore's theories of what may be influencing such accounts. Along the way, some intriguing data and findings emerge.

For example, in the past, sonar has found a series of strange regular prints on the loch bed, nicknamed 'the footprints', whose origin has never been explained, but which may be related to wartime military exercises here. During the two teams' studies, a remote-controlled unmanned craft, the *Sea Owl*, filled with cameras, is sent down to investigate one of these prints, but reveals it to be nothing more startling than a submerged wheelbarrow. Surely, however, as aptly queried by the narrator, submerged wheelbarrows couldn't explain all of these 'footprints', but then he seems to run out of investigative steam, ending with the weak comment that scientists can only speculate. Why can they only speculate? Bearing in mind that the *Sea Owl* had successfully unveiled the identity of one of these prints, how little more time, trouble, and money would it have required simply to have taken this craft along the loch bed a bit further while it was already there, in order to spy on a few more of these prints and find out what they were too? Surely this was a superb opportunity to solve at least one Loch Ness mystery that instead was needlessly lost.

A very notable, unexpected find made by the Natural History Museum team's fish expert, Dr Colin Bean, was that, contrary to a previous estimate, in 1973, that the loch contained 3 tons of fish (and which had been deemed sufficient to support a higher predator), it now appears that a much more realistic estimate is 27 tons. That is, 9 times more fish than hitherto assumed, thereby substantially increasing the possibility that the loch could sustain a large-sized species of top predator - a loch which, incidentally, contains as much water as in all of England and Wales combined.

Following this discovery, the film proceeds to consider the biology of plesiosaurs, deemed the best fit for most Nessie sightings, as well as for Rines's flipper photos. Using the computer enhancement expertise of Brian Reece Scientific Ltd with the original unenhanced photos, the researchers attempt to duplicate the final rhomboid flipper images widely publicised by the Rines team, but are unable to do so. Moreover, when they examine the surgeon's photo, they notice a curious white spot just in front of the neck, which may indicate the presence of something towing the neck along, but equally may just be a blemish on the negative.

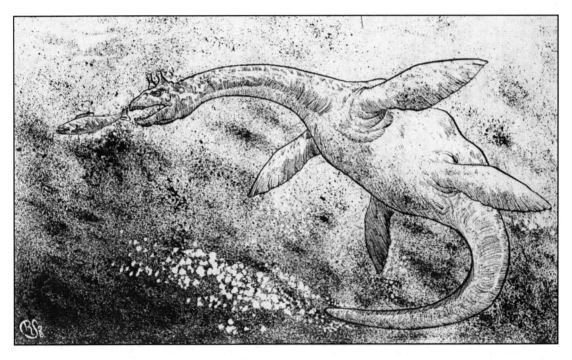

Nessie envisaged as a plesiosaurian creature (Richard Svensson)

The most interesting find made when applying computer enhancement expertise, however, occurs with a frame depicting a very large object moving across the loch from the famous film shot by Tim Dinsdale in 1960. First of all, the team examines not just the frame's positive but also its negative image, and is surprised to see in the negative a shadow behind the object. Furthermore, when the positive is cleaned up by enhancement techniques, a very large underwater shadow directly beneath the object can clearly be seen - implying that whatever this object is, it possesses an extremely sizeable hitherto-

Alien Zoo

unsuspected portion present beneath the water surface, and thereby arguing against the possibility that it is merely a surface vessel such as a boat. (Having said that, Nessie investigator Dick Raynor has opined at http://www.lochnessinvestigation.org/cyberspace.html#seeunderwater that perhaps the underwater shadow is merely a reflection of the shore behind the object.)

The climax of the film, however, comes with the FBA's sonar work aboard their research vessel *Calanus*. During the evening of 19 July 1993, a massive underwater storm is recorded by their sonar equipment as it rages beneath the vessel, an event rarely witnessed before, and guaranteed to disturb the loch's fauna. The following day, while examining the sonar traces recorded during that storm, Dr Colin Bean and other members spot a very large, unidentified sonar trace deep in the water with a second one close by (and perhaps even a third and fourth), which do not appear to be shoals of fish because they are followed by quite a pronounced wake (whereas shoals of fish do not cause wakes). The team members are perplexed, unable to explain these anomalous traces.

The film ends in celebration - what appears to be a totally new species of microscopic nematode worm

Ichthyosaurs and mosasaur, painted by Heinrich Harder

Alien Zoo

has been discovered during the research work. Cryptozoologists, however, may wish that the unexplained sonar traces had elicited as much interest and attention.

Film #2, entitled 'PaleoWorld: The Loch Ness Secret', and lasting 25 minutes, is probably of more direct cryptozoological pertinence, as it attempts to uncover the possible identity of Nessie, by examining three supposed contenders from prehistory - ichthyosaurs, mosasaurs, and plesiosaurs. British palaeontologist Dr Michael Benton discusses the anatomy and lifestyle of each one, supplemented by various specialists from elsewhere around the world and some stunning film of preserved fossils, as well as a reconstruction of pioneering fossil hunter Mary Anning's discovery at Lyme Regis, Dorset, during the 1800s of the first complete ichthyosaur and plesiosaur skeletons.

Personally, I found the ichthyosaur segment superfluous, as this remarkably fish-like or even dolphin-like reptile bore little if any resemblance to eyewitness accounts of Nessie. Indeed, the most memorable part of it came at the very end, with the narrator's chilling closing line - noting that if ichthyosaurs do indeed exist in Loch Ness, it could be the most dangerous place in the world to go fishing! Other than Lake Champlain, perhaps?

Sandwiched between the ichthyosaur and mosasaur segments is a reconstruction of the mystifying land sighting by chauffeur Alfred Cruickshank, which occurred at dusk one evening in summer 1934 according to this film (but normally given by other sources as early morning in April 1933) as he was driving along the north bank of the loch.

At the crest of a hill, his car's headlights picked out a big animal crossing the road. It had a large humped body, estimated at 4 ft high and around 25 ft long, and waddled away on two pairs of legs, its belly on the ground, and its head close to its body, with very little neck. Later, summing up the mosasaur section, the novel question is posed as to whether Cruickshank's mystery beast was a female mosasaur that had come on land to lay her eggs and was now returning to the loch, just as sea turtles come ashore to lay their eggs before going back into the sea.

The third, and most popular, reptilian contender for Nessie is then discussed - the plesiosaur. Included here is an eyewitness reconstruction from 1 June 1994, when, after seeing a mysterious object above the water surface while driving alongside the loch, Fiona Mackay and her friend Errol David jumped out of their car and ran along the bank for a clearer view. The object had a long tall neck and moved swiftly in the water, then suddenly dived, creating such a splash that its two observers had to jump back to avoid being soaked. Moreover, other eyewitnesses saw it that night. However, the film ends with no firm suggestions as to what Nessie may be, always assuming that such a creature does exist.

Films #3 and #4 are no more than a few minutes long. The first of these is a brief interview with Adrian Shine at the onset of Operation Deepscan back in 1987, and the second, less than 2 minutes long, is a montage of film clips of early Nessie expeditions, and images as to what it may look like.

All in all, this DVD is an interesting survey not just of Loch Ness as a famous 'monster' lake, but also as a body of water that is actually as puzzling to mainstream zoology as it is to cryptozoology (though it should be borne in mind that as these films were made during the 1990s, their findings are not current). If you are hoping for an exclusively cryptozoological package, you may be disappointed, but worthy of note here is that the cryptozoological coverage is presented in a relatively optimistic, open-minded manner - in stark contrast to the depressing tendency by so many of the more recent LNM documentaries to rule out of hand with smug self-assurance even the faintest possibility of a cryptozoological mystery existing here.

Alien Zoo

CRYPTOZOOLOGY: SCIENCE AND SPECULATION.
BY CHAD ARMENT.

Coachwhip Publications: Landisville, 2004. Pb, 393pp, appendices, bibliography, index. ISBN 1-930585-15-2

For all its relatively high media profile nowadays, in the eyes of mainstream publishers cryptozoology generally remains a specialised subject that is not expected to sell in vast quantities. Consequently, books on this subject that dare to venture too far beyond the familiar - Nessie, bigfoot, sea monsters, and suchlike - and certainly any that seem overtly scholarly or 'deep' (as immediately suggested if there is not an abundance of eyecatching illustrations, or if the bibliography is more than a few pages long), face the fate of never seeing the light of day. (Yes, I know that there have been exceptions, but those were precisely that - exceptional.)

Thank Heavens, therefore, for Desk Top Publishing - at least, that is, when it is in the hands of genuinely talented, thoughtful writers like Chad Arment. No, there aren't any pictures. Instead, this extraordinarily interesting book is brimming with the kinds of topics and ideas appertaining to cryptozoology that have until now been predominantly confined to the archives of some of the better online cryptozoology chatgroups (such as Arment's own, invitation-only cz@yahoo.groups.com).

Indeed, Part 1 of his book (roughly a third of its total length) is devoted entirely to the science of cryptozoology, and discusses such subjects as: what is cryptozoology?; a scientific foundation for an investigative methodology; a logical foundation for an investigative methodology; an ethnozoological foundation for an investigative methodology; the rationale for cryptozoology; the methodology of cryptozoology; the feasibility of cryptozoology; and the credibility of cryptozoology.

Never before has any hard-copy cryptozoological publication offered such a detailed analysis of these areas - desperately needed if cryptozoological credibility is to continue and increase, yet something that mainstream publishers would simply not have understood or accepted. Tragically, containing swathes of cryptozoological text devoted to "investigative methodology", "rationale", and "logical foundation" but without a single Nessie snapshot or photos of some bigfoot spoor inserted here and there to make it appear (in the eyes of non-cryptozoological editors) more palatable and 'entertaining', this very worthy and extremely important work may never have seen mainstream print. And there are copy editors that might well have been in urgent need of smelling salts if confronted with the likes of "ethnozoological"! Sad, but true.

However, thanks to Arment's bravery in going it alone and publishing his book independently, he has done cryptozoology a great service, and has given us all a valuable insight into what the future of cryptozoological publishing and writing may hold. Namely, books that delve into areas and concepts far too specialised to gain mainstream publishing approval, but which must see print if cryptozoology is ever to advance and not stagnate.

Part 2 of his book is equally rewarding, surveying a range of mystery beasts obscure even by cryptozoological standards but all quite fascinating - from long-tailed bobcats, giant orang utans, and enormous snakes in the Everglades, to dwarf seals, the great naked bear...and you'll have to buy this book if you want to find out what the others are! It ends with three appendices, devoted respectively to suggestions for the obtaining of larger zoological specimens; North American terrestrial mammal description dates;

Alien Zoo

and wildlife genetics laboratories; plus a 17-page bibliography, and a 6-page index.

All in all, a compelling publication that is very different indeed from any previous cryptozoology book. So if you're looking for a light read, this is not for you - but if you're searching for an understanding of the mechanics of cryptozoology, what lies beneath its surface, rather than just another parade of cryptids great and small, you need this book - now.

TASMANIAN TIGER.
BY DAVID OWEN.

Johns Hopkins University Press: Baltimore, 2004. Hb, 228pp, illus, notes, bib, ind, $25.00 in USA, ISBN 0-8018-7952-3

The largest modern-day species of Australian carnivorous marsupial, the thylacine or Tasmanian wolf *Thylacinus cynocephalus* - although most commonly known Down Under as the Tasmanian tiger on account of its stripes (and in spite of its strikingly canine morphology!) - remains one of the world's most readily recognised creatures. This is no mean feat, bearing in mind that it officially died out in 1936. But there is much more to the thylacine than just its singular appearance, or its tragic disappearance, or even its much-hoped-for reappearance. This extraordinary animal and its no-less-exceptional history have become both a symbol of past mistakes on the part of humanity and a warning that must be heeded for humanity to avoid making similar ones now and tomorrow.

This is the message that clearly runs throughout David Owen's absorbing chronicle of the once and future(?) thylacine, recounting in fact-filled but never dry or wordy detail the discovery and destruction of this remarkable species, and documenting the amazingly sparse extent of knowledge concerning it that was gleaned before it was lost to science, and the world, almost 70 years ago. Despite being no bigger than a standard field guide, *Tasmanian Tiger* certainly offers the reader value for money - containing 15 chapters, a sizeable selection of colour and (mostly) b/w photos and engravings (and even a recurring thylacine motif on each right-hand page), 16 pages of detailed notes, a bibliography, and an index.

The first four chapters concentrate upon the thylacine's discovery and naming by European explorers and settlers, its evolution, its biology and lifestyle, and its extinction. The next eight deal with the increasingly hostile attitude of the settlers to the thylacine and their merciless yet largely unwarranted persecution of it, culminating in the lonely and pitiful death of Benjamin (actually most probably a female, not a male) - the world's last captive (and confirmed living) thylacine - in Hobart's Beaumaris Zoo on 7 September 1936. With bitter irony, this was just two months after the Tasmanian Parliament finally granted the thylacine full protection. These chapters also document the major searches that took place in the aftermath of Benjamin's demise, initially confident of locating wild specimens but later fired by the ever-increasing, desperate realisation that the thylacine may have truly gone.

The last three chapters deal respectively with the thylacine in commerce and art, the exciting possibility that the thylacine will be resurrected one day by cloning (though plans by Dr Michael Archer to pursue this dream using DNA extracted from three preserved thylacine pups in the Australian Museum have recently been abandoned, at least for the time being), and sightings and the science of survival. This final chapter is likely to be the single most interesting chapter for cryptozoologists and other *FT* readers, documenting various recent sightings and the likelihood of thylacine survival.

Alien Zoo

However, in view of the extensive coverage given earlier in the book to this species' history, those who consider it plausible that the thylacine does still exist may be disappointed by Owen's very slim, lightweight treatment of this important, tantalising issue in his closing chapter. For example, no mention whatsoever of dobsegna reports from Irian Jaya (Indonesian New Guinea) is included, nor of the persistent sightings emanating from mainland Australia. Even Hans Naarding's classic 1980s encounter in Tasmania's Arthur River region - popularly deemed to be the most credible post-1936 eyewitness account of a thylacine ever recorded - receives only a few lines here.

Clearly, therefore, those seeking thylacine cryptozoology need to look elsewhere for data, but for historical, biological, and non-speculative information this book is a highly valuable addition to any naturalist's library. True, it cannot - and indeed, does not attempt to - compete with the lavishly-illustrated, sumptuously-produced likes of Eric Guiler and Philippe Godard's magnificent, definitive tome *Tasmanian Tiger: A Lesson To Be Learnt* (1998). But as a compact yet rewardingly comprehensive survey of one of modern-day natural history's most enigmatic and also emblematic of beasts, Owen's book has few rivals. And if its tale of tragedy can inspire readers to contribute to the ongoing conservation of our world's unique fauna, then perhaps the thylacine might not have died wholly in vain after all.

BIG BIRD!
BY KEN GERHARD.

CFZ Press: Bideford, 2007. Pb, 96 pages, illus., appendices, bibliog., notes, £10.99 in UK, ISBN 978-1-905723-08-9

In terms of media popularity and general reader familiarity, reports of gigantic flying birds and pterodactylian lookalikes – collectively dubbed Big Birds - in North America rate almost as highly over there as claims for bigfoot and assorted lake monsters. Yet whereas these afore-mentioned cryptids have received countless book-length treatments over the years, similar degrees of documentation for Big Birds are significantly fewer. Indeed, until now the only notable exception to this has been Mark A. Hall's well-received book *Thunderbirds!*, which has passed through a number of revised, expanded editions. However, with the publication of this present book by Ken Gerhard, a second, wholly independent investigation on the subject is now available, and promises to become as classic a work in the cryptozoological literature as Hall's pioneering, ongoing study. Despite being a relatively slim volume at only 91 pages of main text, a remarkable amount of information is packed inside, together with a number of the author's own photos plus some relevant illustrations by acclaimed cryptozoological artist William Rebsamen.

Based as he is in Texas, much of Gerhard's book concentrates upon his own investigations of Big Bird flaps in his home state, which has traditionally been a centre for such sightings. However, there is also a fascinating chapter on neo-pterosaurs reported from elsewhere in the world, including such disparate localities as Zambia (the kongamato), New Guinea (the ropen), Japan, Crete, Namibia, the Amazon Basin, Western Australia, and...West Yorkshire! Also recorded are odd accounts of atypical winged anomalies from North America, including giant mystery bats and an alleged flying snake in Texas.

Traditional Amerindian thunderbird lore is also well-documented and quite fascinating. Did you know, for instance, that some thunderbirds were deemed in native legends to be shape-shifters? Not so much airborne cryptids, in fact, as were-birds, who had only to tilt their heads upwards for their avian beaks

Alien Zoo

Front cover of Ken Gerhard's *Big Bird!* (CFZ Press)

to detach like masks, before casting off their plumaged skins like feathered blankets, to become wholly human.

I documented the North American Big Bird phenomenon in detail within one of my own books, *In Search of Prehistoric Survivors* (1995), and while acknowledging that some descriptions certainly recalled pterosaurian creatures, in my view most of the cryptids referred to as Big Birds seemed to be unequivocally of the ornithological kind, albeit featuring wingspans far in excess of any species known to exist today. Consequently, I was somewhat surprised to find that Gerhard favours a pterosaurian identity as the predominant explanation for his continent's Big Birds, although in fairness the examples that he has concentrated upon in his book are mostly drawn from the leathery-winged contingent, whereas in mine I principally featured the feathered examples. In relation to his theory, he also includes a short appendix by Leland Hayes discussing prehistoric pterosaur lifestyle and diversity.

Having said that, I do agree with Gerhard that certain known types of bird that have been offered in the past as explanations for pterosaur-type Big Bird sightings fall far short of the mark. These include eagles, vultures, condors, and those long-winged maritime species the frigate or man o' war birds. At the time of writing my own prehistoric survivors book, I had never seen a living, breathing frigate bird aloft in the sky, so I could not offer any insight drawn from personal experience as to whether this avian form could indeed be mistaken for a pterosaur in flight.

In 2007, however, I was fortunate enough to find myself standing on Mount Corcovado, beside the base of the stupendous Christ the Redeemer statue overlooking the bay of Rio de Janeiro in Brazil. Gazing up into the sky, I was delighted to see not only the customary circling of vultures above the statue but also, further out, a number of soaring magnificent frigate birds *Fregata magnificens*, more of which were also later spotted by me winging over Copacabana Beach and again above the summit of Sugar Loaf Mountain – and I can confirm that none of them bore even a passing resemblance to pterosaurs. Ditto, incidentally, for the exceedingly rare California condor *Gymnogyps californianus*, two of which I was extremely privileged to see in 2004 while visiting the Grand Canyon where some free-living captive-bred specimens now exist. Completing Gerhard's book is a useful appendix containing a chronological listing of Big Bird sightings, a short bibliography, and a more extensive Notes section. Sadly, there is no index, which I would have greatly appreciated when attempting to locate passages in the main text that I had previously read and wished to re-read while preparing this review. Perhaps, however, one can be included in a subsequent edition, as I feel sure that this handy and very interesting book will prove more than adequately successful to warrant updating on a regular basis in the future, and I happily recommend it to anyone wishing to learn more about North America's highly intriguing aerial cryptids.

Alien Zoo

MONSTER OF GOD:
THE MAN-EATING PREDATOR IN THE JUNGLES OF HISTORY AND THE MIND.
BY DAVID QUAMMEN.

Hutchinson: London, 2004. Hb, 515pp, bib, notes, ind, £25.00, ISBN 0-09-179957-0

Today, humanity stands aloft at the very apex of the planet's foodchain, viewing the numerous lower levels imperiously, secure in the knowledge that we have nothing to fear on a grand scale or even a very regular basis from any other carnivorous species sharing our world. But it was not always so. Lurking like malevolent shadows within the darkest recesses of our collective, inherited psyche are deeply-etched genetic memories of a far-distant time when the likes of feline, ursine, and even crocodiline devourers customarily preyed upon our less-evolved, poorly-protected species. Today, we have won the battle with such monsters, which have been ruthlessly decimated by our kind, persisting within what were once the mere hinterlands of their mighty domain, and even then only by the grace of modern-day humanity's generally more enlightened, ecologically-minded outlook. But the fear that they once engendered has not been eliminated, continuing to prowl in ceaseless fury within the caged jungles of the human mind.

This is the core of Yale/Oxford University-educated David Quammen's latest, superbly-written book. So if you are expecting a straightforward tome surveying the fundamental lifestyle and biology of a selection of contemporary predatory species, look elsewhere. The contents of this volume are much more complex, esoteric...and fascinating. They consist principally of eight hefty chapters, beginning with an examination of our own uneasy and multi-faceted relationship, including mythological and spiritual aspects (featuring notable appearances by the seven-headed Phoenician dragon Lotan and the Holy Bible's Leviathan) with certain dangerous, man-eating species - species that he terms alpha predators.

The next six chapters offer a truly dazzling, even phantasmagorical survey of four such creatures - the lions of India's Gir Forest, the brown bears of Romania's Carpathian Mountains, the saltwater crocodiles of Northern Australia, and the mighty tigers of Siberia. Within each such survey, Quammen deftly combines first-hand fieldwork, snippets of folklore, and compelling psychology, with the scientific contributions of foremost researchers in the field, political backgrounds, and even digressions into the kind of memorable trivia that make this kind of book such a treasure, to yield a breathtaking journey of the very best 'magical mystery tour' variety. At no point in the book is the reader ever quite sure where the next twist or turn will take them, but they know full well that they are only too eager to find out. Do not pick this book up to start reading just before bedtime, because you will still be there, bleary-eyed but totally hooked, the next morning!

Here, then, are just a few of the numerous highlights on offer for each of Quammen's ever-lively quartet of alpha predators. He begins with the near-maneless lions of the Gir Forest *Panthera leo persica* - the alpha predator that inspired him to write this book - opening with a detailed examination of the lion's former geographical glory, extending well inside Europe's boundaries, followed by its inevitable decline at the hands (or hand-weapons) of mankind. Its modern-day history in India receives weighty coverage, but Quammen's light, quirky use of contemporary language ensures that the reader's interest is

Alien Zoo

held - ably assisted with periodic digressions into such diverse areas as leopard biology, and why there are no tiger-devouring monsters around or super-beasties that chomp merrily on great white sharks.

Later portions meticulously probe via direct communication with various representative individuals the deeper, more intense considerations of how various livestock-farming tribes interact with the Gir lions, especially as the Gir Forest is a sanctuary. What is the ultimate destiny of the tribes, and the fate of the lions? It may be 'right-on' for Westerners to bang the drum loudly for lion conservation from their cosy homes well away from these creatures' fangs and claws, but where does that leave the Gir Forest's poverty-stricken tribes living in far closer proximity who still suffer depredation from their leonine neighbours?

Moving onto Quammen's documentation of Australia's toothy crocodile alpha predator, I freely confess that as a biker my favourite chunk of this section was the surprising discovery related in compelling detail that Australian Hell's Angels gangs traditionally adorn their club-houses with the mounted heads of crocodiles, and even send them to one another. Quammen chats to a Darwin-based taxidermist who specialises in preserving such striking exhibits, but don't try this at home - a single scratch acquired from a single tooth in the jaws of a crocodile while preserving its head contains more than enough gangrene-inducing bacteria to separate you from a finger in an alarmingly short space of time. In this section, Quammen carefully examines the totemic power of the saltwater crocodile *Crocodylus porosus* among aboriginal - and biker - communities, as well as modern-day crocodile trade and farming, the continuing danger posed by these ferocious reptiles, and, inevitably, contemporary crocodile hunts.

Quammen's coverage of the Carpathian brown bear, belonging to the Eurasian brown bear subspecies *Ursus arctos arctos*, is no less eclectic. I even found myself perusing with keen interest his potted review of the rise and fall of former Romanian dictator (and very formidable bear-hunter) Nicolae Ceausescu, a figure from European history who had previously failed to engage more than passing attention from me - as good an example as any of how successfully alluring Quammen's digressions can be. Another one is how dragon myths may owe their origin to inherited memories of predation by leopards *Panthera pardus* and other terrifying meat-eaters when our ancestors lived on the plains of East Africa, and also, in Romania's part of the world, to nightmarish confrontations with the gigantic cave bear *Ursus spelaeus* - invoking such legends and folk-epics as the dragon of Klagenfurt, Beowulf's confrontation with Grendel, and even (further east) the mistaken identification of *Protoceratops* dinosaur fossils by early humans as the remains of griffins.

And as before, the ever-present conflicts between alpha predator and human are rigorously explored throughout this section, including an extensive consideration of the brown bear's place in this region's forest management, and how the local shepherds accept its unpredictable omnipresence.

Last but by no means least is Quammen's voyage of discovery in search of the modern world's largest living felid - the magnificent Siberian tiger *Panthera tigris altaica*. Having said that, this section actually opens with a survey of a very different category of 'tiger' - examining the world's erstwhile sabretooths or machairodontids, a distinctive taxonomic subfamily of feline carnivores discrete from bona fide tigers and other true felids, but no less extraordinary. Passing on to the real tiger, Quammen was shocked to learn that its traditional near-deification in this area of the world was being rapidly replaced by the deification of another, very different item - money. Despite being fully protected by law, the Siberian tiger has an extremely valuable pelt that ensured its worth alive as a god was far less than its worth dead as a means of purchasing a new snowmobile. True, the earlier beliefs and myths survive, but economic practicalities faced by local hunters have torn away their potency.

The history of the Siberian tiger's fall, and attempts to engineer at least a partial rise again in recent years thanks to Russian conservation efforts and the studying of wild tigers here, is well documented. And with the news that as no human has been killed by a Siberian (or, as Quammen prefers to call it, Amur) tiger since 1997, this may signal a decrease in tiger poaching, let us hope that this most spectacular big cat will indeed survive.

The final chapter is certainly the most surprising, but also, at least for me, perhaps the most entertaining, of all. Its two highlights are Quammen's coverage of the unexpected discovery of France's Chauvet Cave and the truly amazing cave paintings disclosed inside (including dazzling images of cave lions *Panthera leo spelaea* that lend good support for believing that both male and female in this leonine form were maneless); and his delightful assessment of the rapacious, eponymous space monster that starred in the 'Alien' series of blockbuster sci-fi films as a bona fide alpha predator - a sort of parasitic wasp with attitude, wrought large, bipedal, and photogenic in a gut-exploding, saliva-drooling kind of way. Ably supplementing this bravura performance of a book is an extensive listing of source notes, an equally expansive bibliography, and a very detailed index.

If you prefer your reading material to progress in a steady, predictable, linear manner, this book may not be ideal for you, as it has more side-turns, false starts, and meandering passages than the maze at Hampton Court. If, however, you have the kind of mind that flows or even spurts forward on several fronts at once, you will embrace it and pursue its multitudinous musings with true joy, as I have done. David Quammen did the world of natural history writing proud with his previous offering, *The Song of the Dodo*, and he has done it again with *Monster of God*. Let's hope that we don't have to wait too long for his next masterpiece.

THE LOCALS:
A CONTEMPORARY INVESTIGATION OF THE BIG-FOOT/SASQUATCH PHENOMENON.
BY THOM POWELL.

Hancock House: Blaine, 2003. Pb, 271 pp, ind, $19.95 US, ISBN 0-88839-552-3

Looking across some shelves that quite literally contain dozens of books dealing with bigfoot and other man-beasts, I wonder how many more can this subject sustain, and how can anyone who does choose to write on it hope to find a different approach or slant. I still have no answer to the first of these questions, but I am happy to say from reading this book that its author, Thom Powell, has successfully confronted the second one.

The first welcome change, highlighted in the book's subtitle, is that Powell has concentrated upon contemporary cases instead of merely recycling and reassessing the same hoary old standards. So if you're looking for yet another lengthy re-examination of the Patterson-Gimlin footage, for example, you won't find it here (not even a still of the demure Patty stepping out is included). Indeed, perusing the index (always useful but all-too-often absent from bigfoot books), I only discover a single page-reference to it, p. 25 (although that particular reference is, unfortunately, incorrect - try as I might, I have yet to spot any mention of the P-G footage on p. 25, but it is named on p. 45). Nor are there any photos of casts,

Alien Zoo

Bigfoot (William Rebsamen)

bigfoot statues, eyewitnesses, or any other illustrative material, just one map and some simple chapter-heading line drawings.

Instead, Powell has turned to his own Oregon-based investigations for material, and also to the files of the Bigfoot Field Researchers Organization (BFRO), for whom he was once a curator. Consequently, much of the data presented here will be fresh to many of this book's readers, but there is also coverage of modern headline-hitting cases, most notably the Skookum body cast, which receives an entire chapter.

I also appreciate the clear presentation of this book. Too many bigfoot tomes in the past have consisted of daunting - and sometimes downright off-putting - swathes of solid, closely-spaced text, unbroken by anything more than a few chapter headings, and with only the narrowest of margins. Powell's, conversely, has wide margins, clear paragraph separation, and sub-headings when needed, thereby actively encouraging extended reading, and even dipping. And once readers are enticed, they will be held by Powell's highly accessible, conversational style of writing, avoiding the literary Scylla and Charybdis of dry and sensationalised commentaries that can so often be found in books on this subject.

Perhaps, however, this should come as no surprise - Powell is, after all, a science teacher, and so is professionally versed in presenting data in an attractive, easily-digestible manner. For the same reason, we should expect to find emphasis upon the science behind bigfoot investigation, and this too is amply represented here. An entire chapter is devoted to bigfoot science at the book's beginning, including some very interesting information on hair analyses and DNA, and at its end there are rigorous examinations of the three major explanations currently on offer for bigfoot - as an elusive but real creature, as a paranormal entity, and as an extraterrestrial being. There is also a fascinating Appendix entitled 'Using the Bigfoot Phenomenon to Teach the Scientific Method: A Set of Science Lessons for Students in Grades 8-12', which should surely guarantee attention from even the most jaded student!

All through his book, Powell uses the data gathered to help establish a profile of behaviour exhibited by bigfoot, and mostly this works well, with one exception. As documented in detail by Powell, one of the most baffling bigfoot characteristics reported by eyewitnesses is its apparent ability to vanish into thin air - one second it is there, the next it is gone. Here, I feel, Powell's eagerness to be seen to give all possible solutions, however remote, a fair hearing goes a little too far - especially when he soberly considers the prospect of bigfoot having evolved the extraordinary ability to warp local time, perhaps even to move between dimensions, or at least to use its own mental processes to immobilise temporarily a human eyewitness's perception of local time. All large, forest-inhabiting animals are masters of camouflage and the ability to disappear from sight without needing to resort to such extreme measures, so why should bigfoot need to be made a special case?

Such disagreements aside, however, *The Locals* is an excellent book for the modern-day bigfoot reader. True, it is nothing if not bold for a bigfoot author to all but ignore the Patterson-Gimlin film - rather like writing a book on the Loch Ness monster and not mentioning Tim Dinsdale's film. Nevertheless, Powell's brave gamble pays off, enabling him to haul his hefty subject into the present in terms both of data proffered and of the ways in which such data needs to be examined, and I can certainly recommend his book to anyone seeking a clear, forward-thinking approach to this ever-intriguing subject.

THE NATURAL HISTORY OF UNICORNS.
BY CHRIS LAVERS.

Granta: London, 2009. Hb with dustjacket, 258 pp, b/w illus., bibliog., index, £19.99, ISBN 978-1-84708-062-2

Receiving this book from *FT* to review, it was not without a degree of trepidation that I read its title, *The Natural History of Unicorns* – bearing in mind that I am presently working upon a book of my own entitled *Unicorns: A Natural History*. Happily, there is no conflict, for whereas mine concentrates upon the retelling of famous and not-so-famous unicorn legends from around the world in the same manner as my earlier *Dragons: A Natural History* bestseller, and should be similarly illustrated in full colour, this present book, authored by Chris Lavers (and lacking colour images – see below), focuses upon the origin and history of the unicorn, and for the most part it does so very satisfactorily.

As one entire bookshelf in my library can testify, there have been many books published on the subject of this most fascinating and ethereal of mythological creatures, and include a number of standard works, such as Odell Shepard's *The Lore of the Unicorn* (1930) and Rüdiger Beer's *Unicorn: Myth and Reality* (1977 – yet, strangely, not included in the bibliography of Lavers's book). Consequently, I was curious to see how, if at all, this latest book's coverage would differ from that of its predecessors.

When dealing with the unicorn's early history, somewhat inevitably Clavers does tend to cover much

Alien Zoo

the same ground as earlier works, though with particular emphasis upon Ctesias of Cnidus's contributions in moulding the unicorn into the form familiar to us all today. What I found interesting and entertaining, however, is that Clavers does not shy away from incorporating his own, original ideas and researches, such as proposing, quite reasonably, that instead of the Persian onager *Equus hemionus* being the asinine component of the Indian unicorn's composite identity, for logical geographical considerations it was more likely to have been the kiang *E. kiang*, a large Central Asian wild ass.

That section is followed by other detailed chapters that examine such diverse subjects as the Judaeo-Christian unicorn and the unicorn in Christian imagery, a belligerent if beneficent Asian unicorn of burly form known as the karkadann, the okapi *Okapia johnstoni* as a possible origin for African unicorn lore, Dr Franklin Dove's famous horn-bud grafting experiments of the 1930s to create a dominant central-horned 'uni-bull', and unicorns in ancient civilisations (which, oddly, constitutes the last, rather than the first, chapter in this book).

By far my own favourite, however, is an extensive chapter investigating khutu - the enigmatic horn-like material from which in bygone ages Islamic cutlers would fashion ornate knife handles. This is the first book on unicorns that I have read that documents this subject in detail, and reveals the history of khutu to be every bit as interesting as that of the unicorn, with which it has become almost irrevocably intertwined. Down through the centuries, all manner of identities have been proposed for this mysterious substance, ranging from ivory (variously designated as mammoth, walrus, narwhal, or even hippopotamus), the frontal bones of certain ungulates such as goats or cattle, the teeth of snakes or fishes, and even the roots of a specific tree. Once again, however, after assessing all of these candidates, Claver offers for consideration his own thought-provoking source – the frontal bones and associated horn material of the musk-ox *Ovibos moschatus*, dating back to when this species still existed in Eurasia.

So far, then, so good. However, sometimes Clavers's trawling for unicorn associations spreads a little too wide, in my view. I think it highly unlikely, for example, that Babylon's famous scaly, claw-footed, long-necked sirrush or dragon of the Ishtar Gate lays much claim to unicorn affinity – the mokelembembe maybe, but not the unicorn. Conversely, on certain other occasions, his trawling has not spread anywhere near as wide as it should, or at least could, have done.

In particular, I was disappointed that the remarkable diversity of unicorn types on record was accorded very little coverage here. Indeed, the single unicorn chapter in my own *Dr Shuker's Casebook* (2008) surveys a much wider selection of unicorn varieties than can be found in the whole of Lavers's book. Why no Persian shadhahvar of the flute-like horn and bloodthirsty demeanour, for instance, or the equally rapacious hare-like al-mi'raj, or the web-footed camphor, the woolly-coated pirassoipi, the swivel-horned yale, the Chinese ki-lin, or the Patagonian unicorn depicted in ancient Argentinian rock paintings?

Perhaps the biggest shortcoming is the illustrative content, which consists entirely of 27 mostly small black-and-white photos scattered amongst the book's 258 pages. Although it is clearly intended to be treated primarily as a work of scholarship, I see no reason why colour images could not have been incorporated, if only perhaps as an insert of plates at the centre of the book.

After all, few mythological animals have inspired such an abundance of exquisite artwork as the unicorn – the two sets of justifiably celebrated Unicorn Tapestries instantly come to mind, for example, alongside many beautiful Old Master paintings. It is a great shame that these were not showcased here in their multi-hued glory. Having said that, I did enjoy the composite 'Identikit' unicorn depicted in colour on the dustjacket, but less so the fake foxing effect that was real enough for me to scratch at one spot, unsure of whether or not my review copy had been stored in a damp warehouse before being sent to me!

All in all, quibbles notwithstanding, *The Natural History of Unicorns* is an absorbing read and a worthy addition to the literature of this remarkable creature - surely one of the most famous animals never to have existed.

Some of the books that I've reviewed in *FT* down through the years (Dr Karl Shuker)

UNCONVENTION '94 LECTURE:
FROM PANTHERS TO PLESIOSAURS
- CURRENT AFFAIRS IN CRYPTOZOOLOGY.

Let no one suppose
That the creatures he knows...
Are ALL that the animal kingdom can show:
NO!...

At the corners of dreams,
Round the edges of sleep,
There's a something that seems to be going to creep,
To crawl or to climb, to lumber or leap...
It must be –
PREFABULOUS ANIMILES!...

Then some day he'll see them lurking in lanes,
Or breaking down hedges and fences and stiles –
He'll see the Prefabulous Animiles.

James Reeves – 'Let No One Suppose', in *Prefabulous Animiles*

Finally, I am delighted to be able to include in this book, after years of fruitless searches for it, the following long-lost *FT*-associated item of mine from my personal archives. Back in 1994, I prepared what in hindsight can now be seen to be a prototype of what would become my long-running Alien Zoo column in *FT*. I had been kindly invited by Bob Rickard and Paul Sieveking to speak at *FT*'s inaugural "UnConvention", to be held in London on 18-19 June 1994. Rather than preparing a lecture on a single subject, however, I felt that it might be more entertaining to speak on a range of different topics, and decided therefore to give a slide-accompanied talk on the major cryptozoological events and news stories from the past 12 months. This I did, on 18 June, and was delighted when it received a very positive response from the audience. So much so, in fact, that during subsequent years I have received many requests from colleagues and correspondents for a typed-out transcript of this lecture.

Yet although I had indeed prepared one shortly after giving the lecture at UnConvention, using a tape-recording that a friend in the audience had prepared for me, when I finally decided to seek out the recording and transcript to send copies off in response to one such request, I was unable to find either of them. Numerous searches through my archives since then have always failed to uncover them, so I eventually assumed that they must have somehow been discarded inadvertently, most probably during my major house move during the early 2000s. While completing the final read-through of this present book's manuscript before submitting it to its publisher, however, I needed to recheck some details; and while going through a relevant folder of data, I came upon, filed away in totally the wrong place, the long-lost typed-out transcript of my UnConvention lecture! Consequently, rediscovered just in time to appear here, this transcript is now reproduced below verbatim, together with some of the images that I used when presenting it plus a couple of later ones. [Please note that because this transcript has not been edited or updated in any way since its original lecture presentation in 1994, some of the information contained in it is now outdated, but I have deliberately chosen to present the transcript here in its

Alien Zoo

Holding a cast of a bigfoot print discovered in Washington State, USA, in 1982 (Dr Karl Shuker)

original form in order to preserve its historical relevance and interest.]

In addition to reports and eyewitness accounts of very famous cryptozoological animals, such as the yeti, Loch Ness monster, sea serpents, bigfoot and its giant footprints (like the one I'm holding here [SLIDE]), and so on, a surprising number of totally new, hitherto-unrecorded mystery beasts have also been hitting the headlines in recent times, here and overseas. What I'd like to do now, therefore, is to spend this time reviewing some of the more interesting and potentially important of these latest recruits to cryptozoology.

VU QUANG OX AND GIANT MUNTJAC

Those of you who've read my latest book, *The Lost Ark: New and Rediscovered Animals of the 20th Century* (published last November by HarperCollins) [SLIDE], will already be aware that the largest new mammal to have been discovered for over 50 years turned up as recently as 1992, in a little-explored area of northern Vietnam called Vu Quang. This is the creature

[SLIDE], the size of a buffalo and distantly related to buffalos, but with long antelope-like legs and horns, which is nowadays known as the Vu Quang ox *Pseudoryx nghetinhensis*. Its closest relatives, however, died out over 4 million years ago, which makes it a quite spectacular cryptozoological discovery.

Vu Quang is clearly the place to be for any self-respecting cryptozoologist, because in March of this year this same region made the headlines for a second time - when evidence of another major new species was uncovered here. This is a muntjac or barking deer [SLIDE], of which there are many different species. None of them, however, is very large, and all have short antlers sprouting from relatively long antler-bases or pedicels. In Vu Quang, however, some skulls were found three months ago that belong to an extremely large type of muntjac, twice as big as any previously recorded, and with unique antlers. The antlers them-

Vu Quang ox (William Rebsamen)

selves are long (rather than short), and sprout from antler-bases that are short (rather than long) - in other words, the exact reverse of the normal condition in muntjacs. Consequently, zoologists who have examined them agree that they are definitely from a dramatically new species - the biggest muntjac in the world, which is now, aptly, being referred to as the giant muntjac.

NEPALESE GIANT ELEPHANTS

Moving from one Oriental giant to another, we come to this creature [SLIDE] - which is a truly enormous Asian bull elephant, one of two photographed in a remote valley in western Nepal, during an expedition led by Colonel John Blashford-Snell in February and March 1992. Standing at around 11 ft 3 in at the shoulder, he is the largest Asian elephant *Elephas maximus* ever recorded, and has these unusual bumps on his brow and trunk. Although not characteristic of Asian elephants, they were typical of a supposedly long-extinct creature called a stegodont (the ancestor of modern-day elephants and also of the mammoths) - as pointed out by Canadian palaeontologist Dr Clive Coy. Since this photo was taken, a team has returned to the valley, to collect droppings and other biological samples for scientific analysis - so it shouldn't be too long before we finally discover just what these huge creatures really are. Mutants or mammoths? We'll have to wait and see.

THREE PERUVIAN MYSTERY CATS

During several years of research in Peru, American zoologist Peter Hocking has received reports of at least three different types of mystery cat. One is particularly distinctive, because it is said to be striped, just like a tiger! There is no known species of striped cat inhabiting South America - but as I pointed out in my book *Mystery Cats of the World* (1989), reports of a large, undiscovered striped cat have previously been documented from Colombia and Ecuador, and these reports are thought by some to describe a living sabre-tooth tiger, presumably related to this creature [SLIDE], the extinct American sabre-tooth, *Smilodon*.

The second Peruvian mystery cat is equally strange - it's as large as a jaguar, but instead of possessing the jaguar's rosettes, its entire body is marked with solid black speckles.

As for the third mystery cat, this allegedly resembles a giant black panther, so this may simply be an extra-large form of black jaguar.

ISNACHI

Since the beginning of the 1990s, no fewer than four totally new species of monkey have been found in South America, but according to Peter Hocking, a much more spectacular example may still await discovery here. If you can imagine a monkey with a face similar to this baboon [SLIDE], a very short tail, but a body as big as that of a chimpanzee, then you should have a fairly accurate picture of the isnachi - one of several different local names given to it by a number of separate tribes in Peru who know of it. According to these tribes, it is very rare but is in any case avoided by them whenever possible because they say that it is extremely fierce. One of its most characteristic activities is to rip apart the tops of chonta palm trees in order to obtain the tender vegetable matter inside - and its presence within a given locality can be swiftly confirmed simply by finding trees damaged in this way, because no other creature is strong enough to do this.

Alien Zoo

MAPINGUARY

This is one of the most famous of all cryptozoological photos [SLIDE]. Taken in 1920 by a team of geologists on the Venezuela-Colombia border, it shows what they claimed to be a 5-ft-tall, tailless, bipedal ape-like beast. Although traditional zoologists dismiss the possibility that an unknown form of ape lives in South America, there are native traditions throughout this continent concerning the existence of these creatures, and most cryptozoologists support such claims.

Recently, however, Goeldi Museum zoologist Dr David Oren has come up with a very different identity for Brazil's version, the mapinguary, because he believes it not to be an ape but a living ground sloth, as shown here [SLIDE]. The last ground sloths officially died out several thousand years ago, with only their much smaller, tree-dwelling relatives surviving into the present. But when Oren studied mapinguary descriptions given by the Mato Grosso Indians, he found that they closely recalled one particular type of ground sloth, the mylodontid.

The mapinguary's footprints, body size, red hair colouration, vocalisations, and even its faecal droppings all correspond with those known from fossilised or mummified mylodontids. Also, it is said to be invulnerable to bullets, and we know that the mylodontids had bony nodules in their skin that would have acted as an effective body armour. Interestingly, according to the Indians the mapinguary has an extra mouth, in the centre of its belly, and when threatened, it releases a hideous stench that suffocates its attackers. This may sound quite bizarre, but Oren suggests that these descriptions might simply refer to some form of gas-secreting gland, used for defence.

Ground sloth, life-sized model at Iowa Museum of Natural History (Bill Whittaker/Wikipedia)

At the moment, Oren is seeking finance for a new expedition in search of a mapinguary. If he finds one, it could well prove to be the largest living mammal native to South America, and, if it really is a ground sloth, it will resurrect from extinction one of the most peculiar groups of mammals ever known.

Alien Zoo

THYLACINES IN NEW GUINEA

This is another animal that may be resurrected one day [SLIDE] - the Tasmanian wolf or thylacine *Thylacinus cynocephalus*. The last confirmed thylacine on Tasmania was a zoo specimen that died in 1936, and it officially died out several thousand years ago on mainland Australia and New Guinea. Even so, reports of living thylacine-like beasts have been regularly emerging from Tasmania and the mainland for several decades - and are now emerging from New Guinea too. During the early 1990s, while in Irian Jaya - New Guinea's Indonesian, western half - grazier Ned Terry was amazed to receive accurate descriptions of thylacines from the local natives, who claimed that such creatures existed in the area's remote highlands, where they were called dobsegna. As you can imagine, it would certainly be very ironic if the thylacine were rediscovered not in Tasmania or even in mainland Australia, but in New Guinea. Yet as this is far less well-explored than those other regions, a population could survive undetected here much more easily than elsewhere.

MIGO

Another new mystery beast from New Guinea is the migo. In January of this year, a Japanese TV team journeyed to Papua, with the renowned American cryptozoologist Prof. Roy Mackal as its scientific advisor, in search of a mysterious lake monster known as the migo. The lake in question, Niugini, was around 1200 ft in diameter, about 30 ft deep, and, surprisingly, did not contain fishes. But this worked in the favour of the investigators, because it meant that the animals had to spend more time at the surface than other lake monsters, in order to feed upon the abundant waterfowl there. This enabled the team to film a migo, estimated to be over 33 ft long, moving at a speed of 4 knots, and extremely serpentine in shape, but producing vertical undulations.

CADBOROSAURUS

This [SLIDE] is an eyewitness sketch of Caddy, also nicknamed Cadborosaurus, the sea serpent of

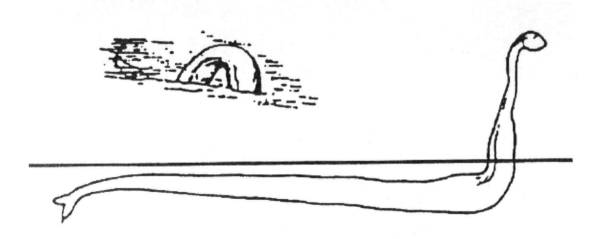

Osmond Fergusson's sketch of Caddy

Alien Zoo

Cadboro Bay in Canada's British Columbia. This particular sketch was made in 1897 by a prospector called Osmond Fergusson, who, with his partner, observed it from just a few feet away, off the Queen Charlotte Islands. They estimated its total length to be about 25 ft.

In July of last year, two pilots, Don Berends and James Wells, actually encountered two 'Cadborosauruses' together in the vicinity of Saanich Inlet, apparently about to begin mating with one another. Further enquiries revealed that locals claim the breeding season for Caddys takes place every year here at this time, with the Caddys giving birth to live young in shallow water or on certain remote beaches. Consequently, if a scientific team could spend some time here this year, we might at last find out just what 'Cadborosaurus' is.

LAKE TIANCHI MONSTER

The most popular water monster identity, especially for long-necked water monsters, is this animal - the plesiosaur [SLIDE], a marine reptile believed to have died out alongside the dinosaurs about 64 million years ago. One Chinese lake currently experiencing monster activity of the plesiosaurian persuasion is Lake Tianchi, in the northeastern province of Jilin, which was once known as Dragon Lake. Back in 1980, one member of a research team investigating the monster unsuccessfully attempted to shoot it, and it has quite wisely kept a low profile since then - until now. Several recent reports have been filed of a golden-coloured water beast with a square head, horns, a long neck, and humps. Consequently, a number of Japanese and Chinese film crews and scientific teams, and even a team from North Korea, are currently poised around the shore, in the hope of bringing to a close this longstanding mystery. No decent monster photos have yet been obtained - certainly nothing as conclusive as this picture would be [SLIDE] - but if you'd like a monster key-ring or teeshirt, the nearby tourist kiosks will be more than happy to sell you one.

INKBERROW PANTHER

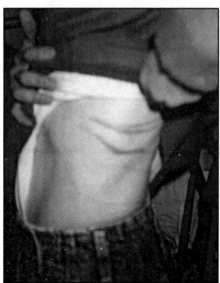

Aside from the Loch Ness monster, Great Britain is best-known, cryptozoologically, for its ABCs or alien big cats. As you can see, I'm holding here a model of the most famous type, the black panther version [SLIDE], which is now also the most infamous - thanks to the dramatic case of the Inkberrow panther. Towards the end of last year, during their investigations of sightings of a panther-like animal near the churchyard of Inkberrow in Worcestershire, former soldier Nick Dyke and his wife Sally went to the churchyard one night after having earlier left some offal there as bait - and experienced a *very* close encounter of the feline kind.

According to Nick Dyke, he suddenly put his foot on something that moved - an enormous black cat that sprang up at him, and then ran towards Sally, leaping through the air and

Sally Dyke exhibiting her injuries (CFZ Collection)

Alien Zoo

swatting her left side with its paw before fleeing. Its claws tore through her clothes, and left behind some deep, 5-in-long scratches in her flesh. Fortunately, there only seem to be a handful of incidents in which a mystery cat has allegedly attacked a human in Britain, but even so these are enough to show that the subject should be treated much more seriously by the media than is normally true.

ISLE OF WIGHT LEOPARD CAT

ABC cases supported by an actual body are far from common, but in January the national newspapers published photos of a large spotted cat identified as either a serval or an ocelot and which seemed to have been shot on the Isle of Wight very recently. When I saw the photos, however, I realised that the cat was actually an Asian leopard cat *Prionailurus bengalensis*, like this animal [SLIDE], and further investigations revealed that it had been shot seven years ago - but it had only now come to light, because the people who had shot it had only been teenagers at the time, and had been afraid that by killing it they had committed a crime. Interestingly, this is the fifth leopard cat corpse discovered in Britain since 1980.

LUDLOW JUNGLE CAT AND JASPER

This specimen [SLIDE], an Asian jungle cat *Felis chaus*, was also found dead in Britain - near to Ludlow, Shropshire, in February 1989, killed by a car. As this species is not native to Europe, it is, once again, clearly an escapee from captivity. What makes this cat uniquely important is that when I investigated its history last year, I discovered that several cats remarkably similar in appearance to crossbreeds between jungle cats and domestic cats have been reported from this same area *since* the discovery of its body. As jungle cats and domestic cats *can* mate and produce fertile offspring, it is possible, therefore, that the Ludlow jungle cat, an adult male, had indeed been interbreeding with local feral domestics.

This cat, Jasper [SLIDE], is an exceptionally large domestic cat that was actually born to one of these mysterious crossbreed lookalikes, and as you can see, except for his shorter legs he does look very like a jungle cat, like these [SLIDE], and these [SLIDE], for example. And I'm not the only person who thinks so. This similarity was commented upon in particular by Gareth Thomas - the Ludlow vet who examined him only a few days after examining the dead body of the Ludlow jungle cat. At present, DNA tests cannot distinguish between jungle cats, domestic cats, and hybrids, but when the technology is sufficiently advanced, these tests may finally be able to discover Jasper's true identity.

WOLVERINES AND BEARS

Since 1992, a number of reports have emerged from Wales, Cornwall, and Devon describing creatures remarkably similar to this animal [SLIDE]. This is a wolverine *Gulo gulo*, a giant weasel as large as a small bear, native to northern North America and Europe, but not to Britain. And there are no wolverines in captivity here either at present, making such reports even more mysterious. Last summer, Joann Crowther, a professional photographer from London, was travelling through Wembworthy in Devon when she saw a dead animal by the roadside. She stopped to look at it, but unfortunately, not being a wildlife photographer, she didn't take a photo. When she later reported her sighting to naturalists, however, the creature was conclusively identified as a wolverine - but by then, inevitably, the body had gone. Wolverines can be extremely savage, so if you do encounter anything that looks like one, leave it severely alone!

Alien Zoo

If bear-sized weasels are not spectacular enough for you, then how about this, the real thing? According to the children's song, if you go down to the woods today, you're sure of a big surprise - which, if those woods happens to be just northwest of Oxford, may well take the form of a large chocolate-coloured bear. Since January of this year, such a creature has allegedly been sighted by several people in the vicinity of Wychwood Forest - including Greg Gilbert, the driver of a bus out in front of which the creature walked, as well as his only passenger, a secretary called Sarah Cooper. One farmer even claims that it has been catching fish in his fish ponds, but all attempts to catch the bear have met with failure.

MAKALALA

I'd now like to mention two very distinctly fortean mystery beasts. Some of the most interesting cryptozoological discoveries initially take place not in the field but in the library, by unearthing some obscure report or article that had been published many years ago yet had never before received any cryptozoological attention. Not very long ago, I came upon one such report - published in the 1878-79 volume of the *Bulletin de la Société Philomatique* by a Count Marschall, who claimed that in a region of Central Africa 8-9 days' journey from the coast of Zanzibar the Wasequa tribe knew of a monstrous bird as tall as an ostrich, with very long legs, the head and beak of a bird of prey, and a favoured diet of carrion torn from animal carcases. They called it the makalala - which means 'noisy' - because its wingtips bore hard horn-like plates, which it would beat together to make a very loud sound.

Life-size model of phorusrhacid at Drayton Manor Park and Zoo (Dr Karl Shuker/reproduction courtesy of Drayton Manor Park and Zoo)

Such a bird would seem totally unbelievable - were it not for this creature [SLIDE]. This is a life-sized model of a terror bird, or, to give it its formal scientific name, a phorusrhacid. These gigantic, flightless flesh-eating birds lived in South America until around 2 million years ago, and one, suitably called *Titanis*, lingered on in North America until about 10,000 years ago and was about 10 ft tall. Although terror bird fossils have never been found in Africa, some scientists are now beginning to believe that they may have a small, *living* relative here - the odd-looking secretary bird *Sagittarius serpentarius*, a stork-like bird of prey named after its crest of feathers that resemble the quill pens a secretary or clerk would push behind his ear in Victorian times.

Needless to say, the makalala may well

be a complete hoax. But if real, could it be a giant, undiscovered form of secretary bird? One interesting point is that very few known birds have the horny wingtips mentioned for the makalala - but the secretary bird is *one* of those few birds, and is the *only one* that is carnivorous. No makalala reports seem to have emerged since Marschall's, more than a century ago - but we may yet find proof of its former existence. Marschall states that the tribal chiefs wore helmets made from the skulls of this giant bird - should one of these helmets pass into the hands of a visiting scientist, he would instantly confirm the onetime reality of the monstrous makalala.

WINGED CATS

Finally: one of the most famous of all fortean animals must surely be the winged cat. This [SLIDE] is one of several perfectly genuine examples of cats with large wing-like outgrowths of fur that have been reported over the years, but never explained. As someone with a longstanding interest in mystery cats, I decided recently to see if I could uncover the answer, and after browsing through some rather obscure sections of the veterinary literature, I finally found what I was looking for. There is, it turns out, a rare and little-known skin disorder of cats, inherited genetically, known as feline cutaneous asthenia, or FCA for short. Cats suffering from FCA have abnormally delicate, flexible skin that will stretch very readily, especially upon their shoulders or along their back, which soon results in the formation of furry wing-like extensions, which can even be moved if they contain muscle fibres. Also, they often drop off, but without causing any bleeding, which thus explains reports of winged cats that have suddenly moulted their wings. If you'd like more details concerning winged cats and FCA, there will be an article of mine dealing with this subject in the next *Fortean Times* issue.

And that, I'm afraid, is all we have time for in this whistle-stop tour of current affairs in cryptozoology. So, thank you for your interest, and I think there's just enough time left for some questions if there are any that you'd like to ask. Thank you very much.

[Due to time limitations, I omitted the following three subjects when I gave my lecture, but I am adding them here in order to present the lecture in its original planned form.]

CLEAR LAKE MYSTERY FISH

This animal [SLIDE] is one of the most famous of all cryptozoological success stories. It was captured alive in the sea off southern South Africa in December 1938, and to the amazement of zoologists worldwide its unique leg-like lobed fins, tripartite tail, armour-plating scales, and other external features conclusively identified it as a coelacanth - a member of an archaic group of fishes called crossopterygians that had supposedly died out alongside the dinosaurs, over 60 million years ago! Other specimens were later captured, mostly around the Comoro Islands near Madagascar, but so far all known coelacanths belong to the same species.

In October 1993, however, fisherman Lyle Dyslin caught a quite bizarre fish in northern California's Clear Lake that in some ways resembled a small coelacanth - for just like this species it had leg-like lobed fins, and large distinctive scales. But *unlike* the coelacanth, its head was strangely dog-like in shape with whisker-like projections around the mouth. The feature that attracted everyone's attention, however, was its tail - for whereas the tail fin of every single fish so far known to science is vertical, the Clear Lake mystery fish's tail was horizontal - like that of aquatic mammals, such as whales, dolphins, and sea cows.

Alien Zoo

Greatly intrigued by this, I've been in contact with a number of ichthyologists regarding its possible identity, and the general consensus is that it is most probably an unusual form of catfish - as indicated by its 'whiskers', which are certainly the barbels that characterise catfishes. As for its unique tail, however, if it really was horizontal, then the fish was either a malformed or mutant specimen [possibly of the channel catfish *Ictalurus punctatus*] - or it represented a dramatic new species never before recorded by science. Tragically, we may never know the answer, because, although Dyslin took photos of the fish, he didn't keep it, but released it back into the lake. His reason for doing so? Because it reminded him so much of his pet dachshund!

PAINTING OF A SEA SERPENT?

Sketch of 'sea serpent' cave painting.

Another aquatic anomaly is this beast [SLIDE]. In 1985, while diving in the Mediterranean Sea near Marseilles, Henri Cosquer discovered a grotto with an entrance 120 ft beneath sea level. Six years later, it was found to lead to a huge underground gallery above sea-level, whose walls were covered in extraordinary paintings dated at 26,000-16,000 BC, and which included this one. Some scientists have attempted to identify it as the now-extinct flightless seabird the great auk *Pinguinus impennis*. Others have likened it to a penguin, but as penguins are primarily Antarctic species that have never been native to this part of the world, this seems somewhat unlikely. French ichthyologist Dr François de Sarre, intriguingly, has recently proposed that it might be a Palaeolithic artist's rendition of a sea monster, of the long-necked sea serpent type, and suggested that it could be some form of long-necked seal. Whatever it is, it doesn't seem to correspond very closely to any animal currrently recognised by science, and is yet another iconographical oddity within the chronicles of cryptozoology.

ISLE OF WIGHT FOOTPRINT

Lastly: in April 1994, Isle of Wight (IOW) naturalist Martin Trippett informed me that a garden in Ride had received an unusual visitor. The garden had been freshly dug on the day in question by its owners, who then placed their garden rubbish in some bin-liners. The garden is completely enclosed by a 3-ft-high wall and its only entrance is via a gate, which they locked that night. The next morning, they found that some unidentified animal had been in their garden, ripping the bin liners to shreds and leaving huge footprints all over the freshly dug soil. This is a cast of one set of the prints [SLIDE], which were later described to me over the phone by some IOW reporters. Measuring 4.5 in long and 4 in across, they had no claw marks at all, which leaned towards a huge cat as an identity, but when the casts were sent to me I could see from the shape of the heel pad and the diverging placement of the toe pads that they were in fact dog prints, albeit from an extremely well-manicured dog - and this is what I told the papers.

Cast of two of the IOW footprints (Dr Karl Shuker)

Inevitably, they were rather disappointed, as this dashed any hopes for them of dramatic headlines concerning giant cats on the loose. Nevertheless, they then confessed that they had actually been informed by the police that a great dane had been loose in this area for the past week.

All of which proves that however tempted you may be to give the media the story that it wants, regardless of your own personal opinion, it is not a good idea to do so. Cryptozoology has a nasty knack of coming back to haunt those who flirt with its favours.

* * * * *

The shadows of evening are drawing near, the gates are closing, and it is time now to leave my Alien Zoo. I hope that you have enjoyed your visit, and you are very welcome to return whenever you wish. Just open this book if ever you need to escape for a time from the jaded world outside, and let its marvels reawaken your curiosity.

I would rather have a mind opened by wonder than one closed by belief.

Gerry Spence – *How To Argue And Win Every Time*

INDEX OF ANIMAL AND PLANT NAMES.

A

Aardwolf, 222
ABC (see Alien big cat),
Adder, 71, 303, 304, 307
Aegotheles savesi, 296
Aepyornis maximus, 46
Aeromys thomasi, 21, 237
Ahool, 128, 331
Ajolote, 280, **280**
Albatwitcher, 13, 174
Alces alces, 19, 95, 153, 221
Alien big cat (ABC),
        Australian, 65, 73, 214
        British, 11, **50**, 51, 52, 53, 54, **54**, 55, 56,
        57, **57**, 58, 59, **59**, 60, **60**, 61, **61,** 62, 63,
        64, 73, 358, **358**, 359, 362, **362**
        French, 73
        Hawaiian, 178, 179
        New Zealand, 73, 96
Allghoi khorkhoi, 141, **141**, **192**, 193, 219, 296
Almas (=almasty), 140, 275, 276, **276**, 296
Alosa alosa, 73
Altamaha-ha, 42
Alux, 293
Amblyornis flavifrons, 282
Ameranthropoides loysi, **108**, 109,135, 263, 299
Ammonite, **302**, 303, **304**, 304, 305, **305**, 306,
306
Amphipod, 120
Amphisbaenian, 218, 280, **280**
Anaconda, giant, 16, 48, 96, **97**, 103
Anodorhynchus glaucus, 296
Anser anser, 88
Ape, Bili, 134, 231
        Loys's, 16, 135, 261, 262, 321, 356
        skunk, 178
Aquila chrysaetos, 122, 204
Ara ararauna, 211
      atwoodi, 211
      erythrocephala, 211

Arachnocampa luminosa, 123
Architeuthis kirkii 100
      longimanus, 35
      sp., **100**, 113, 152
Arctictis binturong, 157, **157**
Arctonyx collaris, 162
Arica Beast, 199
Arocatus roeselii, 276
Athene blewitti, 93
Atlantisia rogersi, 272
Axis porcinus, 73

B

Bacteria, cloud-inhabiting, 118
Badger, hog, 162
Balaenoptera acutorostrata, 319
Balaenoptera musculus, 277
Banakon, 14, 279, 280
Barbastella barbastellus, 141, **141**
Barbastelle, 141, **141**
Barracuda, 143
Bat, Antillean ghost-faced, 169, **169**,
      Australian ghost, 168
      fruit, 168, 169, 170, 171, 246, 331
      mega-, 168, 169, 171
      micro-, 168, 169, 171
      new Foja blossom, 285
Bat-woman, Vietnamese, 167
Beachwalker, Icelandic, 291
Bear, Carpathian brown, 347
      cave, 347
      great naked, 342
      Peruvian mystery pygmy, 71
      tailed, 158
      Wychwood Forest mystery, 360
Beast of Gévaudan, 44, **44**
Bee-eater, European, 160, **160**
Beetle, ant-mimicking tiger, 288
      Maid of Kent, 47
Big Bird, 43, 118, 122, 142, 150, 265, 344, **345**

Alien Zoo

Bigfoot, 16, 17, 18, 19, 43, 92, 99, 101, 137, 152, 157, **172**, 173, 176, 198, 201, 202, **202**, 214, **215**, 219, 221, 257, 258, **258**, 259, 291, 293, 342, 344, 348, 349, **349**, 350, 354, **354**
Binturong, 157, **157**
Bird of paradise, Barraband's mystery twelve-wired, 194, **194**
     Berlepsch's six-wired, 282
     Count Raggi's, 84, **84**
     Goodenough Island mystery, 125
Birdzilla, 14, 265
Blenny, 267, 269
Bobcat, long-tailed, 342
Bombus hypnorum, 141
Bonobo, 70, 157
Bornean mystery mammal, 20, 21, 219, 237
Brno Beast, 109
Brocket, mystery, 118
Bronze, geranium, 44
Bu-rin, 156
Bubo bubo, 19, 112, 118, 213
Bucardo, 121
Bug, millennium (new water strider), 110
     Natural History Museum's mystery, 276
Bullockornis, **186**
Bumblebee, tree, 141
Bunting, Townsend's, 204, **204**
Bunyip, 16, 71, 94, 142, 187
Burnet, New Forest, 122
Bushbaby, mystery giant, 36, 37, **37**
Bustard, great, 184, **184**

C

Cacyreus marshalli, 45
Caddy (=Cadborosaurus), 35, 291, 357, **357**, 358
Cailleach-bheur, 94, **94**, 314
Callaeas cinerea cinerea, 45, **45**, 120, 132, 259
Caluromysiops irrupta, 106
Camel's tail, 219
Campephilus imperialis, 18
     principalis, 18, 93, 296
Canis latrans, 263
Canzanella's mystery animals, 161, **161**, 162, **311**, 312
Capra pyrenaica pyrenaica, 121
Capybara, 48
Carcharhinus leucas, 269

Carcharocles megalodon, 113, **113**
Cat, Abyssinian, 60, 61
     African golden, 197
     African wild, 177
     alien big (see Alien big cat),
     American Kellas, 184, 185
     Asian (Temminck's) golden, 185, 197
     flying, 246, 247
     giant black Peruvian mystery, 355
     Icelandic ghoul, 14, **290**, 291
     jungle, 51, 52, 62, **63**
     Hayling Island, 52
     hybrid, 53, **53**, 55, 56, 58, 59, **59**, 60, **60**, 61, **61**, 62, 359
     Ludlow, 52, 53, **54**, 54, 55, 56, **57**, 58, 359
     Kellas, 184
     leopard, 11, 359
     mystery Madagascan, 177
     scimitar, 177, 197
     speckled Peruvian mystery, 98, 355
     striped Peruvian mystery, 355
     Tennessee multicoloured mystery, 69
     winged, **240**, 241, 242, **242**, 243, 244, 245, **245**, 246, 247, 248, 250, 251, 252, 254, **254**
       Manchester, 247, 248, **249**, 250, **250**, 251, 252, **253**
       Wiveliscombe, 246
Catfish, Asian stinging, 267, **268**
     channel, 362
     transparent Trinidad mystery, 101
     wels, 151, 233, **233**
Catopuma temminckii, 197
Cattle, Gaoligong Shan, 173
     Icelandic sea, **290**, 291
     Swona, 101
Cave fish, St Louis mystery, 120
Cavy, Patagonian, 48
Centipede, giant Sumatran shrieking, 295
     with even-numbered leg pairs, 101
Chameleon, Greek mystery, 71
Champ (see Lake Champlain monster),
Charmosyna diadema, 296
Charybdis japonica, 131
Chaunax suttkusi, 95
Chausie, 62, 63
Chimpanzee, pygmy, 70, 157
Chironectes minimus, 98, 100, **100**, 106, 107, **107**

366

Alien Zoo

Chitra indica, 67, **67**
Chrysochloridae, 326
Chrysopelea ornata, 222
Chupacabras, 66, 263, 309, 313
Cigau, 177, 185, 197
Circus eylesi, 315
Civet, giant palm, 34
        Hose's palm, 19
        Seram mystery, 34
        Vietnamese mystery, 34
Cobra, king, 279, 280
Cockatrice, 175
Cockerel, egg-laying, 175
Coelacanth,
        Indonesian, 15, 115, 116
        silver figurines, 138, 139
Colobus, Miss Waldron's red, 193, 194, **194**
Colugo, 168, 169, 170, **170**
Condor, California, 345
Corvus splendens, 91
Coturnix novaezelandiae, 259, 260, **260**, 311
        ypsilophora, 260, 311
Cougar, eastern, 76
Courser, Jerdon's, 93
Coyote, hairless, 263
Coyote x wolf hybrid, 263
Coypu, 48, 72, **72**
Crab, sponge, 143
Crocidula russula, 275
Crocodile, American, 258
        Cayman Islands, 258, 259
        Cuban, 259
        Johnston's, 190
        Mary River mystery, 189, **189**
        Murua, 190
        Nile, 14, 264, 265
        saltwater, 42, 190, 220, 222, 346, 347
        Siamese, 110
'Crocodile', mystery Welsh, 30
Crocodylus acutus, 258, 259
        johnstoni, 190
        niloticus, 264, 265
        porosus, 42, 190, 347
        rhombifer, 259
        siamensis, 110
Crow, house, 91, 92
        wattled, 45
Cryptoprocta ferox, 111, **111**, 117
        spelea, 111

Cubomedusan, undiscovered, 155, 163, 164
Cumom, 195, **195**
Curelom, 195, **195**
Curlew, eskimo, 296
        slender-billed, 150, **150**, 200
Curupira, 13, 39, 40, **40**

D

'Dark People', Siberian, 291
Dasycyon hagenbecki, 131, 132, **132**
Deer, donkey-eared mystery, 13, 38, 39
        giant, 114
        hog, 73
        Iranian mystery, 262
        musk, 133
Dendrolagus mbaiso, 37, 129, 295
        pulcherrimus, 283
Devil-bird, Sri Lankan, 261, 272
Devil-pig, Papuan, 125
Diatryma, 43, 44, **44**
Dicerorhinus sumatrensis, 93
Dickcissel, 204
Didi, 13, 164, 261, 262
Dingiso, 15, 37, 128, 294
Dinornis, 95, **95**, 271
Dinosaur, Colorado river, 159, 160
        New Mexico mystery, 294
'Dinosaur-bird', Aqeputa, 150
Diplocaulus, 200, 201, **201**
Diplodocus, 294
Diplogale hosei, 19, 237
Dissostichus eleginoides, 175, **175**
Dobhar-chú, 13, 275, 317
Dobsegna, 37, 125, 219, 344, 357
Dodo, 13, 17, 37, 209, 210, **210**, 211, **211**, 212, 273, **273**, **277**, 278, **278**, 348
Dodu, 9, 13, 173, 174
Dog,   Isle of Wight mystery, 363, **363**
        snake-headed, 13, 35, **35**
        Texan hairless, 263
Dolichotis patagonum, 48
Dorcopsulus sp., 284
Dragon, Ishtar Gate, 351
        Klagenfurt, 347
        Mekong River, 13, 121, **121**
        Nepal, 102, **102**, 103
Drepanovelia millennium, 110

Alien Zoo

Driloleirus americanus, 229
Dromaius novaehollandiae, 271
Dryocopus pileatus, 18
Duah, 126, 127, **127**, 128
Duck, pink-headed, **64**, 66, 93
Ducula sp., 284
Duende (=Dwendi), 14, 201, 292, **292**, 293
Dugong, 135, 165
Dugong dugon, 135, 165

E

Eagle, bald, 203, 204, 207
    golden, 122, 204
    Haast's, 315, **315**
    king's, 118, 122
    Malagasy crowned, 47
    Philippine (=monkey-eating), 272
    Washington's, 204, **207**, 207
Ebu gogo, 19
Eel, eunuch, 218
    giant mystery, 35, 66
    long-tailed (slender giant) moray, 268,
268
    Wasdale white mystery, 317
Egernia kintorei, 91
El Sharana mystery bird, 190, 191, **191**
Elephant, African, 140
    African pygmy, 16, 101, 140
    Asian, 123
    Indian pygmy, 19, 312
    Nepalese giant, 355
    pink-tusked black, 92
    water, 159, **159**
Elephant bird, great, 13, 46, **46**, 47, **47**
Elephas maximus, 140, 312, 355
    maximus rubridens, 93
Elk, Irish, 114, 115, **115**
Emela-ntouka, 19
Emu, Australian, 271
Emus hirtus, 47
Ensut-ensut, 135
Entzaeia-yawá, 98
Equus ferus przewalskii, 146
    hemionus, 351
    kiang, 351
    quagga (=*burchelli*) *quagga*, 201
Eremiornis carteri, 91

Esakar-paki, 98
Eskie, 70
Esox lucius, 151
Exmoor Beast, 17, 59
Eyra, 69
Ezitapile mystery beast, 113

F

Fairhaven Lake mystery beast, 310
Felis chaus, 51, 52, 62, 359
    lybica, 177
    lynx, 52
    silvestris, 51, 184
    yagouaroundi, 69
Fish, Clear Lake mystery, 361, 362
    Manaus eel-like mystery, 179
    Shatt al Arab River mystery, 267, 268,
269
Fjorulalli, **290**, 291
Flamingo, 84
Fly, Yates's scuttle, 155
Flycatcher, small-headed, 204
Fossa, 111, **111**, 117
    cave, 111
Fox,
    flying (see Bat, fruit), 167, 168, 169
    platinum, 309, 310, **310**
    red, 52, 55, 56, 109, 232, 310
    Sampson, 232
    silver, 310
Fregata magnificens, 345
Frigate bird, magnificent, 345
Frog, Pinocchio, 284, 285
    pool, **90**, 91

G

Gaper, orangeblotch, 95
Gardener, golden-fronted, 282, **282**
Gasparilla Lake monster, 182
Gavialis gangeticus, 190
    papuensis, 190
Gazeka, 125, 301
Gecko, Delcourt's giant, 312, 313, **313**
    Foja 'gargoyle', 284
Genyornis newtoni, 190, **191**

368

Alien Zoo

George (see Jasper and George),
Gévaudan Beast (see Beast of Gévaudan),
Gharial, 190
Giant, helmeted, 316, 317
       horned, 307, 308
Glider, sugar, 162, **162**, 263
Glironia venusta, 106
Globster, 14, 19, 318, 319
       Chilean, 182
       Florida, 179, **179**
Goat, holy, 103
       Pyrenean wild, 121
Goose, grey lag, 88
Gorilla, Ebo Forest, 178
Grendel, 347
Grey People, 163
Griffin, 347
Grosbeak, blue, 204
Ground sloth, living, 124, **124**, 144, **144**, 158, 262, 356, **356**,
Grouse, Caucasian black, 176
Gryttie, 42
Guan, Peruvian wattleless mystery, 71
Gulo gulo, 158, **158**, 359
Gulus, 272
Gunni, 183
Gymnogyps californianus, 345

H

Haetae (=Haechi), 276, 277
Haliaeetus leucocephalus, 203, 204, 207
Hantu jarang gigi, 135
Harpagornis moorei, 315, **315**
Harrier, Eyles's, 315
Hawk-eagle, Javan, 43
Hawkesbury River monster, 314, 315
Helinaia carbonata, 204
Hessie (see Sea serpent, Aalesund),
Heteralocha acutirostris, 45, 97, **97**
Heteropneustes fossilis, 267
Hierophis viridiflavus, 35
Hobbit, 19
Hokioi, 315
Homo floresiensis, 19
       habilis, 333, 335, **335**
Homotherium latidens, 177, **177**
Honeyeater, wattled smoky, 281, 282, **282**, 285

Honeyguide, Malaysian, 295
Hoplodactylus delcourti, 312, 313, **313**

Horse, blood-sweating, 13, 135
       hairless blue, 200
       Haiti wild, 147
       Nangchen, 147
       Przewalski's, 147, **147**
       Riwoche, 145, 146, **146**, 147, 148
Horse-eel, Irish, 14, 136, 137, 200, 275
Horse-whale, Icelandic, **290**, 291
Howler, Ozark, 68, **68**
Hrosshvalur, **290**, 291
Hucho taimen, 221, 260
Huia, 45, 96, **96**
Humanoid, elephant, 142, **142**
Hyaena, striped, 45, 222, 257
Hyaena hyaena, 45, 222, 257
Hydrochoerus hydrochaeris, 48

I

Ibis, Davison's, 71
Ichthyosaur, **340**, 341
Ictalurus punctatus, 362
Iguanodon, 237, **237**
Indicator archipelagicus, 295
Irtysh fleecy-skinned mystery beast, 324
Isnachi, 355

J

J'ba foti, 96
Jackal, horned, 261
Jackalope, 174, **174**, 183
Jaguar, 48, 69, 98, 99, 195, 267, 355
       white-collared black, 116, 117
Jaguarundi, 69
Jasper and George, 51, 58, 59, **59**, 60, **60**, 61, **61**, 62, 63, 359
Jellyfish, new Great Barrier Reef box, 222
       undiscovered box, 155, 162, 163

K

Kallaana, 19, 312
Kappa, mummified, 199, **199**

Alien Zoo

Kapre, 92
Karkadann, 351
Kawekaweau, 312, 313, **313**
Kayadi, 219, 220
Kha-nyou, 17, 18
Kiang, 351
Kipunji, 17, 223
Koao, 208, **208**
Kokako, South Island, 45, **45**, 46, 120, 132, 259
Kondakovia longimana, 115
Koreke, 259, 260, **260**, 311
Kuil kaax, 14, 293, **293**
Kungstorn, 118, 122

L

Lacerta bilineata, 182, **182**,
Lake Champlain monster, 180, **180**, 341
Lake Gryttjen monster, 42
Lake Kanasi (=Hanas) monster, 220, 260
Lake Khaiyr monster, 321, 322, **322**, 323, 324
Lake Labynkyr monster, 324
Lake Maggiore monster, 20
Lake Manitoba monster, 39, **39**
Lake Mjoesa monster, 143, 155
Lake Murray monster, 42
Lake Okanagan monster, 131, 133, 156, **156**
Lake Pepin monster, 275
Lake Peruca monster, 275
Lake Seljord monster, 17, 66, 138
Lake Somin monster, 323, 324
Lake Tianchi monster, 20, 261, 358
Lake Van monster, 41
Lake Vedlozero monster, 324
Lake Windermere monster, 233, **233**
Lampona cylindrata, 226, 227, 228
Lamprey, giant, 138
Laonastes aenigmamus, 18
Latimeria chalumnae, 139
      menadoensis, 15, 115, 116
Lavellan, 229, 230, **230**
Lemur, flying (see Colugo),
     mainland African, 38, **38**
Leopard, 51, 73, 194, 214, 238, 347
Leptailurus serval, 238, **239**
Leptidea reali, 141
     sinapis, 141
Letayuschiy chelovek, 167

Leviathan, 195, 346
Liasis amethystinus, 188
Li'kela-bembe, 133
Linh duong, 103
Lion, African, 71, 96, 238, 268
     Chilean, 14, 96
     Gir Forest, 346, 347
     marsupial, 33, 65, **65**, 114, 163
     Peruvian jungle, 71
Lionfish, 151, **151**
Litoria sp. 284
Littlefoot, 292
Lizard, hammerhead, 200, 201, **201**,
     monitor, 114, 159, 163, 219, 294
     Trinidad glowing, 13, 101
     western green, 182, **182**
Loch Morar monster, 27, 217, 218
Loch Ness monster, 17, 19, 25, 26, 27, 29, 70, 94, 110, **110**, 111, 138, **138**, 155, 180, 181, **181**, 218, **218**, 236, 291, 297, 321, **336**, 337, 338, **338**, 339, **339**, 340, **340**, 341, 342, 350, 358
Loch Watten monster, **22**, 23, 24, **24**, 25, 26, 27, 28, 29, 30, 31, 32
Longneck, 291
Loof lirpa, 273
Loriciferan, Shuker's, 223, **223**
Lough Eske monster, 70
Lough Ree monster, 136, 137, 138
Loxodonta africana, 140
Loxosceles rufescens, 227
Lutra sumatrana, 114
Lutreolina crassicaudata, 106
Lycosa spp., 227
Lynx, European, 52

M

Macaca fascicularis, 272
Macaque, crab-eating, 272
Macas mystery beast, 98, 99, 100
Macaw, blue-and-yellow, 211
     Dominican green-and-yellow, 211
     glaucous, 296
     Guadeloupe red, 210, **210**
     Hispaniolan red, 210, **210**
     Jamaican green-and-yellow, 211
     St Croix, 212
Machairodontid (see Sabre-tooth),

Alien Zoo

Macroderma gigas, 168
Macrogalidia musschenbroekii, 34
Macropus rufogriseus, 38, 69
Madtsoiid, 188
Maipolina, 266, **266**
Makalala, 360, 361
Mallomys sp., 283
Mami water, 135
'Mammoth', Thailand, 123
Man, Flores, 19
      Piltdown, 185, **185**
Man-beast, Arroyo Salado, 176
      Kyrgyzstan, 140, 141
Man-monkey, Shropshire Union Canal, 264, **264**
Manatee, West African, 122, **122**, 135
Mande burung, 152
Manipogo, 39, **39**
Mapinguary, 13, 124, **124**, 144, **144**, 158, 262, 356, **356**
Mara, 48, **49**
Martin Mere monster, 150
Mau, Egyptian, 60, 61
Megachiroptera, 168
Megalania prisca, 114
Megaloceros giganteus, 114, 115, **115**
Megaselia yatesi, 155
Meiacanthus nigrolineatus, 269
Melipotes carolae, 282, **282**, 285
Merfolk, Japanese, 199
Merhorse, 291
Mermaid, Batroun, 297, 298
      Feejee, 152, 197, 199, 298, **298**, 299, **299**
      Karoo cave painting, 73
Merman, 199, **216**, 217
Merops apiaster, 160, **160**
Messie, 42
Microchiroptera, 168
Midge larvae, Thai mystery glowing, 123
      Waitomo glowing, 123
Migo, 42, 357
Mihirung, 190, 191, **191**
Mini-man, Maine, 231
Mink, American, 243, 313
Mirounga leonina, 165
Mjoesorm (see Lake Mjoesa monster), 135, 136, 143, **154**, 155
Mngwa, 197
Moa, bush, 94, 271

giant, 236, **236**
      living, 16, 94, **94**, 236, **236**, 297
Mokele-mbembe, 14, 16, 17, 18, 37, 69, **69**, 102, 122, 133, 139, 173, 195, 262, **262**, 294, 351
Mole, common, 326, **326**
      golden, 326
      marsupial, 183, 326
      Mesty Croft giant mystery, 325, 327
      Townsend's, 326
Mongolian mystery ungulate, 133
Monitor, giant, 114
Monkey, giant spider, 262
      Mindanao red-faced mystery, 272
      Ozark mountain, 152
Mono rey, 135
Monster, scratch, 14, 160
Moolyewonk, 314, **314**, 315
Moose, New Zealand, 19, 95, 221, **221**, 222
Morag (see Loch Morar monster),
Morelia oenpelliensis, 188
      spilota, 189
Mormoops blainvillii, **169**
Mosasaur, 341
Moschus sp., 133
Mother (see Jasper and George),
Mouse-whale, Icelandic, 15, **290**, 291
Muhnochwa, 14, 160
Muntjac, giant, 15, 354, 355
Müshmurgh, 13, 176
Mushveli, **290**, 291
Musk-ox, 351
'Mutant', Maine, 234
Mycobacterium ulcerans, 226
Myocastor coypus, 48, 72, **72**
Myxas glutinosa, 72

N

Naga, 121, **121**, 123, **123**, 188, 314
Neovison vison, 313
Nesoclopeus woodfordi, 161
Nessie (see Loch Ness monster),
Nessiteras rhombopteryx, 17
Newt, great-crested, **320**, 322, 323
Ngoubou, 195, 197, **197**
Nightgrowler, 13, 33
Ninki-nanka, 17, 18, 20
Nisaetus bartelsi, 43

Alien Zoo

Nittaewo, 14, 260, 261
Notiomystis cincta, 259
Notoryctes spp., 326
Numenius borealis, 297
       tenuirostris, 150, **150**, 200
Nunda, 197

O

Oarfish, 40, **41**, 291
Octopus, new pygmy, 158
Ofuguggi, **290**, 291
Ogopogo, 17, 42, **130**, 131, 132, 156, **156**, 291
Ogre, mummified Japanese, 199
Ogston Reservoir monster, 155
Okapi, 11, 351
Okapia johnstoni, 11, 351
Oliver, 70, **70**, 71
Omajinaakoos, 313
Onager, 351
Onça-canguçú, 118, **119**
Ophiophagus hannah, 279, 280
Ophrysia superciliosa, 93
Opossum, black-shouldered, 106
       bushy-tailed, 106
       thick-tailed, 106, 107
       water, 98, 100, **100**, 106, 107, **107**
Orang bati, 102, 128, 167, 329, 330, 331, 332, 333, 334, 335
Orang pendek, 13, 16, 17, 19, 20, 139, 142, 143, **143**, 177, 180, 182, 185, 261
Orang utan, giant, 220, **220**, 342
Ostrich, 46, 85, 190, 271, 360
Otis tarda, 184, **184**
Otter, Egede's super-, 14, 217
       hairy-nosed, 114
       master, 275
       Omey Island giant, 266
Ovibos moschatus, 351
Owl, European eagle, 19, 112, **112**, 113, 118, 213, **213**
       great grey, 118
       strike owl, 118, 122
       ural, 122
'Owl', Mindanao hairy, 272
Owlet, forest spotted, 92
Owlet-nightjar, New Caledonian, 296
Ox, Vu Quang, 354, **354**

P

Pach-an-a-ho', 43
Pamá-yawá, 98
Pan paniscus, 70, 157
       troglodytes schweinfurthii, 70, **70**, 231
Panther, Angolan mystery giant black, 194
       black, 51, 53, 65, 68, 96, 187, 194, 214, 238, 353, 355, 358, **358**,
       Chinese mystery black, 73
       Inkberrow, 358
Panthera leo, 96, 238
       leo persica, 346
       leo spelaea, 348
       onca, 69
       pardus, 53, 214, 238, 347
       tigris, 238
       tigris altaica, 347
       tigris sondaica, 43, 164, **164**,
Paradigalla, undiscovered, 141
Paradisea raggiana, 84, **84**
Parotia, Berlepsch's, 282
       Queen Carola's, 282
Parotia berlepschi, 282
       carolae, 282
Parrot, Amazon, 209
       Jamaican red mystery, 209, **209**
       night, 296
       paradise, 13, 70, 120, **120**
Passerina caerulea, 204
Pauca billee, 246
Peccary, Ecuadorian mystery, 98
Pelophylax (*Rana*) *lessonae*, 91
Penguin, flying, 274
Pepie (see Lake Pepin monster),
Petaurus breviceps, 162, 263
Pezoporus (=*Geopsittacus*) *occidentalis*, 296
Pharomachrus mocinno, 79
Philesturnus carunculatus, 45, 259
Phorusrhacid, 360, **360**
Physeter macrocephalus, 182
Pigeon, new Foja imperial, 284
Pike, northern, 151
Pili mara, 263
Piliocolobus badius waldronae, 193, 194, **194**
Pithecophaga jefferyi, 272
Plant people, mummified, 197
Platalea leucorodia, 118

Alien Zoo

Plesiosaur, **21**, 181, **181**, 279, 315, 339, **339**, 341, 352, 358
Pliciloricus shukeri, 23, **23**
Pony, Exmoor, 147, 148, **148**
    striped New Guinea mystery, 74, 75
Porphyrio mantelli, 73, 207, 208, **208**, 259
    paepae, 208, **208**
Potto, false, 38
Pouakai, 315
Prionailurus bengalensis, 11, 359
Proctoporus (=Oreosaurus) shrevei, 101
Procyon lotor, 309
Profelis aurata, 197
Propotto leakeyi, 168, 171
Proteles cristatus, 222
Protoceratops, 346
Psephotus pulcherrimus, 70, 120, **120**
Pseudibis davisoni, 71
Pseudonovibos spiralis, 103, **103**
Pseudopotto martini, 38
Pseudoryx nghetinhensis, 354, **354**
Pteranodon, 126
Pterodactyl, preserved Australian, 151
Pterois volitans, 151, **151**
Pteronura brasiliensis, 266
Pteropus, 168, 169, **169**
Pterosaur, 102, 125, 126, **126**, 128, 129, 344, 345, **345**
Puma, Cannich, 75, **75**, 77
    grey (not black), 192
Puma concolor couguar, 76
Python, amesthystine, 188, 189
    carpet (rainbow), **188**, 189
    Oenpelli, 188

Q

Quagga, 200
Quail, Australian brown, 312
    Himalayan mountain, 93, **93**, 297
    New Zealand, 15, 260, **260**, 311
    Tiritiri Matangi mystery, 13, 259, 260, 311
Quetzal, 79, 80, **80**, 83, 84

R

Rabbit, winged, 246, 247
Raccoon, hairless, 309
Raiju, 199
Rail, Inaccessible Island, 272
    Malaita mystery, 161
    Woodford's, 161
Raphia regalis, 86
Raphus cucullatus, 13, 17, 37, 209, 210, **210**, 211, **211**, 212, 273, **273**, 277, 278, **278**, 348
Rat, Beijing giant, 13, 33
    black (ship), 274
    Foja giant woolly, 283, 285
    Laotian rock, 17
Rat king, Otago Museum, 274, **274**
Rattus rattus, 274
Raudkembingur, **290**, 291
Red-crest, Icelandic, **290**, 291
Regalecus glesne, 40, **41**, 291
Rhamphorhynchus, 126, **126**, 128
Rhinoceros, Sumatran, 93
Rhinoptilus bitorquatus, 93
Rhodonessa caryophyllacea, **64**, 66, 93, 202, 296
Ropen, 102, 125, 126, 127, 128, 129, 237, 344
Runan-shah, **216**, 217

S

Sabre-tooth, 13, 197, 347, 355
    aquatic, 266, **266**
    Mexican living, 13, 72, **72**
Sachamama, 13, 46, 48, **48**
Saeneyti, **290**, 291
Sagittarius serpentarius, 360
Saiga, 133
Saiga tatarica, 133
Samotherium, 112, **112**
Sarcophilus harrisii, 66, 139, **139**
Saro, 266
Scapanus townsendii, 326
Schizodactylus monstrosus, 160, **160**
Scratch monster, 14, 160, 161
Sea monster, Masbate, 36
    Yangshashan, 220
Sea serpent, Aalesund, 94, 95, 135
    Bonavista, 112

Alien Zoo

Gambian, 17
Hook Island, 178
Japanese, 232
longneck, 291, 362
Shackleton, 278
Sea serpent cave painting, 362
Sea spider, giant, 178
Seal, dwarf, 342
    monk, 164
    southern elephant, 164, 220
Seal mother, Icelandic, 15, **290**, 291
Secretary bird, 360
Sel'awaa, 257
Selamodir, **290**, 291
Serpent, Garden of Eden, 183
    giant rainbow, 13, 110, 188
Serval, 238, 359
Shad, allis, 73
Shark, bull, 269
    Colombian mystery, 140
    Lake Sentani freshwater, 125
    megalodon, 113, **113**, 114
    prickly, 179
Shelduck, crested, 296
Shell monster, Icelandic, 14, **290**, 291
Shiashia-yawá, 98
Shrew, greater white-toothed, 275
    pygmy, 275
Shrimp, armoured, 179
    pistol, 143
Shunka warak'in, 265
Silurus glanis, 151, 233, **233**
Sinohydrosaurus lingyuanensis, freak two-headed, 237
Siren, two-tailed, 297
Sirrush, 351
Skeljaskaimsli, **290**, 291
Skink, 280
    tjakura, 91
Skoffin, **290**, 291
Skunk, Barbara Woodhouse's pouched, 13, 105, 106, 107
    hog-nosed, **104**, 106, 107, **107**
Slaguggla, 118, 122
Smethwick mystery beast, 264
Snail, glutinous, 73
Snake, Blashers's Dorset mystery, 71
    Everglades giant, 342
    flying, 13

Namibian flying, 128
Navajo flying, 196, **196**
Northern Territory giant mystery, 110
Russian flying, 222
Sarajevo jumping, 35
western whip, 35
Snowman, abominable, 288
Sorex minutus, 275
South Carolina mystery beast, 232, **232**
Spider, fiddleback, 227
    white-tailed, 226, **226**, 227
    wolf, 227
Spinifex bird, 91
Spiza americana, 204
    townsendi, 204, **204**,
Sponge, giant freshwater, 162
Sponge reefs, 133
Spoonbill, European, 118
Sqrat, 95, 113
Squid, flap-armed, 158, 159
    giant, 17, 19, 113, 152, 198, 279
    southern giant, 35, 36, **36**, 100, **100**
'Squid', Carska bara swamp, 137
Squirrel, Thomas's flying, 21, 237
    Wimbledon mystery flying, 262, 263
Stegosaur petroglyph, Cambodian, 230, **230**, 231
Stephanoaetus mahery, 47
Stitchbird, 259
Strix nebulosa, 118
    uralensis, 122
Strophidon sathete, 268, **268**
Struthio camelus, 46, 85, 190, 271, 360
Super-otter, Egede's, 217, **217**
Swift, mystery whiskered, 111
Syconycteris sp., 285
Sylvania microcephala, 204, **204**
Sylvicola montana, 204, **205**

T

Tadorna cristata, 296
Tahina spectabilis, 271
Taimen, 221, 260
Takahe, 73, 74, **74**, 207, 208, **208**, 259
Talpa europaea, 326, **326**
Taniwha, 163, **163**
Tasek Bera monster, 149
Tasmanian devil, **139**

Alien Zoo

mainland Australian, 66, 139
Tasmanian wolf (=Tasmanian tiger), 19, 37, **37**, 41, 44, 116, **117**, 136, **136**, 261, 287, 297, 343, 356
Temeenii suul, 19
Tengu, 199, **199**
Tern, royal, 122
Terror bird, 360, **360**
Tetrao mlokosiewiczi, 176
Thalasseus maximus, 122
Thalattosuchian, 190
Thoracosaur, 190
Thunderbird, 71, 314, 344
Thunnus sp., 118
Thylacine, 13, 19, 20, **37**, 66, 94, 136, **136**, 141, 183, 221, 261, **286**, 287, 291, 297, 343, 344, 356
   mainland Australian, 41,116, **117**, 221, 344
   New Guinea, 37, 44, 45, 125, 219, 344, 356
Thylacinus cynocephalus (see above),
Thylacoleo carnifex, 65, **65**
Thyrsoidea macrura, 268, **268**
Tieke, 45, 259
Tiger, 33, 164, 177, 238
   Bornean, 94
   Cambodian, 114
   Ecuadorian water, 98
   Guyanan water, 266
   Javan, 43, 164, **164**,
   Korean, 164
   Peruvian speckled, 48
   Peruvian striped, 48, 355
   Queensland, 16, 65, 114, 219
   rainbow, 99
   Rongwe, 222
   Siberian, 346, 347, 348
   tapir, 98
   Tasmanian (see Thylacine),
Tiggy1 and Tiggy2, 59
Tikis River monster, 149
Titanis, 360
Tizhurek, 291
Toothfish, Patagonian, 175, **175**
Tortoise, giant Vietnamese freshwater, 67, 68
Tragopan, mystery lyre-tailed, 176
   western, 176, **176**
Tragopan melanocephalus, 176
Tree, Indian cow-clutching, 263
   Madagascan man-eating, **269**, 270, 272
   raffia palm, 86

Tahina palm, 271
tiger, 14, 263
Tree kangaroo, golden-mantled, 283
   New Britain mystery, 294
Trichechus senegalensis, 135
Trilobite, living, 197, 198, **198**
Triturus cristatus, **320**, 323
Troll, Cannock Chase, 259, **259**
Trout, reverse-fin, **290**, 291
Trunko, 518, **518**, 519, **519**
Trunko, Glacier Island, 519
Tsere-yawá, 98
Tshenkutshen, 99
Tuna, 118
Turtle, Mongolian mystery, 49
   narrow-headed soft-shelled, 66, 67, **67**
Tygomelia, 153
Tzuchinoko, 199

U

Ukang, 272, **272**
Ulama, 272
Ular tedong, 149
Uma-ay, 272
'Uni-bull', 351
Unicorn, 13, 71, 163, 165, 276, 350, 351, 352
Upah, 295, **295**
Urdarkottur, **290**, 291
Ursus arctos arctos, 347
   spelaeus, 347

V

Veado branco, 118
Vipera berus, 71, 303, 304, 307
Vorompatra, 46, **46**
Vulpes vulpes, 52, 55, 56, 109, 232, 310

W

Wallaby, Foja dwarf, 284, 285
   red-necked, 38, 69
   UK naturalised, 38, 69
Warbler, Blue Mountain, 204, **205**
   carbonated swamp-, 204
Water goblin, Lake Vedlozero, 324

Alien Zoo

Wattie (see Loch Watten monster),
Werewolf, Cannock Chase, 71, 259
Whale, baleen, with unique song, 201
       blue, 257, 277
       minke, 319
       sperm, 100, 182
Wild-man, Chinese, 17
Wildcat, African, 60, 177, 178
       Irish, 275
       Scottish, 51, 57, 58, 184
Windy (see Lake Windermere monster),
Wolf, Achill Island dwarf, 275
       Andean, 131, 132, **132**
Wolverine, 158, **158**, 359
Wonambi, 110, 188
Wonambi naracoortensis, 188
Wood white, common, 141
       Real's, 141
Woodpecker, imperial, 18
       ivory-billed, 17, 18, 92, 93, 296
       pileated, 18
Worm, giant Palouse, 229
       Mid-Atlantic Ridge mystery, 198
       Mongolian death, 13, 17, 18, 114, 139, 140, 194, 201, 217, 218, 219, 296
Worm-lizard (see Amphisbaenian),

X

Xenorhina arboricola, 284

Y

Yapok, 98, 100, **100,** 106, 107, **107**
Yarri, 33, 65, 219
Yeti, 13, 16, 18, 134, **134**, 143, 288, 291, 297, 308, 309, **309**, 354
Yowie, 20, 99, 137, 141, 142, 187

Z

Zeuglodont, 39, 42, 43, **43**
Zygaena viciae, 122

ACKNOWLEDGEMENTS.

During my many years of writing for *Fortean Times*, I have been assisted in my varied researches and cryptozoological chroniclings by numerous people, whose kindness and interest in my work I very greatly appreciate.

First and foremost, I offer my very sincere and grateful thanks to Bob Rickard, Paul Sieveking, Steve Moore, Mike Dash (collectively The Gang of Fort), David Sutton, Ian Simmons, Val Stevenson, Etienne Gilfillan, Hunt Emerson, Jane Watkins, Joe McNally, James Brown, John Brown Publishing, James Brown Publishing, I Feel Good Publishing, Dennis Publishing, and all of the many other members of the *FT* team during my own *FT* tenure for their invaluable assistance, support, encouragement, and enthusiasm down through the years, and without whom my *FT* career, and this present book, would simply not have been possible.

I also wish to thank all of the following persons for their own greatly-valued contributions to the contents of my Alien Zoo and my other *FT* writings republished in this book:

Nick [surname withheld by me]; Chad Arment; Neil Arnold; Robert Baker; Simon Baker; the late Mark K. Bayless; Dr Simon K. Bearder; Matthew Bille; Janet and Colin Bord; Dr Michael W. Bruford; John and Lesley Burke; Juan Cabana; Don Campbell; Guillaume Chapron; Carmen Garcia-Frias Checa; Anthony Cheke; the late Mark Chorvinsky; Prof. John L. Cloudsley-Thompson; Margaret Cockbill; Matthew J. Coefer; Loren Coleman; Andrew Collins; Scott Corrales; Paul Cropper; Gary Cunningham; Adam Davies; John Edwards; David Farrier; Steve Feltham; Angel Morant Forés; Keith Foster; Richard Freeman; Alan Friswell; Errol Fuller; Andrew Gable; Tonio Galea; Dorian Gallant; Ken Gerhard; Bill Gibbons; Prof. Colin P. Groves; Ann Harper; Markus Hemmler; Isabela Herranz; Phil Hide; Dr Paul Hillyard; Shaun Histed-Todd; Chris Holloway; Terry Hooper; Tyson Hughes; Brian Irwin; Todd Jurasek; John Kahila; Prof. Reinhardt Møbjerg Kristensen; Brad LaGrange; Alex Lamprey; Shane Lea; Bryan Long; Tony Lucas; Prof. Roy P. Mackal; Ivan Mackerle; *Manchester Evening News*'s archive staff; Dr Adrienne Mayor; Ruchi Mehta; Ron Mills; Dr Ralph Molnar; Dr Timothy Mowl; Richard Muirhead; Dr Darren Naish; Michael Newton; the late Scott Norman; Mark North - Dark Dorset; Prof. Stephen O'Brien; Mark O'Shea; Dr Charles Paxton; Peter Pesavento; Richard Pharo; Martin Phillipps; Roger Pope; Jeanette and Robert Powell; Alan Pringle; Fred Rackham; Dr Robert J. Raven; Michel Raynal; Nick Redfern; Chris Rehberg; the late Roy Robinson; Dr Marc van Roosmalen; Andy Sanderson; Dr François de Sarre; Marko Sciban; Anna Severson; Malcolm Sewell; Lance Shirley; Mary D. Shuker; Bob Simmons; Martine Smids; Bruce Spittle; Kevin Stewart; Dr George Stoecklin; Nick Sucik; Jan-Ove Sundberg; Gareth Thomas; Lars Thomas; Spencer Thrower; Larry Tribula; the late Martin Trippett; Derek Uchman; Richard Wells; the late Jan Williams and her husband Keith; Mike Williams; Rod Williams; Ted Williams; Ben Willis; Harry Wilson; Michael Woodley; Chris Woodyard; Christopher M. Younger.

In addition, I am especially grateful to Jade Bengco, Markus Bühler, Tim Morris, Andy Paciorek, William M. Rebsamen, Stephanie Sparkman (née Canzanella), and Richard Svensson for so kindly permitting me to utilise their superb artwork in this book and also for specifically creating a number of new illustrations to be included in it.

Last, but definitely not least, I owe a huge debt of thanks to Hunt Emerson, *FT*'s very own longstanding cartoonist-in-residence, for preparing the hilarious image of me for this book's recurring 'Alien Zoo' motif inside and also for so kindly permitting me to utilise as its recurring 'Lost Ark' motif his

Alien Zoo

equally marvellous 'Lost Ark' caricature of me that was originally used as the header for my 'Lost Ark' articles in *FT*; to Bob Rickard and David Sutton once more, specifically this time for penning their delightful forewords to this book; and to Jonathan Downes and the CFZ Press for seeing into print yet another book of mine that I've wanted to prepare for such a long time.

MY WEBSITE AND BLOGS

If you have enjoyed this or any of my other books and articles, and have any information or personal experiences/eyewitness accounts relating to any aspect of cryptozoology, animal anomalies, animal mythology, or any other mysterious phenomenon that you consider may be of interest to me and that you wish to share with me, please email me at: karlshuker@aol.com

For full details concerning my work and publications, please visit my official website at: http://www.karlshuker.com

I also have two online blogs. My ShukerNature blog at http://www.karlshuker.blogspot.com deals with cryptozoology and animal anomalies; and my Starsteeds blog at http://www.starsteeds.blogspot.com is devoted to my poetry, especially those poems of mine contained in my book *Starsteeds and Other Dreams: The Collected Poems* (CFZ Press: Bideford, 2009).

In addition, you can check out my official entry on Wikipedia, the Net's most comprehensive online encyclopedia, which can be accessed at: http://en.wikipedia.org/wiki/Karl_Shuker - and you can also check out my Facebook page and join my Facebook Fan Page., and access my Twitter page.

Many thanks indeed.

Alien Zoo

A LISTING OF MY ORIGINAL ARTICLES

All of the Lost Ark chapters in this book are based upon the following Lost Ark articles of mine (listed below in the same order as this book's corresponding chapters) that were originally published in *Fortean Times*:

SHUKER, Karl P.N. (2009). In search of the missing monster [re Wattie]. *Fortean Times*, no. 253 (September): 52-5.

SHUKER, Karl P.N. (1993). The lovecats. *Fortean Times*, no. 68 (April-May): 50-1.

SHUKER, Karl P.N. (1993). Lovecats: the next generation. *Fortean Times*, no. 72 (December-January): 46.

SHUKER, Karl P.N. (2009). The Gabriel feather. *Fortean Times*, no. 245 (February): 46-9.

SHUKER, Karl P.N. (2009). Kicking up a stink. *Fortean Times*, no. 248 (May): 54-5.

SHUKER, Karl P.N. (2002). Flying graverobbers [re ropen]. *Fortean Times*, no. 154 (January): 48-9.

SHUKER, Karl P.N. (1998). Horseplay in Tibet. *Fortean Times*, no. 86 (May): 42.

SHUKER, Karl P.N. (1997). Are our high-flying kin batty? *Fortean Times*, no. 97 (April): 48-49.

SHUKER, Karl P.N. (2005). The Ark down under [re Australian mystery beasts]. *Fortean Times*, no. 199 (August): 54-5.

SHUKER, Karl P.N. (2007). Crypto-twitching. *Fortean Times*, no. 222 (May): 42-4.

SHUKER, Karl P.N. (1997). Beware: flesh rotter [re Australian spider mystery]. *Fortean Times*, no. 95 (March): 44.

SHUKER, Karl P.N. (1994). Cat flaps [re winged cats explanation]. *Fortean Times*, no. 78 (December-January): 32-3.

SHUKER, Karl P.N. (2009). Winged cat gallery. *Fortean Times*, no. 244 (January): 42-5.

SHUKER, Karl P.N. (1996). Fins, fangs and poison [re Shatt al Arab River's mystery venomous fish]. *Fortean Times*, no. 93 (December): 44.

SHUKER, Karl P.N. (2006). A new Eden in New Guinea. *Fortean Times*, no. 209 (May): 38-40.

SHUKER, Karl P.N. (1997). Headless and petrified [re entombed adder and ammonites]. *Fortean Times*, no. 103 (November): 50.

SHUKER, Karl P.N. (2008). The truth behind the monster [re Lake Khaiyr monster]. *Fortean Times*, no. 232 (February): 58-9.

SHUKER, Karl P.N. (2009). Driven batty by the orang bati. *Fortean Times*, no. 255 (November): 52-3.

DISCLAIMER

I have sought permission for the use of all illustrations known or suspected by me to be still in copyright. Any omission brought to my attention will be rectified in future editions of this book.

Birmingham City Council Presents

Myths and Monsters

INCLUDING LIFESIZE ANIMATRONIC CREATURES!

UNRAVELLING THE TRUTH

24th May - 31st
Tickets: 303 19
www.bmag.org.uk

Birmingham City Council   Natural History Museum

ABOUT THE AUTHOR.

Born and still living in the West Midlands, England, Dr Karl P.N. Shuker graduated from the University of Leeds with a Bachelor of Science (Honours) degree in pure zoology, and from the University of Birmingham with a Doctor of Philosophy degree in zoology and comparative physiology. He now works full-time as a freelance zoological consultant to the media, and as a prolific published writer.

Dr Shuker is currently the author of 16 books and hundreds of articles, principally on animal-related subjects, with an especial interest in cryptozoology and animal mythology, on which he is an internationally-recognised authority, but also including a poetry volume. In addition, he has acted as consultant for several major multi-contributor volumes as well as for the world-renowned *Guinness Book of Records/Guinness World Records* (he is currently its Senior Consultant for its Life Sciences section), and he has compiled questions for the BBC's long-running cerebral quiz 'Mastermind'.

Dr Shuker has travelled the world in the course of his researches and writings, and has appeared regularly on television and radio. Aside from work, his diverse range of interests include motorbikes, the life and career of James Dean, collecting masquerade and carnival masks, quizzes, philately, poetry, travel, world mythology, and the history of animation.

He is a Scientific Fellow of the prestigious Zoological Society of London, a Fellow of the Royal Entomological Society, a Member of the International Society of Cryptozoology and other wildlife-related organisations, he is Cryptozoology Consultant to the Centre for Fortean Zoology, and is also a Member of the Society of Authors.

Dr Shuker's personal website can be accessed at http://www.karlshuker.com, and his mystery animals blog, ShukerNature, can be accessed at http://www.karlshuker.blogspot.com. His poetry blog can be accessed at http://starsteeds.blogspot.com. There is also an entry for Dr Shuker in the online encyclopedia Wikipedia, and a fan page on Facebook.

AUTHOR BIBLIOGRAPHY

Mystery Cats of the World: From Blue Tigers To Exmoor Beasts (Robert Hale: London, 1989)
Extraordinary Animals Worldwide (Robert Hale: London, 1991)
The Lost Ark: New and Rediscovered Animals of the 20th Century (HarperCollins: London, 1993)
Dragons: A Natural History (Aurum: London/Simon & Schuster: New York, 1995; republished Taschen: Cologne, 2006)
In Search of Prehistoric Survivors: Do Giant 'Extinct' Creatures Still Exist? (Blandford: London, 1995)
The Unexplained: An Illustrated Guide to the World's Natural and Paranormal Mysteries (Carlton: London/JG Press: North Dighton, 1996; republished Carlton: London, 2002)
From Flying Toads To Snakes With Wings: From the Pages of FATE Magazine (Llewellyn: St Paul, 1997; republished Bounty: London, 2005)
Mysteries of Planet Earth: An Encyclopedia of the Inexplicable (Carlton: London, 1999)
The Hidden Powers of Animals: Uncovering the Secrets of Nature (Reader's Digest: Pleasantville/ Marshall Editions: London, 2001)
The New Zoo: New and Rediscovered Animals of the Twentieth Century [fully-updated, greatly-

expanded, new edition of *The Lost Ark*] (House of Stratus Ltd: Thirsk, UK/House of Stratus Inc: Poughkeepsie, USA, 2002)

The Beasts That Hide From Man: Seeking the World's Last Undiscovered Animals (Paraview: New York, 2003)

Extraordinary Animals Revisited: From Singing Dogs To Serpent Kings (CFZ Press: Bideford, 2007)

Dr Shuker's Casebook: In Pursuit of Marvels and Mysteries (CFZ Press: Bideford, 2008)

Dinosaurs and Other Prehistoric Animals on Stamps: A Worldwide Catalogue (CFZ Press: Bideford, 2008)

Star Steeds and Other Dreams: The Collected Poems (CFZ Press: Bideford, 2009)

Karl Shuker's Alien Zoo: From the Pages of Fortean Times (CFZ Press: Bideford, 2010)

Consultant and also Contributor

Man and Beast (Reader's Digest: Pleasantville, New York, 1993)

Secrets of the Natural World (Reader's Digest: Pleasantville, New York, 1993)

Almanac of the Uncanny (Reader's Digest: Surry Hills, Australia, 1995)

The Guinness Book of Records/Guinness World Records 1998-present day (Guinness: London, 1997-present day)

Consultant

Monsters (Lorenz: London, 2001)

Contributor

Of Monsters and Miracles CD-ROM (Croydon Museum/Interactive Designs: Oxton, 1995)

Fortean Times Weird Year 1996 (John Brown Publishing: London, 1996)

Mysteries of the Deep (Llewellyn: St Paul, 1998)

Guinness Amazing Future (Guinness: London, 1999)

The Earth (Channel 4 Books: London, 2000)

Mysteries and Monsters of the Sea (Gramercy: New York, 2001)

Chambers Dictionary of the Unexplained (Chambers: Edinburgh, 2007)

Chambers Myths and Mysteries (Chambers: Edinburgh, 2008)

The Fortean Times Paranormal Handbook (Dennis Publishing: London, 2009)

Plus numerous contributions to the annual *CFZ Yearbook* series of volumes.

THE CENTRE FOR FORTEAN ZOOLOGY

So, what is the Centre for Fortean Zoology?

We are a non profit-making organisation founded in 1992 with the aim of being a clearing house for information, and coordinating research into mystery animals around the world. We also study out of place animals, rare and aberrant animal behaviour, and Zooform Phenomena; little-understood "things" that appear to be animals, but which are in fact nothing of the sort, and not even alive (at least in the way we understand the term).

Why should I join the Centre for Fortean Zoology?

Not only are we the biggest organisation of our type in the world, but - or so we like to think - we are the best. We are certainly the only truly global Cryptozoological research organisation, and we carry out our investigations using a strictly scientific set of guidelines. We are expanding all the time and looking to recruit new members to help us in our research into mysterious animals and strange creatures across the globe. Why should you join us? Because, if you are genuinely interested in trying to solve the last great mysteries of Mother Nature, there is nobody better than us with whom to do it.

What do I get if I join the Centre for Fortean Zoology?

For £12 a year, you get a four-issue subscription to our journal *Animals & Men*. Each issue contains 60 pages packed with news, articles, letters, research papers, field reports, and even a gossip column! The magazine is A5 in format with a full colour cover. You also have access to one of the world's largest collections of resource material dealing with cryptozoology and allied disciplines, and people from the CFZ membership regularly take part in fieldwork and expeditions around the world.

How is the Centre for Fortean Zoology organised?

The CFZ is managed by a three-man board of trustees, with a non-profit making trust registered with HM Government Stamp Office. The board of trustees is supported by a Permanent Directorate of full and part-time staff, and advised by a Consultancy Board of specialists - many of whom are world-renowned experts in their particular field. We have regional representatives across the UK, the USA, and many other parts of the world, and are affiliated with other organisations whose aims and protocols mirror our own.

I am new to the subject, and although I am interested I have little practical knowledge. I don't want to feel out of my depth. What should I do?

Don't worry. We were *all* beginners once. You'll find that the people at the CFZ are friendly and approachable. We have a thriving forum on the website which is the hub of an ever-growing electronic community. You will soon find your feet. Many members of the CFZ Permanent Directorate started off as ordinary members, and now work full-time chasing monsters around the world.

I have an idea for a project which isn't on your website. What do I do?

Write to us, e-mail us, or telephone us. The list of future projects on the website is not exhaustive. If you have a good idea for an investigation, please tell us. We may well be able to help.

How do I go on an expedition?

We are always looking for volunteers to join us. If you see a project that interests you, do not hesitate to get in touch with us. Under certain circumstances we can help provide funding for your trip. If you look on the future projects section of the website, you can see some of the projects that we have pencilled in for the next few years.

In 2003 and 2004 we sent three-man expeditions to Sumatra looking for Orang-Pendek - a semi-legendary bipedal ape. The same three went to Mongolia in 2005. All three members started off merely subscribers to the CFZ magazine.

Next time it could be you!

Project Kerinci, Sumatra - 2003
In search of the bipedal ape Orang Pendek

How is the Centre for Fortean Zoology funded?

We have no magic sources of income. All our funds come from donations, membership fees, works that we do for TV, radio or magazines, and sales of our publications and merchandise. We are always looking for corporate sponsorship, and other sources of revenue. If you have any ideas for fund-raising please let us know. However, unlike other cryptozoological organisations in the past, we do not live in an intellectual ivory tower. We are not afraid to get our hands dirty, and furthermore we are not one of those organisations where the membership have to raise money so that a privileged few can go on expensive foreign trips. Our research teams, both in the UK and abroad, consist of a mixture of experienced and inexperienced personnel. We are truly a community, and work on the premise that the benefits of CFZ membership are open to all.

What do you do with the data you gather from your investigations and expeditions?

Reports of our investigations are published on our website as soon as they are available. Preliminary reports are posted within days of the project finishing.

Each year we publish a 200 page yearbook containing research papers and expedition reports too long to be printed in the journal. We freely circulate our information to anybody who asks for it.

Is the CFZ community purely an electronic one?

No. Each year since 2000 we have held our annual convention - the *Weird Weekend* - in Exeter. It is three days of lectures, workshops, and excursions. But most importantly it is a chance for members of the CFZ to meet each other, and to talk with the members of the permanent directorate in a relaxed and informal setting and preferably with a pint of beer in one hand. Since 2006 - the *Weird Weekend* has been bigger and better and held on the third weekend in August in the idyllic rural location of Woolsery in North Devon.

Since relocating to North Devon in 2005 we have become ever more closely involved with other community organisations, and we hope that this trend will continue. We also work closely with Police Forces across the UK as consultants for animal mutilation cases, and we intend to forge closer links with the coastguard and other community services. We want to work closely with those who regularly travel into the Bristol Channel, so that if the recent trend of exotic animal visitors to our coastal waters continues, we can be out there as soon as possible.

We are building a Visitor's Centre in rural North Devon. This will not be open to the general public, but will provide a museum, a library and an educational resource for our members (currently over 400) across the globe. We are also planning a youth organisation which will involve children and young people in our activities.

Apart from having been the only Fortean Zoological organisation in the world to have consistently published material on all aspects of the subject for over a decade, we have achieved the following concrete results:

- Disproved the myth relating to the headless so-called sea-serpent carcass of Durgan beach in Cornwall 1975
- Disproved the story of the 1988 puma skull of Lustleigh Cleave
- Carried out the only in-depth research ever into the mythos of the Cornish Owlman
- Made the first records of a tropical species of lamprey
- Made the first records of a luminous cave gnat larva in Thailand
- Discovered a possible new species of British mammal - the beech marten
- In 1994-6 carried out the first archival fortean zoological survey of Hong Kong
- In the year 2000, CFZ theories were confirmed when an new species of lizard was added to the British list
- Identified the monster of Martin Mere in Lancashire as a giant wels catfish
- Expanded the known range of Armitage's skink in the Gambia by 80%
- Obtained photographic evidence of the remains of Europe's largest known pike
- Carried out the first ever in-depth study of the *ninki-nanka*
- Carried out the first attempt to breed Puerto Rican cave snails in captivity
- Were the first European explorers to visit the `lost valley` in Sumatra
- Published the first ever evidence for a new tribe of pygmies in Guyana
- Published the first evidence for a new species of caiman in Guyana
- Filmed unknown creatures on a monster-haunted lake in Ireland for the first time
- Had a sighting of orang pendek in Sumatra in 2009
- Published some of the best evidence ever for the almasty in southern Russia
- In the year 2010, CFZ theories were confirmed when relict populations of pine martens were found in various parts of southern and central England

EXPEDITIONS & INVESTIGATIONS TO DATE INCLUDE:

- 1998 Puerto Rico, Florida, Mexico *(Chupacabras)*
- 1999 Nevada *(Bigfoot)*
- 2000 Thailand *(Giant snakes called nagas)*
- 2002 Martin Mere *(Giant catfish)*
- 2002 Cleveland *(Wallaby mutilation)*
- 2003 Bolam Lake *(BHM Reports)*
- 2003 Sumatra *(Orang Pendek)*
- 2003 Texas *(Bigfoot; giant snapping turtles)*
- 2004 Sumatra *(Orang Pendek; cigau, a sabre-toothed cat)*
- 2004 Illinois *(Black panthers; cicada swarm)*
- 2004 Texas *(Mystery blue dog)*
- Loch Morar *(Monster)*
- 2004 Puerto Rico *(Chupacabras; carnivorous cave snails)*
- 2005 Belize *(Affiliate expedition for hairy dwarfs)*
- 2005 Loch Ness *(Monster)*
- 2005 Mongolia *(Allghoi Khorkhoi aka Mongolian death worm)*
- 2006 Gambia *(Gambo - Gambian sea monster , Ninki Nanka and Armitage's skink*
- 2006 Llangorse Lake *(Giant pike, giant eels)*
- 2006 Windermere *(Giant eels)*
- 2007 Coniston Water *(Giant eels)*
- 2007 Guyana *(Giant anaconda, didi, water tiger)*
- 2008 Russia *(Almasty)*
- 2009 Sumatra *(Orang pendek)*
- 2009 Republic of Ireland *(Lake Monster)*
- 2010 Texas *(Blue Dogs)*

HOW TO START A PUBLISHING EMPIRE

Unlike most mainstream publishers, we have a non-commercial remit, and our mission statement claims that "we publish books because they deserve to be published, not because we think that we can make money out of them". Our motto is the Latin Tag "Pro bona causa facimus" (we do it for good reason), a slogan taken from a children's book `The Case of the Silver Egg` by the late Desmond Skirrow.

WIKIPEDIA: "The first book published was in 1988. `Take this Brother may it Serve you Well` was a guide to Beatles bootlegs by Jonathan Downes. It sold quite well, but was hampered by very poor production values, being photocopied, and held together by a plastic clip binder. In 1988 A5 clip binders were hard to get hold of, so the publishers took A4 binders and cut them in half with a hacksaw. It now reaches surprisingly high prices second hand.

The production quality improved slightly over the years, and after 1999 all the books produced were ringbound with laminated colour covers. In 2004, however, they signed an agreement with LightningSource, and all books are now produced perfect bound, with full colour covers."

Until 2010 all our books, the majority of which are/were on the subject of mystery animals and allied disciplines, were published by `CFZ Press`, the publishing arm of the Centre for Fortean Zoology (CFZ), and we urged our readers and followers to draw a discreet veil over the books that we published that were completely off topic to the CFZ.

However, in 2010 we decided that enough was enough and launched a second imprint, `Fortean Words` which aims to cover a wide range of non animal-related esoteric subjects. Other imprints will be launched as and when we feel like it, however the basic ethos of the company remains the same: Our job is to publish books and magazines that we feel are worth publishing, whether or not they are going to sell. Money is, after all - as my dear old Mama once told me - a rather vulgar subject, and she would be rolling in her grave if she thought that her eldest son was somehow in `trade`.

Luckily, so far our tastes have turned out not to be that rarified after all, and we have sold far more books than anyone ever thought that we would, so there is a moral in there somewhere…

Jon Downes,
Woolsery, North Devon
July 2010

Other Books in Print

The Mystery Animals of Ireland by Gary Cunningham and Ronan Coghlan
Monsters of Texas by Gerhard, Ken
The Great Yokai Encyclopaedia by Freeman, Richard
NEW HORIZONS: Animals & Men *issues 16-20 Collected Editions Vol. 4* by Downes, Jonathan
A Daintree Diary -
Tales from Travels to the Daintree Rainforest in tropical north Queensland, Australia by Portman, Carl
Strangely Strange but Oddly Normal by Roberts, Andy
Centre for Fortean Zoology Yearbook 2010 by Downes, Jonathan
Predator Deathmatch by Molloy, Nick
Star Steeds and other Dreams by Shuker, Karl
CHINA: A Yellow Peril? by Muirhead, Richard
Mystery Animals of the British Isles: The Western Isles by Vaudrey, Glen
Giant Snakes - Unravelling the coils of mystery by Newton, Michael
Mystery Animals of the British Isles: Kent by Arnold, Neil
Centre for Fortean Zoology Yearbook 2009 by Downes, Jonathan
CFZ EXPEDITION REPORT: Russia 2008 by Richard Freeman *et al*, Shuker, Karl (fwd)
Dinosaurs and other Prehistoric Animals on Stamps - A Worldwide catalogue by Shuker, Karl P. N
Dr Shuker's Casebook by Shuker, Karl P.N
The Island of Paradise - chupacabra UFO crash retrievals,
and accelerated evolution on the island of Puerto Rico by Downes, Jonathan
The Mystery Animals of the British Isles: Northumberland and Tyneside by Hallowell, Michael J
Centre for Fortean Zoology Yearbook 1997 by Downes, Jonathan (Ed)
Centre for Fortean Zoology Yearbook 2002 by Downes, Jonathan (Ed)
Centre for Fortean Zoology Yearbook 2000/1 by Downes, Jonathan (Ed)
Centre for Fortean Zoology Yearbook 1998 by Downes, Jonathan (Ed)
Centre for Fortean Zoology Yearbook 2003 by Downes, Jonathan (Ed)
In the wake of Bernard Heuvelmans by Woodley, Michael A
CFZ EXPEDITION REPORT: Guyana 2007 by Richard Freeman *et al*, Shuker, Karl (fwd)
Centre for Fortean Zoology Yearbook 1999 by Downes, Jonathan (Ed)
Big Cats in Britain Yearbook 2008 by Fraser, Mark (Ed)
Centre for Fortean Zoology Yearbook 1996 by Downes, Jonathan (Ed)
THE CALL OF THE WILD - Animals & Men issues 11-15
Collected Editions Vol. 3 by Downes, Jonathan (ed)

Ethna's Journal by Downes, C N
Centre for Fortean Zoology Yearbook 2008 by Downes, J (Ed)
DARK DORSET -Calendar Custome by Newland, Robert J
Extraordinary Animals Revisited by Shuker, Karl
MAN-MONKEY - In Search of the British Bigfoot by Redfern, Nick
Dark Dorset Tales of Mystery, Wonder and Terror by Newland, Robert J and Mark North
Big Cats Loose in Britain by Matthews, Marcus
MONSTER! - The A-Z of Zooform Phenomena by Arnold, Neil
The Centre for Fortean Zoology 2004 Yearbook by Downes, Jonathan (Ed)
The Centre for Fortean Zoology 2007 Yearbook by Downes, Jonathan (Ed)
CAT FLAPS! Northern Mystery Cats by Roberts, Andy
Big Cats in Britain Yearbook 2007 by Fraser, Mark (Ed)
BIG BIRD! - Modern sightings of Flying Monsters by Gerhard, Ken
THE NUMBER OF THE BEAST - Animals & Men issues 6-10
Collected Editions Vol. 1 by Downes, Jonathan (Ed)
IN THE BEGINNING - Animals & Men *issues 1-5 Collected Editions Vol. 1* by Downes, Jonathan
STRENGTH THROUGH KOI - They saved Hitler's Koi and other stories by Downes, Jonathan
The Smaller Mystery Carnivores of the Westcountry by Downes, Jonathan
CFZ EXPEDITION REPORT: Gambia 2006 by Richard Freeman *et al*, Shuker, Karl (fwd)
The Owlman and Others by Jonathan Downes
The Blackdown Mystery by Downes, Jonathan
Big Cats in Britain Yearbook 2006 by Fraser, Mark (Ed)
Fragrant Harbours - Distant Rivers by Downes, John T
Only Fools and Goatsuckers by Downes, Jonathan
Monster of the Mere by Jonathan Downes
Dragons:More than a Myth by Freeman, Richard Alan
Granfer's Bible Stories by Downes, John Tweddell
Monster Hunter by Downes, Jonathan

Fortean Words

The Centre for Fortean Zoology has for several years led the field in Fortean publishing. CFZ Press is the only publishing company specialising in books on monsters and mystery animals. CFZ Press has published more books on this subject than any other company in history and has attracted such well known authors as Andy Roberts, Nick Redfern, Michael Newton, Dr Karl Shuker, Neil Arnold, Dr Darren Naish, Jon Downes, Ken Gerhard and Richard Freeman.

Now CFZ Press is launching a new imprint. Fortean Words is a new line of books dealing with Fortean subjects other than cryptozoology, which is - after all - the subject the CFZ is best known for. Fortean Words is being launched with a spectacular multi-volume series called *Haunted Skies* which covers British UFO sightings between 1940 and 2010. Former policeman John Hanson and his long-suffering partner Dawn Holloway have compiled a peerless library of sighting reports, many that have never been made public before.

Other books include a look at the Berwyn Mountains UFO case by renowned Fortean Andy Roberts and a forthcoming series of books by transatlantic researcher Nick Redfern.

CFZ Press is dedicated to maintaining the fine quality of its work with Fortean Words. New authors tackling new subjects will always be encouraged, and we hope that our books will continue to be as ground breaking and popular as ever.

Lightning Source UK Ltd.
Milton Keynes UK
UKOW021826121112

202090UK00008B/3/P